Lecture Notes in Civil Engineering

Volume 214

Lecture Notes in Civil Engineering (LNCE) publishes the latest developments in Civil Engineering - quickly, informally and in top quality. Though original research reported in proceedings and post-proceedings represents the core of LNCE, edited volumes of exceptionally high quality and interest may also be considered for publication. Volumes published in LNCE embrace all aspects and subfields of, as well as new challenges in, Civil Engineering. Topics in the series include:

- Construction and Structural Mechanics
- Building Materials
- Concrete, Steel and Timber Structures
- Geotechnical Engineering
- Earthquake Engineering
- Coastal Engineering
- Ocean and Offshore Engineering; Ships and Floating Structures
- Hydraulics, Hydrology and Water Resources Engineering
- Environmental Engineering and Sustainability
- Structural Health and Monitoring
- Surveying and Geographical Information Systems
- Indoor Environments
- Transportation and Traffic
- Risk Analysis
- Safety and Security

To submit a proposal or request further information, please contact the appropriate Springer Editor:

- Pierpaolo Riva at pierpaolo.riva@springer.com (Europe and Americas);
- Swati Meherishi at swati.meherishi@springer.com (Asia - except China, and Australia, New Zealand);
- Wayne Hu at wayne.hu@springer.com (China).

All books in the series now indexed by Scopus and EI Compendex database!

More information about this series at https://link.springer.com/bookseries/15087

Norazian Mohamed Noor · Sung Ting Sam ·
Aeslina Abdul Kadir
Editors

Proceedings of the 3rd International Conference on Green Environmental Engineering and Technology

IConGEET 2021, Penang, Malaysia

Springer

Editors
Norazian Mohamed Noor
Faculty of Civil Engineering Technology
Universiti Malaysia Perlis
Arau, Perlis, Malaysia

Aeslina Abdul Kadir
Faculty of Civil Engineering and Built Environment
Universiti Tun Hussein Onn
Parit Raja, Johor, Malaysia

Sung Ting Sam
Faculty of Chemical Engineering Technology
Universiti Malaysia Perlis
Arau, Perlis, Malaysia

ISSN 2366-2557 ISSN 2366-2565 (electronic)
Lecture Notes in Civil Engineering
ISBN 978-981-16-7919-3 ISBN 978-981-16-7920-9 (eBook)
https://doi.org/10.1007/978-981-16-7920-9

This Springer imprint is published by the registered company Springer Nature Singapore Pte Ltd.
The registered company address is: 152 Beach Road, #21-01/04 Gateway East, Singapore 189721, Singapore

Preface

The 3rd International Conference on Green Environmental Engineering and Technology (IConGEET 2021) took place virtually on 8–9 September 2021 in joint with the 7th International Conference on Green Design and Manufacture (IConGDM 2021). The conference was organized by the Sustainable Environmental Research Group (SERG), Centre of Excellence Geopolymer and Green Technology (CEGeoGTech) and the Faculty of Civil Engineering Technology, Universiti Malaysia Perlis. The conference received international and local collaboration, support and participation of the National Institute for Research and Development in Environmental Protection (INCDPM) Romania, the Department of Civil and Environmental, Faculty of Engineering at Rabigh, King Abdulaziz University as well as the Center of Excellence, Water Research and Environmental Sustainability Growth (WAREG), Universiti Malaysia Perlis.

It brought together researchers, practitioners, engineers and students worldwide in hope to extend networking, integrate knowledge and exchange ideas among participants. This year, the theme "Enhancing Sustainable Environment for Humanity: Challenges and opportunities in the Post-Covid Era" has been in point with the COVID19 situation faced by the world impacting us economically, socially and environmentally. Over 100 distinguish researchers, practitioners, engineers and students from across the country have presented their current and intriguing research in the scope of civil and environmental technologies and management, which evolved around five major areas: water and wastewater technology, green technology, air pollution and climate change, environmental management and protection as well as environmental sustainability and development.

The conference has gained tremendous success due to collective efforts from the organizing committees, the co-organizers, the reviewers, presenters and all participants. Special acknowledgment goes to the keynote and invited speakers who assent our invitation and successfully delivered interesting talks to catch all the participants' attention. The keynote speeches were delivered by Prof. Dr. Joel Oliveira Correla Vasco (Polytechnic of Leiria, Portugal), Prof. Dr. Fauziah Ahmad (Universiti Sains Malaysia) and Associate Prof. Dr. Andriana Surieva (University of Chemical Technology and Metallurgy, Bulgaria), while the invited speakers involved were Dr.

Marwan Mustafa Kheimi (King Abdulaziz University), Ir. Ts. Gs. Dr. Chow Ming Fai (Monash University Malaysia) and Dr. Mohd. Faridh Ahmad Zaharuddin (Universiti Teknologi Malaysia).

We would like to thank all the reviewers involved in ensuring each contributed paper were refereed and accepted based on interests, relevance, innovation and pure contribution towards the civil and environmental engineering fields. Acknowledgement gratitude also goes to Associate Prof. Dr. Aeslina Abdul Kadir, visiting researcher of SERG from Universiti Tun Hussien Onn, Malaysia for her invaluable work as one of the editor of this proceeding. Finally, special thanks go to the Universiti Malaysia Perlis for the continuous support throughout the conference event.

Arau, Malaysia

Norazian Mohamed Noor
Chairman
The 3rd International Conference on Green Environmental Engineering and Technology 2021 (IConGEET 2021)

Contents

Air Pollution and Climate Change

Appropriate Community-Based Emergency Response Elements for Toxic Gas Release in Pasir Gudang, Johor

Noor Afzan Ahmad, Kamarizan Kidam, and Rahmat Mohsin

Abstract The number of harmful gas-using industries has expanded dramatically, with the majority of them located near populated areas. Consequently, this presents a significant challenge for government officials to reduce hazards in the vicinity of hazardous gas facilities and offer safe and effective community emergency response team, where serious accidents may occur. The objective of this study is to determine the community-based emergency response elements for toxic gas release in Pasir Gudang. In this work, questionnaire survey was distributed to the community for data collection. A total of 384 communities in Pasir Gudang were chosen as respondents for the study. The performance of the communities in emergency response was utilised as quantitative data and analysed using the Rasch model. The four main constructs evaluated are coordination, collaboration, communication, and capability, with a total of 32 elements of emergency response that were evaluated of its measure mean (logit) and score mean value. Overall, the findings show that the construct of the emergency response has a measure mean (logit) of 0.00 and a score mean of 3.67–5.00. This demonstrates that the level of agreement is high on the elements found in the main constructs.

Keywords Toxic gas release · Community · Emergency response · Construct · Rasch model

N. A. Ahmad (✉) · K. Kidam · R. Mohsin
School of Chemical and Energy Engineering, Universiti Teknologi Malaysia (UTM), 81310 Johor Bahru, Johor, Malaysia
e-mail: nafzan9@graduate.utm.my

K. Kidam
e-mail: kamarizan@cheme.utm.my

R. Mohsin
e-mail: rahmat@utm.my

N. Mohamed Noor et al. (eds.), *Proceedings of the 3rd International Conference on Green Environmental Engineering and Technology*, Lecture Notes in Civil Engineering 214,
https://doi.org/10.1007/978-981-16-7920-9_1

1 Introduction

Fire and release of toxic material are common in industrial accidents. These accidents are caused by human errors or technological failures during storage, processing, or transportation activities. The total death in such accident gradually increases with industrial development. The mortality rate in such accident is higher in India, Mexico, and Brazil, compared with that in other industrialised countries [1]. Toxic gas release is among the most potentially dangerous and far-reaching risks in oil and gas operations. Gas leak causes explosions. Fires, in certain weather conditions, quickly spread deadly toxins farther and wider, causing harm or death to people, damaging the environment, and disrupting operations, leading to financial losses and tarnished reputations. Hence, once a toxic gas release accident occurs, an efficient emergency response is of utmost importance to save human lives and protect the environment [2].

Pasir Gudang is an industrial area where heavy industries operate, such as shipbuilding, petrochemicals manufacturing, and other logistics and transportation. The increasing population, along with over 45 high-risk industries have led to several safety issues towards the surrounding communities. Pasir Gudang City Council [3] reported that a scenario of emergency occurs due to major incidents involving hazardous materials, explosions, transportation accidents, uncontrollable fires, chemical spillage, or combinations thereof, resulting in injuries, evacuations, and fatalities. This paper aimed to determine Pasir Gudang community-based emergency response elements for toxic gas release. The researcher has identified four constructs in emergency response: coordination, collaboration, communication, and capability, as moderators that can influence the community to produce an effective emergency response based on document analyses and expert interviews. These four constructs cover all emergency response elements and can help the community response to toxic gas release more efficiently and effectively.

2 Methodology

This study creates a quantitative data collection. This phase illustrates a study to determine Pasir Gudang community-based emergency response elements for toxic gas release. The researcher describes the findings of the study through a questionnaire that has been developed as a result of the pilot test. Subsequently, 34 elements of emergency response have been analysed using Fuzzy Delphi method and determined using the four main constructs, namely, coordination, collaboration, communication, and capability, based on expert agreement and consensus. This quantitative data was obtained based on a questionnaire that was distributed to respondents from a community in Pasir Gudang, Johor. Before the questionnaire was distributed to the respondents in the actual study, the questionnaire was put through a pilot study process first, where the researcher conducted content validity and face validity tests, determined item reliability and item polarity, performed identification of fit item and value measurement with standardised residual correlations in determining the dependent items.

The researcher conducted this pilot study by distributing a set of questionnaires to 50 respondents from community in Pasir Gudang. The number of respondents in this pilot study, which is 50 people, is deemed adequate because according to Cooper et al. [4], the appropriate number of respondents in the pilot study ranged from 25 to 100 people. Meanwhile, Johanson and Brooks [5] suggested a minimum number of 30 people for a pilot study, which is intended for initial study or scale development. Pilot study was conducted prior the actual distribution of questionnaire to test the reliability of the questionnaire contents, where the result of a pilot study was analysed using Winsteps software with Rasch measurement model approach.

The Rasch model was used because it is well suited to the construction of instruments using the Likert scale [6]. Based on the Rasch measurement model approach, the acceptable Cronbach's Alpha (α) value for reliability is between 0.71 and 0.99 (best level). The results of the analysis of the pilot study found that the reliability value obtained based on the Cronbach's Alpha (α) value is 0.95. The results of the pilot study also show that the value of the person reliability is 0.96, where the value of reliability is considered acceptable. The value of the separation index obtained also shows a high result of above 2.0. Linacre [7] found that the value of a good separation index is more 2.0.

For quantitative method of data collection, the population in this study is made up of a group of Pasir Gudang communities. Population selection is based on the results of the study, as well as on the table for sample sizing [8]. Therefore, samples from the community are selected using a cluster sampling method. To obtain quantitative data, the researcher has extracted 384 out of 533,868 population from the Pasir Gudang communities as the sample size, based on Krejcie and Morgan [9] requirement for sample sizing. The questionnaire survey was conducted using Google Form and Winsteps software, with the Rasch measurement model approach used for data analysis. The questionnaire survey consisted of a total of 34 elements and the five-point Likert scale was used, as suggested by [9] is 1 = Strongly Disagree, 2 = Disagree, 3 = Strongly Agree, 4 = Agree, and 5 = Strongly Agree. This Likert scale is capable of measuring the level of agreement of respondents' perceptions through constructed items.

3 Result

3.1 Demographic Results

There were 384 communities from Pasir Gudang who participated in the study, showing a response rate of 100%. The majority (80.4%) were Pasir Gudang residents, and (78%) were between 20 and 55 years of age. Most of the respondents (60%) have

Table 1 Respondents demographic

Profile	Frequency	Percentage
Resident		
Pasir gudang resident	337	88
Working/business/studying in Pasir Gudang	47	12
Age		
<20 years	33	9
20–55 years	301	78
55 years and above	50	13
Academic qualification		
MCE/MVC	123	32
Diploma/degree	231	60
Master/Ph.D.	29	8
Career/profession		
School student/university	36	9
Government staff	80	21
Private staff	147	38
Business/self-employed	74	19
Retiree/housewife	47	12

a diploma and/or a degree, and (38%) are private staff. Table 1 shows the complete profile of the respondents.

3.2 Rasch Analysis Results

Table 2 shows the summarize of the complete analysis of respondent's agreement for emergency response element in each of constructs. To ensure that the elements of

Table 2 The level of respondent's agreement for emergency response element

Construct	Measure mean (logit)	Score mean	Level
Part B (Coordination)	0.00	4.48	High
Part C (Collaboration)	0.00	4.59	High
Part D (Communication)	0.00	4.67	High
Part E (Capability)	0.00	4.69	High

emergency response found in the construct of coordination, collaboration, communication and capabilities formed have a significant relationship, the researcher uses a quantitative approach where mean score and mean measure are used to see the level of agreement on the elements of emergency response obtained in all constructs. The method used by the researcher in analysing the data obtained from the questionnaire is based on the mean scale interval of [10], obtained a mean scale interval of 1.33 (Low 1.00–2.33, Middle 2.34–3.66, and High 3.67–5.00).

Based on the Table 2, the overall constructs in part B (Coordination) recorded a measure mean value of 0.00 and a score mean of 4.48. While in part C (Collaboration), the measure mean value is 0.00 and the score mean is 4.59. Part D (Communication) shows the measure mean value is 0.00 and the score mean is 4.67 followed by Part E (Capability), the measure mean value is 0.00 and the score mean is 4.69. Mahmud et al. [11] claimed that if the logit value is larger, this shows that it is easier for the respondent to certify the items presented. On the other hand, if the logit value is smaller, it shows that the respondent is more difficult to certify the items provided and answering difficult items. While for negative logit values, respondents showed low ability on the given items and the items were easy to answer. Overall shows the level of agreement is at a high level and they agree with the elements found in the main constructs in parts B (Coordination), C (Collaboration), D (Communication) and E (Capability).

4 Discussion

In this study, it was revealed that the analysis of the Pasir Gudang community-based emergency response for toxic gas release in each construct showed a high level of agreement on the emergency response elements based on the community. The construct on the emergency response element from the aspects of coordination, collaboration, communication, and capability recorded a high level of agreement between the mean value and mean score. Bond et al. [12] reported that a positive respondent's logit value indicates that individuals are more capable of answering difficult items, while for a negative logit value, respondents show low ability in answering a given item, where the item is easy to answer. Overall, a high level of agreement is demonstrated, with agreement with the elements found in the main constructs of emergency response elements in terms of coordination, collaboration, communication, and capability.

5 Conclusion

Based on the discussion and findings of the study, it is evident that the community-based emergency response elements for toxic gas release covered by four constructs: coordination, collaboration, communication, and capability, are accepted by Pasir

Gudang community. The findings of the study are in line with purpose of establishing the Pasir Gudang community-based emergency response for toxic gas release. Indirectly, the findings of this study can be used by Pasir Gudang community to strengthen their knowledge about the elements of emergency response to the release of toxic gases. The emphasis on community-based emergency response elements will help the community to be better prepared when encountering toxic gas emission emergencies, which could occur suddenly and unexpectedly.

References

1. de Souza Porto MF, De Freitas CM (2003) Vulnerability and industrial hazards in industrializing countries: an integrative approach. Futures 35(7):717–736. https://doi.org/10.1016/S0016-3287(03)00024-7
2. Duke AC (2012). Emergency response for toxic gas release incidents. In: SPE middle east health, safety, security, and environment conference and exhibition. OnePetro. https://doi.org/10.2118/152611-MS
3. Pasir Gudang City Council (2019). Department of Planning and Landscape
4. Cooper DR, Schindler PS, Sun J (2006) Business research methods, vol 9, pp 1–744. Mcgraw-Hill, New York
5. Johanson, GA, Brooks GP (2010) Initial scale development: sample size for pilot studies. Educat Psychol Measur 70(3):394–400. https://doi.org/10.1177/0013164409355692
6. Bond TG, Fox CM, Lacey H (2007) Applying the Rasch model: fundamental measurement. In: In the social sciences, 2nd edn
7. Linacre JM (2005). Measurement, meaning and morality. In: Pacific Rim objective measurement symposium and international symposium on measurement and evaluation, Kuala Lumpur, Malaysia
8. Krejcie RV, Morgan DW (1970) Determining sample size for research activities. Educ Psychol Measur 30(3):607–610. https://doi.org/10.1177/001316447003000308
9. Hair JF, Black WC, Babin, BJ, Anderson RE and Tatham R (2006) Multivariate data analysis. Uppersaddle River
10. Levin IL, Rubin DS (2000) Statistics for management, 7th edn
11. Mahmud Z, Rosmini I, Othman J, Masodi MS (2010) Overcoming measurement hurdles in statistical education research using Rasch measurement model. In: Proceedings of the 9th WSEAS international conference on education and educational technology (EDU10), pp 95–101
12. Bond TG, Fox CM. Lacey H (2007) Applying the Rasch model: fundamental measurement. In: In the social sciences, 2nd edn

Carbon Footprint Assessment from Purchased Electricity Consumption and Campus Commute in Universiti Malaysia Perlis (UniMAP): Pre- and During COVID19 Pandemic

Nor Atiqa Baharom, Sara Yasina Yusuf, Siti Khadijah Za'aba, Norazian Mohd Noor, Nor Ashikin Ahmad, Wan Amiza Amneera Wan Ahmad, and Madalina Boboc

Abstract Most institutions and organizations nowadays have been taking responsibility in reducing their carbon footprint (CF) to curtail the global warming impact to at least 20–25% reduction by 2030. Universities and higher learning institutions are starting to invest in becoming greener and carbon-free. Current COVID19 communicable disease has swayed the routine and concurrently influenced regular trends of greenhouse gases (GHG) emissions throughout the world. This study explored the possible GHG emissions (calculated as CO_2e) from internal campus commute and purchased electricity consumption from the year 2018–2020 at Universiti Malaysia Perlis main campus to analyze the influence of COVID19 pandemic on its CO_2e emission. The average amount of CO_2e emitted during pre-COVID19 period (n = 26)

N. A. Baharom · N. A. Ahmad
Faculty of Civil Engineering Technology, Universiti Malaysia Perlis, 02600 Arau, Perlis, Malaysia
e-mail: ashikinahmad@unimap.edu.my

S. Y. Yusuf (✉) · N. Mohd Noor
Sustainable Environment Research Group (SERG), Centre of Excellence Geopolymer and Green Technology (CEGeoGTech), Universiti Malaysia Perlis, 01000 Kangar, Perlis, Malaysia
e-mail: sarayasina@unimap.edu.my

N. Mohd Noor
e-mail: norazian@unimap.edu.my

S. K. Za'aba
Center for Energy Management and Sustainable Campus (COSCEM), Universiti Malaysia Perlis, 02600 Arau, Perlis, Malaysia
e-mail: khadijah@unimap.edu.my

W. A. A. Wan Ahmad
Centre of Excellence, Water Research and Environmental Sustainable Growth (WAREG), Universiti Malaysia Perlis, 02600 Arau, Perlis, Malaysia
e-mail: amneera@unimap.edu.my

M. Boboc
National Institute for Research and Development in Environmental Protection Bucharest (INCDPM), 294, Splaiul Independentei Street, 6th District, 060031 Bucharest, Romania

N. Mohamed Noor et al. (eds.), *Proceedings of the 3rd International Conference on Green Environmental Engineering and Technology*, Lecture Notes in Civil Engineering 214,
https://doi.org/10.1007/978-981-16-7920-9_2

was 1,518.8 tCO_2e/year while during COVID19 period, it was 1,071.5 tCO_2e/year (n = 10), marked as 29.5% reduction. Due to completeness and quality of data for contracted bus (monitoring period of years 2018, 2019 and 2020 as 12 months, 12 months, and 2 months, respectively), year 2019 was determined as the appropriate baseline year for setting the CO_2e reduction target due to COVID19 pandemic precedented year. In comparison to pre-COVID19 pandemic, almost 95%/year and 7%/year reductions of CO_2e were recorded for both Scope 1 and Scope 2, respectively. Comparing Scope 1 and 2, it was obviously observed that the purchased electricity consumption (Scope 2) was the predominant contributor to GHG emission at UniMAP campus by 78% despite of current pandemic influence and its reduction was indistinct (7%/year reduction). Thus, the reduction target in future should be venturing in energy savings and energy auditing in addition to carbon offsetting.

Keywords Carbon footprint · Greenhouse gases · SDG 13 · CO_2 equivalent

1 Introduction

Climate change and global warming have become worsen over the decades due to greenhouse gas (GHG) emissions build upon the Earth's atmosphere. There are a lot of GHG that are made up in the atmosphere. The main six greenhouse gases, which are the significant culprits of global warming, are carbon dioxide (CO_2), methane (CH_4), nitrous oxide (N_2O), and fluorinated gases such as chlorofluorocarbons (CFCs), hydrofluorocarbon (HFCs), and halon. Mainly, the implications of inaction are tremendous in enhancing global warming and result in the release of excessive greenhouse gases that could lead to increasing sea levels, melting glaciers, and climate change [1].

Malaysia is also one of the countries with the highest economic, social, and land development growth. In line with that, electricity demand, materials purchasing and procurement as well as waste generation have also increased for transport, industry, and domestic resources. This is driven by the high population and high demand in Malaysia [2]. According to the BP Statistical Review of World Energy 2019 research, Malaysia's CO_2 emissions amounted to 250.3 million tons for 2018, up from 241.6 million tons in 2017. The primary sources of the emissions were energy consumption, transportations, and municipal solid waste that ends up in landfills [3].

Considering the advantages and relevance of energy efficiency to reduce GHGs emission, many institutions give excellent examples of how to fight climate change by committing to zero carbon emissions. These include behavioral or attitude changes in management, green procurement, and waste management. Goal 13 of the Sustainable Development Goals (SDG) specifically focuses on addressing climate change, in which one of its objectives is to increase awareness among institutions and individuals about climate change mitigation and improve their potential to do so [4]. Universities have been at the forefront of solving environmental sustainability and global warming challenges through sustainable energy use, energy conservation initiatives and clean energy technologies [5]. COVID19 pandemic however has affected the normal operations of most of organizations in Malaysia whereby the energy and transportation consumption has marked a 20–45% reduction in electricity consumption at workplace and 10–20% curtailment in transportation usage. Due to this, baseline year setting shall be relooked to represent the actual scenario, waste, and energy consumptions at institutions. Current GHG and energy audits in response to COVID19 new norms shall also be offsetting this anomaly as it may underestimate the monitoring results.

University Malaysia Perlis (UniMAP) is committed to reduce the GHG emission associated with campus activities as aligned to combat climate change and global warming. To support the government's aim, this research is to manage and develop university greenhouse gas inventory at UniMAP from the year 2018 until 2020, as to set the baseline year for monitoring its carbon footprint as well as planning for carbon offsetting programs. This will be due by assessing the carbon emission from two scopes, which were scope 1 and scope 2. The aim of the study is clearly to reduce the environmental effects from campus activities which could potentially contribute towards high GHG emissions.

2 Materials and Method

2.1 Greenhouse Gases Quantification

The GHG inventory was assessed using four basic steps, GHG inventory design, data collection, emissions calculation using appropriate emission factors (EF) as well as data analysis and result interpretation. To help describe the sources of direct and indirect emissions, GHG emissions from university's activities are categorized into two scopes, which are scope 1 (direct emission) and scope 2 (indirect emission) as outlined by the ISO14064 and Department for Environment, Food and Rural Affairs, DEFRA documents [6]. Activity under scope 1 was recognized as the transportation owned and controlled by the university. While scope 2 covers GHG emissions from energy consumption (purchased electricity). Table 1 shows the data collection activities, unit and sources for each scope. For Scope 1 (direct emission), the fuel consumption data for UniMAP owned vehicles were sourced from the manual logbook at the Department of Development UniMAP (Transportation and Operation Unit) while the data collection of contracted bus was supplied as travelling distance, obtained from the online tracking system developed by the Centre of Excellence for Advanced Sensor Technology (CEASTech) UniMAP (https://www.fleet.ceastech.com/). Purchased electricity consumption for Scope 2 (indirect emission) was obtained in unit kWh from the Tenaga Nasional Berhad (TNB) National Grid Source, conserved by the Department of Development UniMAP.

Estimation of carbon emission regarding transportation usage applies different emission factors and unit either fuel-based or distance based on data availability. As in this study, the emission factors for the transportation type and electricity are

Table 1 Data collection details and provider

Scope	Activity	Collected data (unit)	Data provider
1	Direct emission from activities transportation owned by UniMAP	Fuel consumption (L)	Department of development UniMAP (transportation and operation unit)
	Direct emission from activities transportation controlled by UniMAP	Travelled distance if UniMAP Third Party Contracted Buses (km)	Tracking system developed by CEASTech
2	Indirect emissions associated with electricity purchased	Purchased electricity consumption from TNB National Grid Sources (kWh)	Department of development UniMAP/COSCEM

Table 2 Emission factors for different activities

Activity data	Unit	Emission conversion factor 2018 ($kgCO_2e$)	Emission conversion factor 2019 ($kgCO_2e$)	Emission conversion factor 2020 ($kgCO_2e$)
Petrol[a]	Litres	2.3053	2.31495	2.31467
Diesel[a]	Litres	2.68779	2.68697	2.68787
Bus[a]	Passenger.km	0.12007	0.12076	0.1195
Purchased electricity[b]	kWh	0.694	0.694	0.694

[a]Department of Food and Rural Affair (DEFRA)
[b]Sustainable Energy Development Authority Malaysia (SEDA)

provided by the Department of Food and Rural Affair (DEFRA) as it a common factor for several countries, including in Asia [6, 7] and Sustainable Energy Development Authority Malaysia (SEDA), respectively, as shown in Table 2.

To assess the differences in CO_2e emissions during pre-COVID19 and COVID19 pandemic, data was calculated in rate unit (tCO_2e/year), averaging from January 2018–February 2020 for pre-COVID19 period ($n = 26$) and March 2020–December 2020 ($n = 10$) for COVID19 period.

All conversion factors shown in Table 2 are expressed in 'kilograms of carbon dioxide equivalent of *Y* per *X* (kg CO_2e of *Y* per *X*), where *Y* is the emitted gas and *X* denotes the unit activity. A unit of CO_2e is a universal unit of measurement for the global warming potential (GWP) of greenhouse gases (GHGs), which is expressed as the GWP of a single unit of carbon dioxide [6]. The global warming potentials (GWPs) used in the computation of CO_2e are based on the Fourth Assessment Report (AR4) of the Intergovernmental Panel on Climate Change (IPCC) over a 100-year timeframe. The general quantification of GHG emissions (unit) is given by the following Eq. (1) [6]:

$$\text{GHG emissions} = \text{Activity data} \times \text{Emission Conversion Factor} \tag{1}$$

While for the Scope 1, the following equations were used for both UniMAP owned vehicles (2) and contracted bus (3), extended from [6]:

$$\text{GHG emissions, } kgCO_2e = \sum_i^n V_{fuel} \times EF_{1y} \tag{2}$$

$$\text{GHG emissions, } kgCO_2e = \sum_i^n d \times EF_{2y} \times number\ of\ passenger \tag{3}$$

where,

V_{fuel}	diesel's or petrol's volumes (litre)
d	travelling distance (km)
EF_1	UniMAP owned transportation's emission factor ($kgCO_2e$/litre)
EF_2	contracted bus' emission factor ($kgCO_2e$/km passenger)
y	year of emission factors
$i \ldots n$	types and number of transport

The purchased electricity on was quantified applying the following Eq. (4) as extended from [6]:

$$\text{GHG emissions, } kgCO_2e = \sum_i^n E \times EF_{3y} \tag{4}$$

where,

E	purchased electricity consumption (kWh)
EF_2	purchased electricity emission factor ($kgCO_2e$/kWh)

3 Result and Discussion

3.1 Annual Greenhouse Gas Emissions Generated

Figure 1a illustrates annual GHG emissions for years 2018, 2019, and 2020 which marked as 13,772.06 tCO_2e, 22,925.92 tCO_2e, and 14,190.3 tCO_2e, respectively. In comparison to pre-COVID19 pandemic event, a significant decrease in Scope 1 was observed from 414.21 tCO_2e/year to 17.96 tCO_2e/year (Fig. 1b) due to Movement Control Order (MCO) announced by the Malaysia government starting from March 2020 until end of the year. During the MCO, all transportation within UniMAP was interrupted. However, as seen in Scope 2, a slight reduction during COVID19 (1,053.49 tCO_2e/year) was observed as 7%/year decrement compared to during pre-COVID19 (1,130.71 tCO_2e/year) situation. However, purchased electricity was the

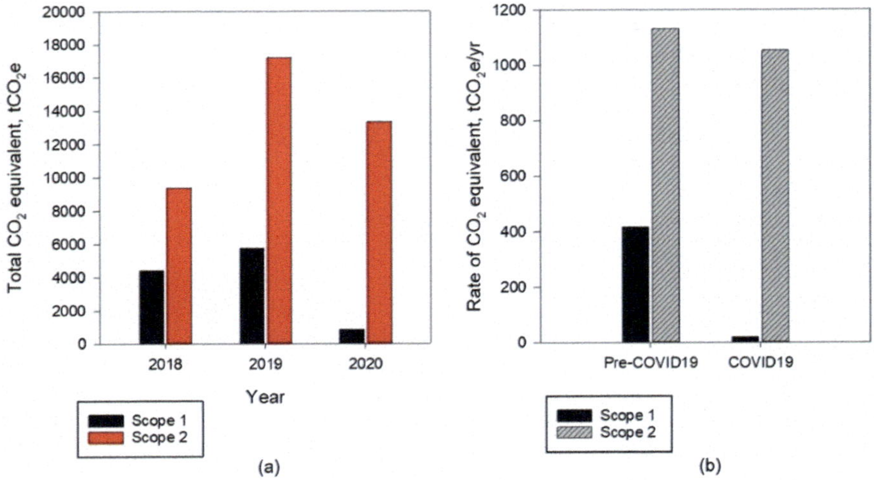

Fig. 1 **a** Cumulative GHG emissions of Scope 1 and 2 for the year 2018 (n = 12), 2019 (n = 12) and 2020 (n = 2); **b** Annual rate of CO_2e pre- and during COVID19 occurrences

highest contributor to GHGs emissions in UniMAP despite of COVID19 outbreak due to the nonstop services besides accommodating lots of UniMAP population and buildings unlike for transportation, where it was only operated to cater the students and staffs mobility.

3.2 Greenhouse Gas Emissions from Transportation Expenditure

The types of transportation used by university employees were divided into two categories: (a) UniMAP owned transportation and (b) UniMAP contracted buses, as the quantification of these means of transportation were applying different emission factors (Table 2). The transportation's inventory consisted of different type of fuel (diesel and petrol) used by the university's owned transportation fleet and mileages of contracted buses for student commuting (Table 3). The total cumulative GHG emission for year 2018, 2019 and 2020 were quantified as 4407.15 tCO_2e, 5713.09 tCO_2e, and 828.77 tCO_2e, respectively (Fig. 2).

This shows that the emissions from 2019 were higher than in years 2018 and 2020, and the lowest was in the year 2020. The primary factor affecting GHG emissions from transportation is the transportation's mileage and fuel consumption itself [7]. The fluctuating trends by months were influenced by the academic calendar, which in linear with the presence of students within the university. The months of January until March marked the semester's study weekdays while April until June was the

Table 3 Cumulative total CO_2e emissions for different fuel types and means of transportation

Fuel	Description	Emission for 2018 (tCO_2e)	Emission for 2019 (tCO_2e)	Emission for 2020 (tCO_2e)
Petrol	UniMAP owned transportation	277.58	297.91	164.19
Diesel	UniMAP owned transportation	216.77	219.96	93.51
Mileage	UniMAP contracted buses	3912.8	5173.22	571.07

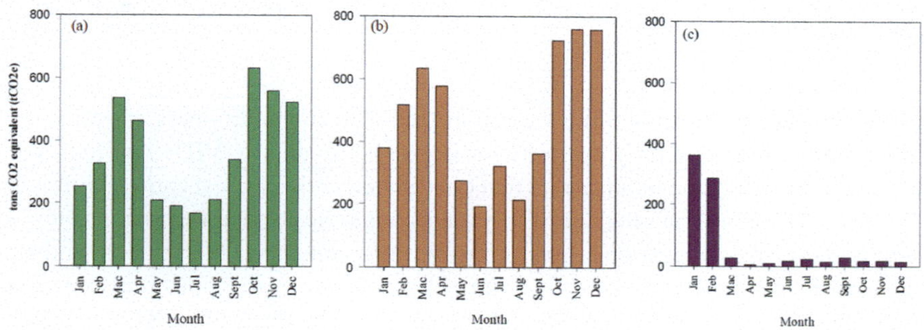

Fig. 2 GHG emissions from transportation for the year **a** 2018, **b** 2019 and **c** 2020

semester break. Table 3 shows that mileages from buses was the highest GHG emissions compared to fuel consumption. The emission of GHGs steadily increased from January until March 2018 and 2019 as shown in Fig. 2. During that month, transportation, especially buses, commuted students to their faculties at different places. The trends were starting to decrease in April. The emissions decreased in 2020 due to the COVID-19 outbreak. The government's new policies to work from home during the outbreak have reduced transportation and directly reduced fuel consumption and mileages.

3.3 *Greenhouse Gas Emission from Purchased Electricity Consumption*

Figure 3 shows the trend of GHG emissions from electricity consumption for the year 2018–2020. Total GHG emission for the year 2018 is 9364.9 tCO2e, while GHG emission for 2019 is 17212.83 tCO2e. The higher emission in 2019 was the result of the higher electricity consumption in the year 2019, influenced by the COVID19 situations. From the graph, the GHG emissions are high mainly in March, April, Mei, October, November and December in both years 2018 and 2019. The use of electricity increased as the classroom, hostel, café and other building were fully utilized by

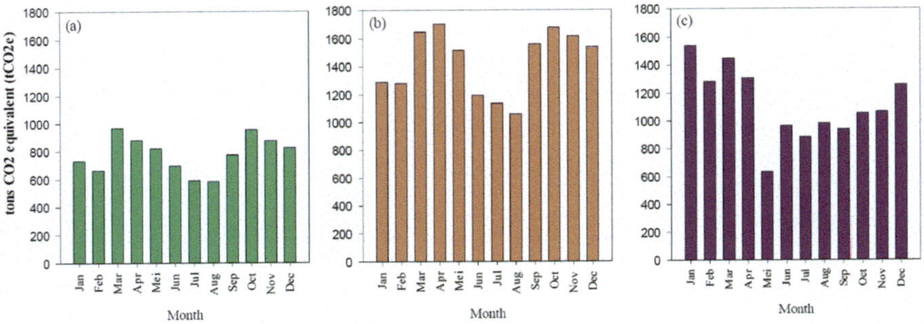

Fig. 3 GHG emissions from purchased electricity consumption for the year **a** 2018, **b** 2019 and **c** 2020

students and staffs. Most facilities were equipped with air conditioners, in which the chiller will require 24 h fully operated, during both pre- and COVID19 periods. Air conditioner is one of the main causes that contributed to the highest carbon emission [7]. This result also shows that electricity emissions were the main emission sources since it shows the highest emission compared to commuting. In previous studies, purchased electricity was found to be the main contributor to greenhouse gas emission on campus especially in hot and humid regions [7]. The trend of GHGs emissions for 2020 shows almost the same pattern as 2019, but the total emissions were less than the previous years due to the COVID19 movement control order. UniMAP has closed all the non-essential building and started implementing working from home. The emission begins to drop from March, April until July.

4 Conclusion

The result will be a baseline for the university to reduce GHG emissions, expressed by CO_2 equivalent emission. The total amount of CO_2e emitted over three years was measured to be 22,925.9 tCO2e. This total GHG emissions consist of scope 1 and scope 2 with percentages of 22% and 78%. From the data that has been analyzed, purchased electricity is the main contributor to the GHGs emissions in UniMAP that generated 39,933.24 tCO2e for those three years. It is followed by transportation for 10,955.02 tCO2e. There were increasing GHG emissions from year 2018 to 2019 and decreased in 2020 due to the Covid-19 crisis. Therefore, this study will support and encourage the campus community to be more sustainable as the concerns towards environmental issues and achieve the green campus initiatives. Thus, energy-saving management teams should be established to reassess the energy consumption and possibilities to reduce the CO_2e. Besides, the study on existing tree planting as carbon storage also need to be hastened.

Acknowledgements The author would like to acknowledge the Center of Sustainable Campus and Energy Management, Department of Development and Centre of Excellence for Advanced Sensor Technology (CEASTech), UniMAP for their support in data collection and collation.

References

1. Kamyab H, Lim JS, Khademi T, Ho WS, Rahmalan A, Haslenda H, Ho CS, Keyvanfar A (2015) Greenhouse gas emission of organic waste composting: a case study of Universiti Teknologi Malaysia green campus flagship project. J Teknol 74(4):113–117. https://doi.org/10.11113/jt.v74.4618
2. Sharif SSA, Muhammad RR, Azahan A (2019) Tren Penggunaan Tenaga Elektrik dan Pembebasan Gas Rumah Hijau di Malaysia (The Trend of Electricity Consumption and Greenhous Gases Emissions in Malaysia). Asian J Environ 3(1):39–45
3. BP, Full report—BP Statistical Review of World Energy 2019
4. Pablo Y, Arijit S, Marcia V (2020) Carbon footprint estimation in a university campus: evaluation and insights. Sustain 12(1). https://doi.org/10.3390/su12010181
5. Abdul- I, Ho C (2015) Realizing low carbon emission in the university campus towards energy sustainability. Open J Energy Eff 4:15–27. https://doi.org/10.4236/ojee.2015.42002
6. Department for Environment, Food and Rural Affairs (DEFRA) (2009) Guidance on How to Measure and Report Your Greenhouse Gas Emissions
7. Nor Izana MS, Mohd S, Wan NA, Wan Ali@Yaacob, Norizan MA, Siti RMS (2017) Carbon footprint assessment at Universiti Teknologi Mara Seri Iskandar Campus, Malaysia. Malaysian J Sustain Environ (MySE) 2(1):59–72. https://doi.org/10.24191/myse.v2i1.5576

Changes of Carbon in a Hardwood Forest by Forecasts Using a Forest Model

Norbert Bara, György Deák, Lucian Laslo, Anda Rotaru, Monica Matei, Madalina Boboc, Natalia Enache, and Sara Yasina Yusuf

Abstract Carbon sequestration proved to be one of the best strategies in climate mitigation processes and forest have the capacity to sequester and store important amounts of carbon. In this study the carbon sequestration capacity of a hardwood forest in Eastern Romania is analyzed by using the European Information Scenario Model through a business-as-usual perspective. The results showed that under the actual management practices the forested area will increase in time. The volume of harvested wood will fluctuate, negatively affecting the increment and growing stocks. Although, the hardwood will store a total of 939.16 Gg C during the entire simulation, the sequestration capacity will decrease with 11%. It is important to mention that the input data is used from the management plan of the researched area. This method increases the confidence of the results, giving an insight for future research to define the best management practices in the studied area.

Keywords Carbon sequestration · Forest · EFISCEN · Biomass · Management practices

1 Introduction

In the context of a greenhouse gas emissions balancing strategy, carbon sequestration by forests is one of the most cost-effective method [1]. Forests is one of the largest terrestrial carbon pools, storing approximately 40% of belowground and 80% of aboveground carbon [2, 3]. Sustainable forestry can stimulate the forests to store more carbon [4]. In order to develop strategies for sustainable management of forests different methods were developed and used. The European Information Scenario

N. Bara (✉) · G. Deák · L. Laslo · A. Rotaru · M. Matei · M. Boboc · N. Enache
National Institute for Research and Development in Environmental Protection Bucharest (INCDPM), 294, Splaiul Independentei Street, 6th District, 060031 Bucharest, Romania

S. Y. Yusuf
Sustainable Environment Research Group (SERG), Centre of Excellence Geopolymer and Green Technology (CEGeoGTech), Universiti Malaysia Perlis, 01000 Kangar, Perlis, Malaysia
e-mail: sarayasina@unimap.edu.my

N. Mohamed Noor et al. (eds.), *Proceedings of the 3rd International Conference on Green Environmental Engineering and Technology*, Lecture Notes in Civil Engineering 214,
https://doi.org/10.1007/978-981-16-7920-9_3

Model (EFISCEN) was developed for such a purpose [5–7]. It is used to provide forecasts, especially for European forests, by generating scenarios.

In this study are presented the results of a simulation developed with EFISCEN model under a business-as-usual premise for the management of a hardwood forest. Therefore, the aim of this research is to contribute in developing a methodology for refining and overseeing Romanian forest management plans, as well as to enhance reporting methodologies for National Greenhouse Gas Inventories and Forest Reference Levels.

2 Materials and Method

The C stocks fluctuate in forests due to the several processes that are happening in these ecosystems and for understanding the carbon dynamics it is crucial to use accurate data [8]. The input data is built based on the information from the management plan of the studied area. Therefore, the data was stratified as having one region: the forest district with five subunits of production (SUP) and one tree group of species composed of 18 hardwood tree species. The area, volume, increment, rotation lengths as well as the extracted timber volume were used based on the data from the aforementioned source. The climatic parameters were obtained using the cumulative degree days (DD) and drought index (DI), both of which were calculated during the vegetative period, from the same source. The DD was calculated by adding the average temperatures for the months of May, June, July, August, and September of 2017. The quantity of precipitation was subtracted from the amount of potential evapotranspiration. The simulation timeframe has 50 years, starting from 2015 and ending in 2065.

3 Result and Discussion

The results of the simulation showed that the current management practices are stimulating forest area growth (Fig. 1). In order to understand the changes that will be presented in this paper, it is essential to observe the volume of wood that is extracted trough the management practices (Fig. 2).

There are presented two basic management practices: thinning and final felling. It can be observed that the highest volume extracted through thinnings is at the beginning of the simulation and that the volume from fellings is increasing by the end of it. This is because the fellings are conducted among the oldest trees. The aforementioned practices influenced the area development. Thus, the area covered by hardwood tree species at the beginning of the simulation was 2.675 ha and increased throughout the entire period, reaching by the end of it at 3.692 ha. Although the increment is used as input parameters, the model simulates the changes of it. Thus, the increment volume was observed to increase during the first ten years of the

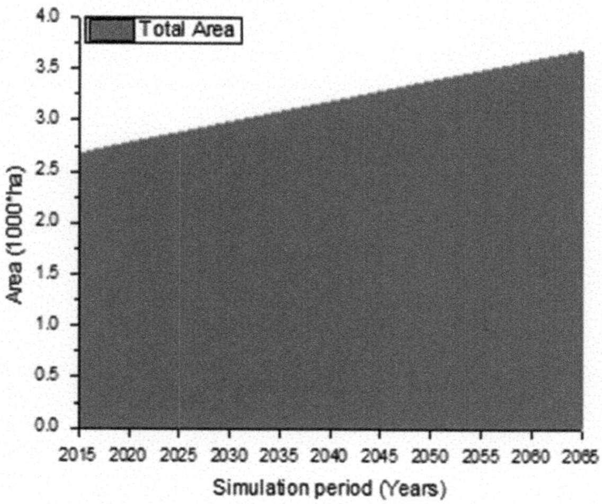

Fig. 1 Area changes in the simulation

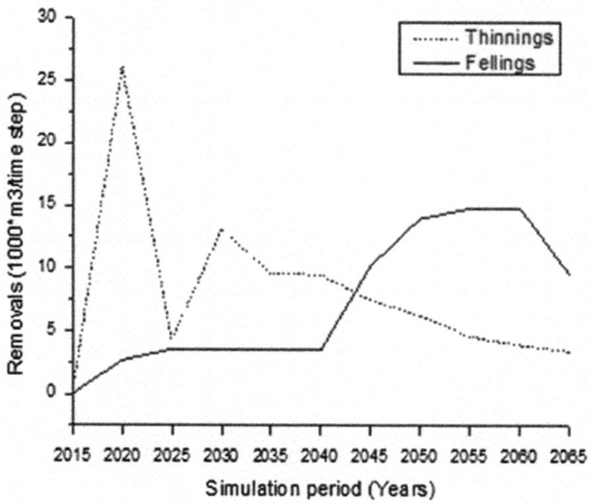

Fig. 2 Simulated volume of harvested wood from thinnings and fellings

simulation, reaching its maximum of 1.42 m3/ha/year. After that it decrease steadily and by the end of the simulation decreased to 0.81 m3/ha/year (Fig. 3). Despite fluctuations can be observed in the growing stock, its volume will decrease by the end of the simulation (Fig. 4).

The evolution of the total carbon content in hardwood biomass, as well as in each SUP was forecasted (Fig. 5). The highest volume of carbon, 89.44 Gg C, was observed to be sequestered in the first-time step of the simulation, afterwards is fluctuant with a decreasing trend. By the end of the simulation the carbon content decreased to 80.75 Gg C. The total sequestered carbon is 939.16 Gg C.

However, differences were observed between the SUPs. The highest volume of carbon sequestered by the trees is in SUP Q (409.89 Gg C), followed by SUP A

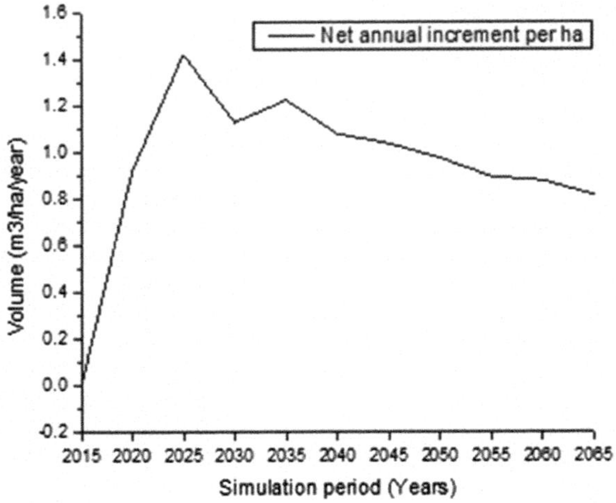

Fig. 3 Changes in the net annual increment

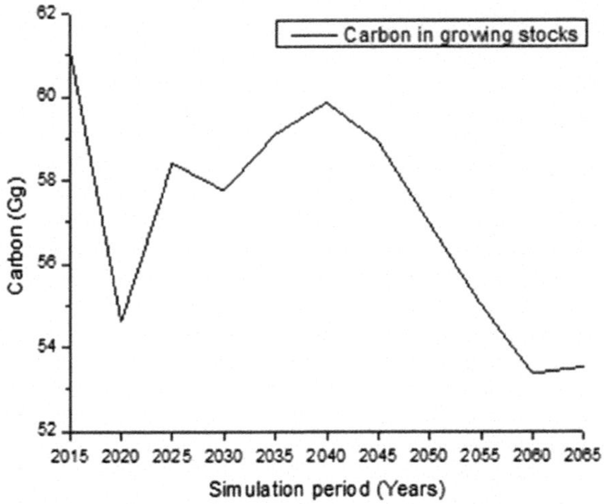

Fig. 4 Simulated changes in the growing stock

(330.17 Gg C), SUP M+E (102.6 Gg C), SUP O (62.78 Gg C) and SUP K (33.69 Gg C). Although management practices are also conducted in SUP Q, the carbon content is increasing due to the decreased volume of harvested wood, which is lower than in other SUP. This suggests that decreasing the intensity of management practices ca stimulate the carbon sequestration capacity of the hardwood types of trees.

By conducting management practices carbon is removed from the forest (Fig. 6). The total carbon removed from thinnings was 29.45 Gg C, while in the final fellings was 26.72 Gg C.

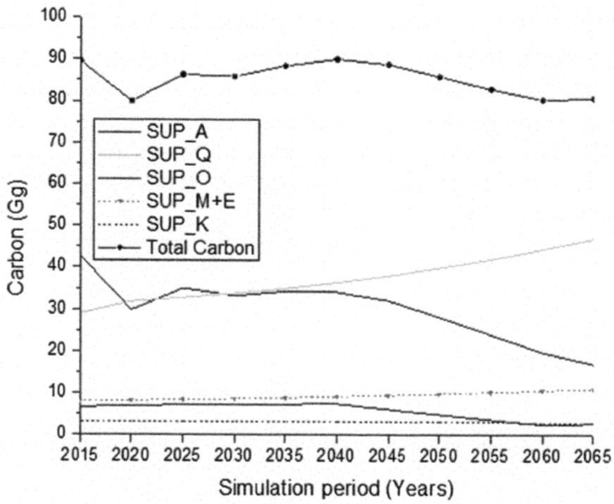

Fig. 5 Changes in carbon stocks. Comparison between subunits of production and the total volume

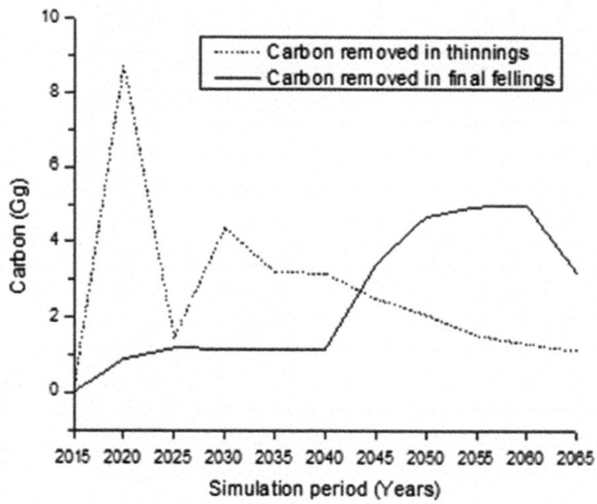

Fig. 6 Carbon removed by management practices

4 Conclusions

This study examines the changes occurred among hardwood trees as a result of the management techniques used in the study region. The results of the simulations were obtained by using the European Information Scenario Model. The input data was used from the management plan of the researched area. The results revealed that under current management techniques, the wooded area will grow over time. The amount of harvested wood will vary, while increment and growing stock volumes will decrease. Although, there are two subunits of production where the carbon content is increasing, the total volume of sequestered carbon is decreasing by the end of the simulation. Therefore, it is crucial to develop accurate scenarios in order to compare the outcomes for the identification of the most sustainable management practices.

Acknowledgements This paper was developed through Nucleu Programme—Contract number 39N/2019 carried out with the Romanian Ministry of Research and Innovation support, within the project "Evolution of sturgeon populations and Danube ichthyofauna in the context of the Danube course alterations from recent decades".

References

1. Richards KR, Stokes C (2004) A review of forest carbon sequestration cost studies: a dozen years of research. Clim Change 63(1):1–48
2. WeiY YuD, Lewis BJ, Zhou LI, Zhou W, Fang X, Zhao W, Wu S, Dai L (2014) Forest carbon storage and tree carbon pool dynamics under natural forest protection program in northeastern China. Chin Geogra Sci 24(4):397–405
3. Bousquet P, Peylin P, Ciais P, Le Quéré C, Friedlingstein P, Tans PP (2000) Regional changes in carbon dioxide fluxes of land and oceans since 1980. Science 290(5495):1342–1346. https://doi.org/10.1126/science.290.5495.1342
4. Sedjo RA (2001) Forest carbon sequestration: some issues for forest investments. No. 1318-2016-103352
5. European Forest Institute (2016) EFISCEN: European Forest Information SCENario model, version 4, European Forest Institute, Joensuu, Finland. http://efiscen.efi.int. Accessed 15 Nov 2020
6. Sallnäs O (1990) A matrix growth model of the Swedish forest. No. 183
7. Bostedt G, Mika M, Peichen G (2016) Increasing Forest biomass supply in Northern Europe-Countrywide estimates and economic perspectives. Scand J For Res 31(3):314–322. https://doi.org/10.1080/02827581.2015.1089930
8. Bara N, Rotaru A, Laslo L, Matei M, Boboc M, Coman V, Voicu M, Enache N, Deak Gy (2020) Carbon Dynamics Projection for the Quercus petraea and Robinia pseudoacacia Species-Case Study. IOP Conf Ser Earth Environ Sci 616(1):012007

Comparative Study of Sulfur Dioxide Removal Using Mesoporous Silica KCC-1 and SBA-15

Muhammad Adli Hanif, Naimah Ibrahim, Khairuddin Md Isa, Masitah Hasan, Tuan Amran Tuan Abdullah, and Aishah Abdul Jalil

Abstract Sulfur dioxide (SO_2) emitted into the atmosphere by fossil fuel burning in the industries posed significant negative effects on humans and the environment. SO_2 removal performance of two mesoporous silica: KCC-1 and SBA-15, are compared through breakthrough experiments on a lab-scale fixed bed reactor. The mesoporous silicas were characterized via nitrogen (N_2) adsorption-desorption isotherm and field emission scanning electron microscopy (FESEM). KCC-1 demonstrates characteristics of capillary condensation and non-uniform slit-shaped pores while SBA-15 displays characteristic of a narrow range of mesopores with minimal network effects. Surface area, total pore volume and average pore diameter of KCC-1 are significantly greater than SBA-15 due to the presence of dendrimeric fibrous morphology. Under tested conditions, SO_2 adsorption capacities of KCC-1 and SBA-15 are 614.01 mg/g and 274.64 mg/g, respectively. Superior performance by KCC-1 can be attributed to better accessibility of SO_2 towards the active sites due to higher surface area provided by the dendrimer fibers.

Keywords SO_2 removal · Dry flue gas desulphurization · Mesoporous silica · KCC-1 · SBA-15

M. A. Hanif · N. Ibrahim (✉) · K. Md Isa · M. Hasan
Universiti Malaysia Perlis, 02600 Arau, Perlis, Malaysia
e-mail: naimah@unimap.edu.my

M. A. Hanif
e-mail: adlihanif@studentmail.unimap.edu.my

K. Md Isa
e-mail: khairudin@unimap.edu.my

M. Hasan
e-mail: masitah@unimap.edu.my

T. A. Tuan Abdullah · A. A. Jalil
Universiti Teknologi Malaysia, 81310 UTM Johor Bahru, Johor, Malaysia
e-mail: tuanamran@utm.my

A. A. Jalil
e-mail: aishahaj@utm.my

N. Mohamed Noor et al. (eds.), *Proceedings of the 3rd International Conference on Green Environmental Engineering and Technology*, Lecture Notes in Civil Engineering 214,
https://doi.org/10.1007/978-981-16-7920-9_4

1 Introduction

Despite current enhancement of green and renewable technology, the majority of industries are still dependent on the usage of fossil fuels as a source of energy due to its high energy density. The combustion of fossil fuels resulted in the release of sulfur dioxide (SO_2) gas into the atmosphere. SO_2 emitted can be carried over long distances due to its long residence time (3–5 days), depending on the meteorological conditions. SO_2 is known as a precursor of acid rain which consequently resulted in acidification of water bodies, damage of crops, corrosion of buildings etc. [1]. The effects towards human beings have also been reported which include respiration difficulties, pulmonary function changes and worsened cardiovascular diseases [2].

Siliceous materials are classified as materials containing silicon dioxide (SiO_2) as the principal constituent with low composition of metallic oxides and alkaline materials. Waste materials such as fly ash, rice husk ash and oil palm ash; and natural materials like quartz, pumice, clinoptilolite and metakaolin are some of the examples of siliceous materials. Meimand et al. [1] and Czuma et al. [3] studied the performance of commercial clinoptilolite and coal combustion fly ash zeolites in SO_2 removal where adsorption capacities of 21.9 mg/g and 107.62 mg/g were obtained, respectively. Compared to other sorbents, these performances can be deemed as weak which can be majorly ascribed to the small pore size of these sorbents.

Mesoporous silica (MS) is a synthetic siliceous material with porosity in the range of 2–50 nm. MS possesses unique properties of ordered pore structures, variable morphology and high surface area contributed by nano-templating approach during its fabrication. MS such as MCM-41, SBA-15 and KIT-6 have been successfully employed in a wide range of applications such as catalysis, air cleaning, waste water treatment, drug delivery system etc. [4]. KCC-1 is a recent addition to the MS family where unlike its predecessors, KCC-1 has a fibrous morphology with an open and flexible framework which results in easily accessible high surface areas [5]. The utilization of KCC-1 in air cleaning is seldom reported and is only focused on carbon dioxide methanation or adsorption [6, 7].

This study is focused on comparing the performance of two of these MS, i.e., laboratory-synthesized KCC-1 and commercial SBA-15 in removing SO_2 from the atmosphere. The study was conducted on a lab-scale fixed-bed reactor using synthetic gas mixtures. The properties of these two silica such as surface area, porosity and morphology were characterized to determine their importance in the SO_2 removal.

2 Materials and Method

2.1 Materials

Fibrous KCC-1 was prepared using microwave (MW) assisted hydrothermal methods. Cetylpyridinium bromide and urea (0.22: 0.83) was mixed in a distilled water and added into a solution mixture of tetraethyl orthosilicate, cyclohexane and 1-pentanol (1: 23.08: 1.08) which was previously stirred in a Teflon bottle. The resulting mixture was stirred at room temperature for 30 min and exposed to intermittent MW irradiation at 400 W for 4 h. The mixture was cooled, centrifuged and rinsed with water, oven-dried overnight at 383 K and finally calcined at 823 K for 6 h. Commercial SBA-15 was obtained from Sigma-Aldrich and used as received.

The surface area and porosity of the samples was determined via BET N_2 adsorption–desorption using Tristar 3000 instrument at 77.35 K. The samples were degassed at 393 K for 3 h prior to characterization. Surface morphology of the samples pre-coated with platinum were determined using JEOL JSM-7600F Field Emission Scanning Electron Microscopy (FESEM) at magnification of 20,000 times.

2.2 Method

SO_2 breakthrough experiment was conducted on a lab-scale fixed bed quartz reactor (length = 25 cm, internal diameter = 8.8 mm) using special-mix gas of 0.3% SO_2/N_2. Temperature and inlet SO_2 concentration were controlled by a tubular furnace with programmable controller and mass flow controller, respectively. Prior to each experiment, the sample was pre-treated with nitrogen (N_2) gas at 150 °C for 1 h to remove any impurities and the reactor is then cooled down to 313 K. Upon achieving stable temperature, the simulated flue gas (SO_2 concentration = 1500 ppm, flow rate = 200 mL/min) was passed through the reactor and the outlet concentration was continuously measured using a portable gas analyzer. Due to the limitation of the instruments, the breakthrough experiments were only conducted up to 2 h.

The sorption bed is considered to have achieved breakthrough if 5% of the inlet SO_2 concentration was detected by the analyzer in the outlet stream. The adsorption capacity of the sample was calculated at $C/C_0 = 0.80$ using the following equation [8], where q is adsorption capacity (mg/g), C_0 and C_A is SO_2 concentration at inlet and at specific time (mg/L), respectively; Q_f is gas flow rate (L/min), y_t is gas molar fraction and m_c is mass of the activated carbon (g).

$$q = \frac{C_0 Q_f y_t}{m_c} \int_0^{\infty} \left(1 - \frac{C_A}{C_0}\right) dt \tag{1}$$

3 Result and Discussion

3.1 Characterization of Sample

N_2 adsorption-desorption isotherm and the surface area and porosity of KCC-1 and SBA-15 mesoporous silica are shown in Fig. 1 and Table 1, respectively. Both KCC-1 and SBA-1 show Type IV (a) isotherm confirming their mesoporous characteristics. KCC-1 exhibits H3 hysteresis demonstrating a characteristic of capillary condensation and non-uniform slit-shaped pores while SBA-15 displays H1 hysteresis which is a characteristic of a narrow range of mesopores with minimal network effects [9]. KCC-1 displays greater total pore volume and average pore size compared to SBA-15. Unlike SBA-15 where its surface area is dependent on the pores, KCC-1 possesses significantly higher surface area due to the presence of dendrimeric silica fibers and their respective channels [5].

FESEM micrographs of KCC-1 and SBA-15 are shown in Fig. 2. KCC-1 possesses well-defined fibrous morphology of colloidal nanospheres made by three-dimensional arrangement of the dendrimeric fibers with diameter of 240–1160 nm. Commercial SBA-15 used in this study displays a cluster of monodisperse longitudinal short rods morphology with length between 370–650 nm.

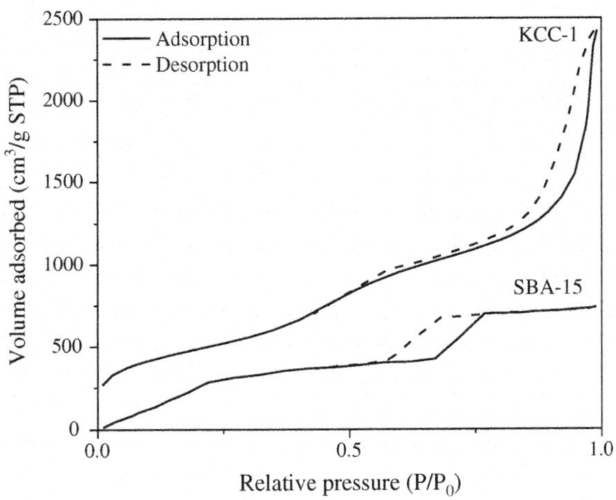

Fig. 1 N_2 adsorption-desorption isotherm of KCC-1 and SBA-15

Table 1 Surface area and pore properties of KCC-1 and SBA-15

Sample	KCC-1	SBA-15
Surface area (m^2/g)	1765.98	778.09
Total pore volume (cm^3/g)	3.86	0.84
Average pore size (nm)	7.7	4.3

Fig. 2 FESEM micrographs of KCC-1 and SBA-15

3.2 SO_2 Removal

SO_2 adsorption performance of KCC-1 and SBA-15 are shown in Fig. 3 and the details of the breakthrough experiment are summarized in Table 2. Both sorbents

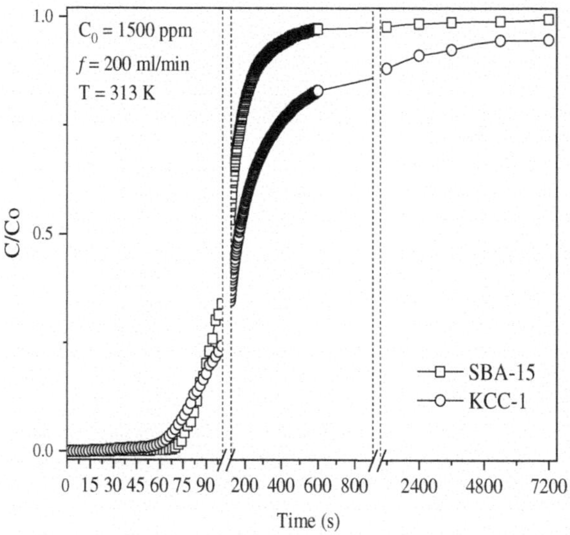

Fig. 3 SO_2 breakthrough curves of KCC-1 and SBA-15

Table 2 Breakthrough time and adsorption capacity of KCC-1 and SBA-15

Sample	KCC-1	SBA-15
Breakthrough time (s)	70	77
Time (s) at calculated adsorption capacity $C/C_0 = 0.8$	500	200
Adsorption capacity (mg/g) at $C/C_0 = 0.8$	614.01	274.64

achieved breakthroughs at similar times, approximately 70 s. Beyond this point, KCC-1 exhibits gradual increase of SO_2 outlet concentration while a steep increase was observed for SBA-15, resulting in saturation of SBA-15 after 4800 s. The saturation was not observed for KCC-1 with $C/C_0 = 0.95$ until nearly 7200 s.

The adsorption capacity of KCC-1 is 614.01 mg SO_2/g sorbent, which is 2.24 times higher compared to the adsorption capacity achieved by SBA-15. Based on the characterization of both samples, superior performance of KCC-1 can be ascribed to its higher surface area due to their dendrimeric fibrous morphology which resulted in enhanced accessibility for SO_2 gas towards the active sites [10]. The adsorption capacity attained can be further improved by introduction of additives e.g. metal; which may provide greater number of active sites lacking in a typical MS [11].

4 Conclusion

Comparative study of mesoporous silica KCC-1 and SBA-15 performance on SO_2 removal was performed on a lab-scale fixed bed reactor. Characterization of the sorbents exposed that KCC-1 possesses greater surface area, total pore volume and average pore size compared to SBA-15 due to the presence of dendrimeric silica fibers and their respective channels. These superior properties lead to enhanced accessibility of SO_2 towards the active sites. SO_2 adsorption capacity obtained by KCC-1 is 614.01 mg SO_2/g sorbent, 2.24 times higher than SBA-15. Introduction of additives may result in further enhancement of desulfurization performance by mesoporous silica KCC-1.

References

1. Meimand MM, Javid N, Malakootian M (2019) Adsorption of sulfur dioxide on clinoptilolite/nano iron oxide and natural clinoptilolite. Health Scope 8(2):e69158
2. WHO (2005) WHO Air quality guidelines for particulate matter, ozone, nitrogen dioxide and sulfur dioxide
3. Czuma N, Zarębska K, Baran P (2016) Analysis of the influence of fusion synthesis parameters on the SO_2 sorption properties of zeolites produced out of fly ash. E3S Web Conf 10:23–26. https://doi.org/10.1051/e3sconf/20161000010
4. Hanrahan JP, O'Mahony TF, Tobin JM, Hogan JJ (2017) Mesoporous silica and their applications. https://www.sigmaaldrich.com/technical-documents/articles/materials-science/renewable-alternative-energy/mesoporous-silica.html. Accessed 17 July 2021
5. Bayal N, Singh B, Singh R, Polshettiwar V (2016) Size and fiber density controlled synthesis of fibrous nanosilica spheres (KCC-1). Sci Rep 6:1–11. https://doi.org/10.1038/srep24888
6. Shahul Hamid MY, Triwahyono S, Jalil AA, Che Jusoh NW, Izan SM, Tuan Abdullah TA (2018) Tailoring the properties of metal oxide loaded/KCC-1 toward a different mechanism of CO_2 methanation by in situ IR and ESR. Inorg Chem 57(10):5859–5869. https://doi.org/10.1021/acs.inorgchem.8b00241

7. Nasir MSRM, Khairunnisa MP, Jusoh NWC, Jalil AA (2020) Characteristics and carbon dioxide adsorption performance of amine-impregnated KCC-1 with different loading ratio. IOP Conf Ser Mater Sci Eng 808:12031. https://doi.org/10.1088/1757-899X/808/1/012031
8. Hanif MA, Ibrahim N, Md Isa K, Tuan Abdullah TA, Abdul Jalil A (2021) Sulfur dioxide removal by mesoporous silica KCC-1 modified with low- coverage metal nitrates. Mater Today Proc 47(6):1323–1328. https://doi.org/10.1016/j.matpr.2021.02.807
9. Thommes M et al (2015) Physisorption of gases, with special reference to the evaluation of surface area and pore size distribution (IUPAC Technical Report). Pure Appl Chem 87:1–19. https://doi.org/10.1515/pac-2014-1117
10. Polshettiwar V, Cha D, Zhang X, Basset JM (2010) High-surface-area silica nanospheres (KCC-1) with a fibrous morphology. Angew Chemie Int Ed 49(50):9652–9656. https://doi.org/10.1002/anie.201003451
11. Li X, Zhang L, Zheng Y, Zheng C (2015) SO_2 absorption performance enhancement by ionic liquid supported on mesoporous molecular sieve. Energy Fuels 29(2):942–953. https://doi.org/10.1021/ef5022285

Monsoonal Impact of the Precursors Toward Tropospheric Nocturnal Ozone O_3 (Night-Time Ozone) During a Decadal Observation in Kuching, Sarawak, Malaysia

Mohd Hakkim Firdaus Hamzah, Ku Mohd Kalkausar Ku Yusof, Nurul Adyani Ghazali, Samsuri Abdullah, Amalina Abu Mansor, and Azman Azid

Abstract Malaysia Borneo is known for diverse rainforests and has a lower population density. Changes in pollutant profiles are anticipated, as cities grow, and commercial-industrial activities expend. The aim of this study is to compare the temporal variation of nocturnal O_3 and understanding the relationship between O_3 and the precursor (NO_2, NO_x, CO and CH_4) between Northeast Monsoon (NEM) and Southwest Monsoon (SWM) in Kuching, Sarawak. The data sources for O_3 and its precursors (NO, NO_2, NO_x, CO, NmHC) which is consisted for 108 months consisting of 216 h (24 h × 9 years) of observation from January 2006 until December 2015 were acquired from the Air Quality Division, Malaysian Department of Environment (DOE). Principal component analysis (PCA) was conducted in this study. From the analysis, the minimum value recorded as much as 62.13% increment since 2015. However, the maximum values showed different as the maximum of O_3 level decreased from 20.50 ppbv (2006) to 11.18 ppbv (2015). Nocturnal O_3 showed similarities pattern between both monsoon as the eclipse overlapped within each other. Nevertheless, SWM has significant impact on nocturnal O_3 compared to NEM. The O_3 have linear relation with NO_2, NO_x, CO, and NO in descending sequence however, CH_4 was the only parameter which is does not influence on O_3 in this study.

M. H. F. Hamzah · K. M. K. K. Yusof (✉) · N. A. Ghazali · S. Abdullah · A. A. Mansor
Universiti Malaysia Terengganu, 21300 Terengganu, Malaysia
e-mail: kukausar@umt.edu.my

N. A. Ghazali
e-mail: nurul.adyani@umt.edu.my

S. Abdullah
e-mail: samsuri@umt.edu.my

A. Azid
Universiti Sultan Zainal Abidin, Gong Badak Campus, 21300 Terengganu, Malaysia
e-mail: azmanazid@unisza.edu.my

N. Mohamed Noor et al. (eds.), *Proceedings of the 3rd International Conference on Green Environmental Engineering and Technology*, Lecture Notes in Civil Engineering 214,
https://doi.org/10.1007/978-981-16-7920-9_5

Keywords Tropospheric ozone · Principal component analysis · Borneo · Monsoon · Ozone precursor

1 Introduction

Ozone (O_3) consists of two layers known as stratosphere (primary production) and troposphere (secondary production). Stratosphere ozone consists approximately 90% which are crucial in filter ultraviolet (UV) radiation, especially UV-b and UV-c as it is harmful towards humans, living organisms, and ecosystems. Besides, the absorption of UV-b by ozone in the stratosphere will regulate and maintain the global temperature of Earth's surface and atmosphere. Tropospheric ozone produced by pollution consists only of 10% of the ozone layer. In addition, it plays a central role in oxidation via trace gases that are chemically and climatically important, thus regulating its lifetime in the atmosphere. In tropospheric chemistry, numerous studies stated that tropospheric O_3 plays an important role due it is one of the key precursors of hydroxyl radical (OH), which regulates the oxidizing capacity in the lower atmosphere [1]. A variety of complex photochemical reactions between ozone precursors, such as nitrogen oxides (NO_x) or volatile organic compounds (VOCs) and solar radiation (photochemical), accelerated the O_3 production [2, 3]. During the day, UV-b is vital for initiating interactions among the O_3 precursors that lead to the formation of tropospheric O_3 [4]. At night, on the other hand, there is no UV-b to make ozone. Instead, nocturnal tropospheric O_3 is depleted through titration reactions with nitric oxide (NO), which produces nitrogen dioxide (NO_2) and oxygen (O_2). The NO_2 generated can react with O_3 to form nitrate (NO_3) and oxygen (O_2) [5]. As a result, the presence of nitrogen oxides (NO_x) at night should have a large reduction in the concentrations of nocturnal tropospheric O_3 in a given area. However, due to the inefficiency of the current O_3 removal mechanism, it has been observed that the existing O_3 removal mechanism during the evening could be greater than typical [5].

In Malaysia, there is lack of study in tropospheric O_3 due to limited data collection (missing data) for certain ozone precursors with tropospheric O_3 concentration [6]. Usually, the majority of missing data were located in rural, and Borneo (Sabah and Sarawak) areas compared to Peninsular Malaysia. Furthermore, most of the study was conducted in short period and in specific areas, places, or states which not covered- up whole Malaysia (Peninsular Malaysia and Borneo). The study also does not emphasize the major monsoonal seasons known as Northeast Monsoon (NEM) and Southwest Monsoon (SWM). In this study, ten years (2006–2015) of hourly dataset of nocturnal O_3 level (night-time) and its precursors were used for the investigation the influenced towards nocturnal O_3 concentration.

2 Methodology

2.1 Study Areas

Kuching, one of rapid developing city situated at N01° 33.734, E110° 23.329 is selected in this study. The understanding of nocturnal O_3 pattern in this area is crucial in determining long term trend to have a better perspective what nocturnal O_3 will in the near future.

2.2 Data Sources

Ozone and its precursors (NO, NO_2, NO_x, CO, NmHC), consisted of 108 months consisting of 216 h (24 h × 9 years) of observation starting from January 2006 until December 2015 were acquired from the Air Quality Division, Malaysian Department of Environment (DOE). Tropospheric O_3 concentration data was recorded by using a system based on the Beer-Lambert Law that will measure the low ranges of ozone in ambient air. It was manufactured by Teledyne Technologies Incorporated (Model 400E). It works by UV light (254 nm) passed through the sample cell which will absorb a proportion of ozone present [7]. The concentration of NO_x (NO and NO_2) ozone precursors was calculated using the theory of chemical luminescence calculation, coupled with state-of-the-art microprocessor technology for monitoring high and medium nitrogen oxide levels (Teledyne Models 200A). All monitoring instruments are frequently calibrated for quality control and quality assurance of air monitoring data by DOE's vendor, Alam Sekitar Malaysia (ASMA).

2.3 Data Analysis

The dataset was rearranged accordingly in time (100–2400 h) and year (January 2006–December 2015). Next, the dataset was averaged as input in Principal Component Analysis. From all datasets, year 2012 was removed since there was no O_3 level recorded. No amputation method was applied in this study. The goals of the PCA are: (1) to investigate the influenced of NEM and SWM onto nocturnal O_3 and its precursor and, (2) to identify the relationship between the precursors towards ozone during NEM and SWM.

The analyses were performed using XLSTAT Software by Addinsoft (Perpetual Ver. 2019, USA). Prior to PCA, Bartlett's sphericity test and Kaiser-Meyer-Olkin (KMO) was performed to identify the adequacy of the dataset and multicollinearity between variables. To investigate the relationship between observed data, PCA is utilized to extract significant elements (principal components; PCs) [8]. Factors with substantial eigenvalues (>1.0) were selected from a previous study by [9] to achieve a better understanding of the correlation relationship between the variables. The percentage of variation is accounted for by each PC, which has a diminishing trend as the rank of the PCs increases. In general, the number of PCs represents the number of independent variables. In most modifications in the data set, the original observations are frequently attributed to the top few PCs. The first PC consolidated the majority of data variances, followed by the second PCs, which matches the remaining of the PCs in the PCAS. All PCA results in this study were produced using the varimax rotation to extract the powerful information from the factor loading (FL). The eigenvalues extracted from eigenfactors and the variable communalities resulting from the retention of a given number of factors were used to determine the number of factors retained in each of the analyses [10, 11]. PCA is good in dealing with independent variables, but it limited to a regression equation with predictors, especially when the independent variables are highly correlated.

3 Result and Discussion

3.1 Temporal Variation

The fluctuation of nocturnal ozone from 0100 to 0600 and 1900 to 2400 h since January 2006 until December 2015 are illustrated in Fig. 1a and b. From the Fig. 1a, there was and increment of minimum values of nocturnal O_3. As much as 62.13% increment for a decadal study whereby the minimum O_3 values in 2006 was 3.67 ppbv, yet it was recorded as high as 5.91 ppbv in 2015. However, the maximum values from 2006 to 2015 showed a contrast pattern with the O_3 has decreased from 20.50 ppbv to 18.42 ppbv (Fig. 1b). In between 0100 and 0600 h (Fig. 1b), the O_3 level relatively similar 6.20 ppbv in average. Starting 1900 h, the O_3 level abruptly increased up to 12.00 ppbv showing a decreasing trend pattern over time to 2400 h.

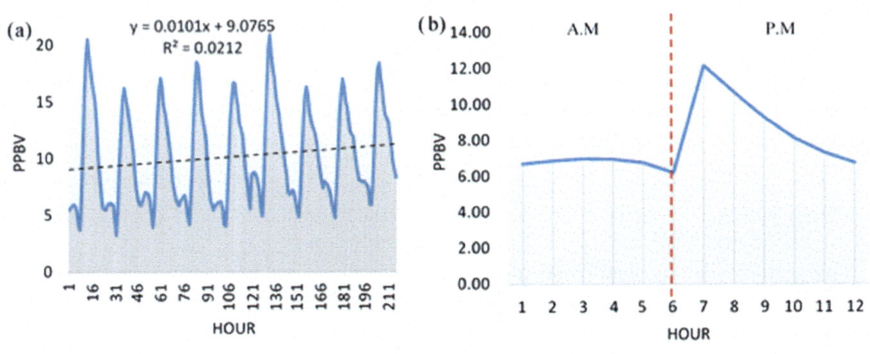

Fig. 1 **a** The overall observation of O_3 level trend (0100–2400) since January 2006 until December 2015. **b** The comparison of nocturnal O_3 between 0100 and 0600 am and 0700 and 1200 pm

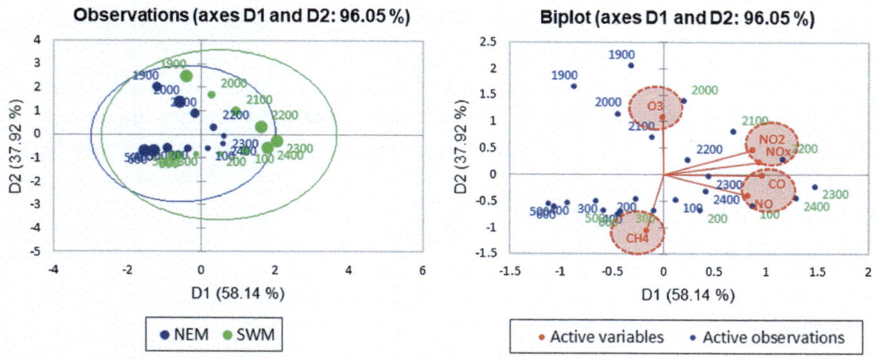

Fig. 2 **a** The segregation of eclipses between NEM and SWM. **b** Relationship between noc- turnal O_3 with its precursor (NO_2, NO_x, CO and CH_4

3.2 *The Relationship Between O_3 and the Precursor*

The monsoonal clustering between NEM and SWM of nocturnal O_3 is demonstrated by different set of color's eclipses in Fig. 2a. Generally, nocturnal O_3 in Kuching depicted the similarities pattern between both monsoons since the eclipse overlapped into each other. To prove the finding from PCA, t-test study was conducted to prove if there is a significant different between NEM and SWM or vice versa. Based on the result (two-tailed), there was no significant differences between monsoons, $t(11) = -0.175, p = 0.86$). However, SWM has a substantial impact on nocturnal O_3 compared to NEM. Figure 2b explained the relationship between O_3 and its precursor. Each component of precursor has been group based on their unique characteristic. From the figure, the O_3 has a linear relation between NO_2, NO_x, CO, and NO in descending order. In contradiction, CH_4 was the only parameter that does not have any influence on O_3 in Kuching for a decade study.

The seasonal dispersion of ozone concentration was mostly driven by seasonal wind direction and sample station locations. First, prior to its creation, ozone is closely linked to NO. NO is emitted by cars and is commonly seen in heavily populated areas. Second, it is possible that the high UV-b and associated meteorology factors of the locations during SWM are to blame.

4 Conclusion

Nocturnal O_3 shown increment of minimum values in this decadal study begin from 2006 (3.67 ppbv) until 2015 (5.91 ppbv) that increase by 62.13%. Contradict with maximum values as it decreases significantly from 2050 ppbv (2006) to 11.18 ppbv (2015). Comparison between nocturnal O_3 were relatively similar (6.20 ppbv) in average started between 0100 and 0600 h. However, for 1900 h, O_3 level peak at highest level to 12.00 ppbv and decrease gradually to 2400 h. Profile similarities pattern for nocturnal O_3 eclipse were overlapped with both monsoons. Nevertheless, SWM show substantial impact on nocturnal O_3 compared to NEM which explained the relationship of each component of O_3 precursors group based on the characteristics. Only CH_4 was the only parameter that does not have any influence on O_3 in Kuching for a decade study. Besides, the O_3 precursor also played an important role as to form tropospheric O_3 concentration thus, affecting the ecosystem in the area.

References

1. Monks PS, Archibald AT, Colette A, Cooper O, Coyle M, Derwent R, Fowler D, Granier C, Law KS, Mills GE, Stevenson DS, Tarasova O, Thouret V, von Schneidemesser E, Sommariva R, Wild O, Williams ML (2015) Tropospheric ozone and its precursors from the urban to the global scale from air quality to short-lived climate forcer. Atmos Chem Phys 15(15):8889–8973. https://doi.org/10.5194/acp-15-8889-2015
2. Tan KC, San Lim H, Jafri MZM (2016) Prediction of column ozone concentrations using multiple regression analysis and principal component analysis techniques: a case study in peninsular Malaysia. Atmos Pollut Res 7(3):533–546. https://doi.org/10.1016/j.apr.2016.01.002
3. Ghazali NA, Ramli NA, Yahaya AS, Yusof NFFM, Sansuddin N, Al Madhoun WA (2010) Transformation of nitrogen dioxide into ozone and prediction of ozone concentrations using multiple linear regression techniques. Environ Monit Assess 165(1):475–489. https://doi.org/10.1007/s10661-009-0960-3
4. Kulkarni PS, Dasari HP, Sharma A, Bortoli D, Salgado R, Silva AM (2016) Nocturnal surface ozone enhancement over Portugal during winter: Influence of different atmospheric conditions. Atmos Environ 147:109–120. https://doi.org/10.1016/j.atmosenv.2016.09.056
5. Awang NR, Ramli NA, Yahaya AS, Elbayoumi M (2015) High nighttime ground-level ozone concentrations in Kemaman: NO and NO_2 concentrations attributions. Aerosol Air Qual Res 15(4):1357–1366. https://doi.org/10.4209/aaqr.2015.01.0031
6. Yusoff MF, Latif MT, Juneng L, Khan MF, Ahamad F, Chung JX, Mohtar AAA (2019) Spatio-temporal assessment of nocturnal surface ozone in Malaysia. Atmos Environ 207:105–111. https://doi.org/10.1016/j.atmosenv.2019.03.023
7. Turnipseed AA, Andersen PC, Williford CJ, Ennis CA, Birks JW (2017) Use of a heated graphite scrubber as a means of reducing interferences in UV-absorbance measurements of atmospheric ozone. Atmos Measur Tech 10(6):2253–2269. https://doi.org/10.5194/amt-10-2253-2017
8. Latif MT, Dominick D, Ahamad F, Khan MF, Juneng L, Hamzah FM, Nadzir MSM (2014) Long term assessment of air quality from a background station on the Malaysian Peninsula. Sci Total Environ 482:336–348. https://doi.org/10.1016/j.scitotenv.2014.02.132
9. Mun'im Mohd Han N, Latif MT, Othman M, Dominick D, Mohamad N, Juahir H, Tahir NM (2014) Composition of selected heavy metals in road dust from Kuala Lumpur city centre. Environ Earth Sci 72(3):849–859. https://doi.org/10.1007/s12665-013-3008-5
10. Du KL, Swamy MNS (2019) Principal component analysis. In: Neural networks and statistical learning, pp 373–425. Springer, London. https://doi.org/10.1007/978-1-4471-7452-3-13
11. Jolliffe IT, Cadima J (2016) Principal component analysis: a review and recent developments. Philos Trans R Soc A Math Phys Eng Sci 374(2065):20150202. https://doi.org/10.1098/rsta.2015.0202

Spatio-Temporal Variation of Particulate Matter (PM_{10}) During High Particulate Event (HPE) in Malaysia

Nursyaida Amila Mohammad Ridzuan, Norazian Mohamed Noor, Nur Alis Addiena A. Rahim, Izzati Amani Mohd Jafri, and Deak Gyeorgy

Abstract Particulate matter (PM_{10}) is the key indicator of air quality index in Malaysia and Southeast Asia's main haze-related pollutant. PM_{10} emanation is believed to cause the strongest harm to public health and the environment. Therefore, it is very important to study the temporal and spatial characteristics of PM_{10} and the weather parameters, hence the relationship between them can be identified. A database with hourly PM_{10} concentration and weather parameters were obtained from Department of Environment (DOE) Malaysia from the period of 2012–2016 at two study areas that are located in Klang Valley, namely, Petaling Jaya and Shah Alam. The temporal analysis for PM_{10} concentration was observed by using descriptive statistics, boxplot and time series plot whereas the spatial analysis was conducted using windrose diagram. The finding shows that the highest average concentration of PM_{10} at Petaling Jaya and Shah Alam in 2015 exceeded the Malaysia Ambient Air Quality Standard, which were 60.13 $\mu g/m^3$ and 66.22 $\mu g/m^3$ respectively. It was due to high particulate event (HPE) that had affected Malaysia during the period of Southwest Monsoon, where the massive land and forest fires came from Sumatra and Kalimantan, Indonesia. According to the wind rose rose diagram, the wind mostly blew from northeast in January until February as Malaysia experienced northeast monsoon where rainfall happened. Shah Alam received stronger wind compared to the Petaling Jaya because the topography of city.

Keywords Air quality · Air quality modeling · Haze · Particulate matter

N. A. M. Ridzuan · N. M. Noor (✉) · N. A. A. A. Rahim · I. A. M. Jafri
Faculty of Civil Engineering Technology, Universiti Malaysia Perlis, 02600 Arau, Perlis, Malaysia
e-mail: norazian@unimap.edu.my

Sustainable Environment Research Group (SERG), Centre of Excellence Geopolymer and Green Technology, Universiti Malaysia Perlis, 02600 Arau, Perlis, Malaysia

D. Gyeorgy
National Institute for Research and Development in Environmental Protection Bucharest (INCDPM), 294, Splaiul Independentei Street, 6th District, 060031 Bucharest, Romania

N. Mohamed Noor et al. (eds.), *Proceedings of the 3rd International Conference on Green Environmental Engineering and Technology*, Lecture Notes in Civil Engineering 214,
https://doi.org/10.1007/978-981-16-7920-9_6

1 Introduction

PM_{10} is particulate matter with a diameter of less than 10 μm. It is a key indicator of the air quality index and the main pollutant in the high particulate event (HPE) episodes. PM_{10} concentration can rise up tremendously due to smoke discharges from biomass burning. Biomass burning has played a recognizable part in unforeseen extreme haze episodes, known as high particulate event (HPE) that covered with the essential and auxiliary pollutants determined from motor debilitates and coal combustion [1]. Deficient combustion, especially from huge ranges of biomass burning ordinarily supplies high quantity of smoke and fine particles to the atmospheric conditions. Due to territorial winds, these fine particles can be transported across boundaries from their primary causes [2].

Air pollutants emitted during HPE in one country will have an effect on that country and the near and regional countries further [2]. These events, mainly due to the very dry weather associated with the incidence of robust El Niño situations, resulted in the international transportation of unsafe pollutants from the original pollution sources from Indonesia's Borneo and Sumatra to the highly populated Malaysian Peninsula [3]. This HPE events have been suffered by country at Southeast Asia (SEA), such as Singapore, Brunei Darussalam, Thailand, Indonesia and Malaysia. The creation of haze is particularly related to weather patterns, air pollutant emissions and conversion of gas to particles. HPE is structured under stable weather conditions such as low winds and reversed thermodynamic elements due to high particle concentrations and gas-to-particle conversion. Meteorological variables such as atmospheric pressure, precipitation, humidity levels, atmospheric stabilization, physical and chemical relationships will determine the destiny and composition of suspended particles when they are transported [4, 5]. Due to the perseverance of fine particles in the air and the ability of particles to be transferred over wider areas, the effect of HPE is commonly geographic in scale.

HPE events and its associated deterioration of air quality also affected Malaysia's economy, not to mention the agricultural production and biological diversity of the country. Episodes of haze have resulted in a decrease in production and development output and a decrease in tourist industry profit, particularly because of cancellation of flights. A study by [6] economic losses due to ambulatory expenses during haze episodes in this country during 2005, 2006, 2008 and 2009 resulted in losses of USD 91,000/year with an average loss of USD 4,789 per haze day occurrence.

This research was carried out at two specific air monitoring stations in Klang Valley, which were Petaling Jaya and Shah Alam. These air monitoring stations have different landuse and have been designed to determine the pattern and status of PM_{10} at various stations that are located at Sek. Ren. Sri Petaling, Petaling Jaya and Sek. Keb. TTDI Jaya, Shah Alam during high particulate event (HPE). Petaling Jaya was chosen as the study area because the station is the closest station to the city of Kuala Lumpur, where the location is encircled by industrial, residential and commercial locations and as well as congested roads. In a meanwhile, Shah Alam is situated in a residential district, while it is close to major streets, the density of congestion is

Table 1 Two geographical air quality monitoring stations in Klang Valley

Location ID	Station location	Coordinates		Background
		Latitude (E)	Longitude (N)	
CA016	Sek. Ren. Sri Petaling, Petaling Jaya	E 101 38.322	N 03 06.553	Industrial areas
CA025	Sek. Keb. TTDI Jaya, Shah Alam, Selangor	E 101 33.375	N 03 06.281	Urban and residential areas

lower compared to Petaling Jaya and generally only notable in the morning and late afternoon after the hours of rush.

2 Materials and Method

2.1 Monitoring Dataset

Secondary dataset from Department of Environment (DOE), Malaysia consists of hourly data of PM_{10} concentration and weather parameters such as wind speed and wind direction from 2012 until 2016 were used in this study.

2.2 Study Location

The selected study areas are located in Peninsula Malaysia i.e. in Klang Valley, Malaysia that includes Kuala Lumpur and its subsurbs, neighboring cities and towns in the state of Selangor. Klang Valley is geographically outlined by Titiwangsa Mountains and the Straits of Malacca with 6.9 million of populations recorded in 2013. Klang Valley is now recognized as an economic city in Malaysia with all of extensive physical development of public infrastructure, industrialization and rapid urbanization that has gradually deteriorated the air quality [7]. The specific location of selected study areas are in Petaling Jaya and Shah Alam. The description of the selected location for this project are tabulated in Table 1.

2.3 Data Pre-Treatment and Graphical Presentation

Dataset of traces gases and meteorological parameters obtained from DOE were analyzed by using descriptive statistics. In this research, the missing value of the data set were replaced by expectation maximization (EM) where the EM method calculates the means, the covariance matrix and the correlation of quantitative (scale) variables with missing data using SPSS software [8]. Descriptive statistics were utilized in order to analyse and summarize the air pollutant and weather parameters data in systematic manner specifically for huge dataset.

Time series plot or time series graph represented the hourly PM_{10} concentrations in Petaling Jaya and Shah Alam against annually from 2012 until 2016. Time series plot is used to observe the trend of PM_{10} concentration over time hence the duration of high particulate event can be detected.

2.4 *Wind-Rose Diagram*

Wind rose were constructed from the wind speed and wind direction data obtained from Department of Environment (DOE) over a given time period, from 2012 until 2016. By using average daily data over 5 years, monthly wind rose diagrams were developed for both study areas. Wind rose diagram is constructed using a suitable scale to portray the percentage frequencies of wind directions and suitable indices shades, lines, etc. to depict different wind speed. The software used to construct the wind rose diagram is Hydrognomon. Hydrognomon is a modern software tool that can analyse the hydrological data. It is an open source tool that running on a standard Microsoft Windows platforms, and as a part of the openmeteo.org framework [9].

3 Result and Discussion

3.1 *Descriptive Analysis*

The descriptive analysis of hourly PM_{10} concentration at Petaling Jaya and Shah Alam from 2012 to 2016 were provided in Tables 2 and 3, respectively. PM_{10} concentration at Petaling Jaya revealed the highest mean of concentration in 2015 with the value of 60.13 $\mu g/m^3$ whereas the lowest PM_{10} concentration was recorded in 2016,

Table 2 Statistical test of PM_{10} concentration at Petaling Jaya

Year	Mean ($\mu g/m^3$)	Mode ($\mu g/m^3$)	Median ($\mu g/m^3$)	Maximum ($\mu g/m^3$)	Minimum ($\mu g/m^3$)	Range	Standard deviation
2012	49.35	38.00	45.00	338.00	5.00	333.00	25.48
2013	48.43	38.00	43.00	373.00	17.00	355.00	29.31
2014	54.84	52.00	46.00	322.00	17.00	305.00	31.52
2015	60.13	54.00	49.00	390.00	12.00	378.00	43.94
2016	42.84	38.00	40.00	142.00	13.00	129.00	15.50

Table 3 Statistical test of PM_{10} Concentration at Shah Alam

Year	Mean ($\mu g/m^3$)	Mode ($\mu g/m^3$)	Median ($\mu g/m^3$)	Maximum ($\mu g/m^3$)	Minimum ($\mu g/m^3$)	Range	Standard deviation
2012	48.36	34.00	44.00	272.00	5.00	267.00	27.04
2013	46.39	52.00	42.00	421.00	12.00	409.00	30.04
2014	54.87	39.00	46.00	282.00	9.00	273.00	32.07
2015	66.22	54.00	52.00	426.00	17.00	409.00	49.51
2016	50.33	52.00	47.00	160.00	15.00	145.00	20.52

with the concentration of 42.84 $\mu g/m^3$. Next, the highest mean of PM_{10} concentration at Shah Alam had a peak value of 66.22 $\mu g/m^3$ during 2015 while 2013 had the opposite mean value of PM_{10} concentration, 46.39 $\mu g/m^3$. According to Malaysian Ambient Air Quality Standard (MAAQS), the averaging time for PM_{10} for 1 year is 40 $\mu g/m^3$. It is shown that the concentration exceeded the standard at both areas in 2015 as the concentration at Petaling Jaya and Shah Alam where the concentrations were 60.13 $\mu g/m^3$ and 66.22 $\mu g/m^3$, respectively.

From the Tables 2 and 3, it indicated that the mean values were higher compared to median value that proven there were extreme event occurred during that year [10] in both locations. The highest PM_{10} concentration that can be classified as high particulate event (HPE) occurred in September–October in 2015 for both study areas where the Air Pollutant index (API) were recorded above the unhealthy level. The haze phenomenon occurred due to fire burning over forests and peatlands in Sumatra and Kalimantan, Indonesia, during that year.

3.2 *Variation of PM_{10} Concentration*

Figures 1 and 2 represented the time series plot of daily PM_{10} concentration at Petaling Jaya and Shah Alam from 2012 until 2016, respectively. Solid line indicates the Malaysia Ambient Air Quality Standard (MAAQS) for 24 h averaging time which is 100 $\mu g/m^3$.

From Figs. 1 and 2, it can be concluded that most of the time, the peak concentration of PM_{10} occurred in June specifically in 2012 and 2013 at both study areas. In 2012, Malaysia experienced several short haze event due to the transboundary air pollution from the forest clearing from neighboring country, Indonesia and occurred during dry season that occurred in June hence contributed to slight degradation in term of air quality in Malaysia. In addition, [11] stated that Malaysia also experienced severe haze event in June 2013 due to the transboundary atmosphere pollution from Indonesia. However, in 2014, the peak concentration of PM_{10} at both study areas occurred in March. This can be compared to [11] as stated that this country experienced moderate haze episode that affected the declination of air quality level and the dry period occurred between February and March in 2014. Next, in 2015 the

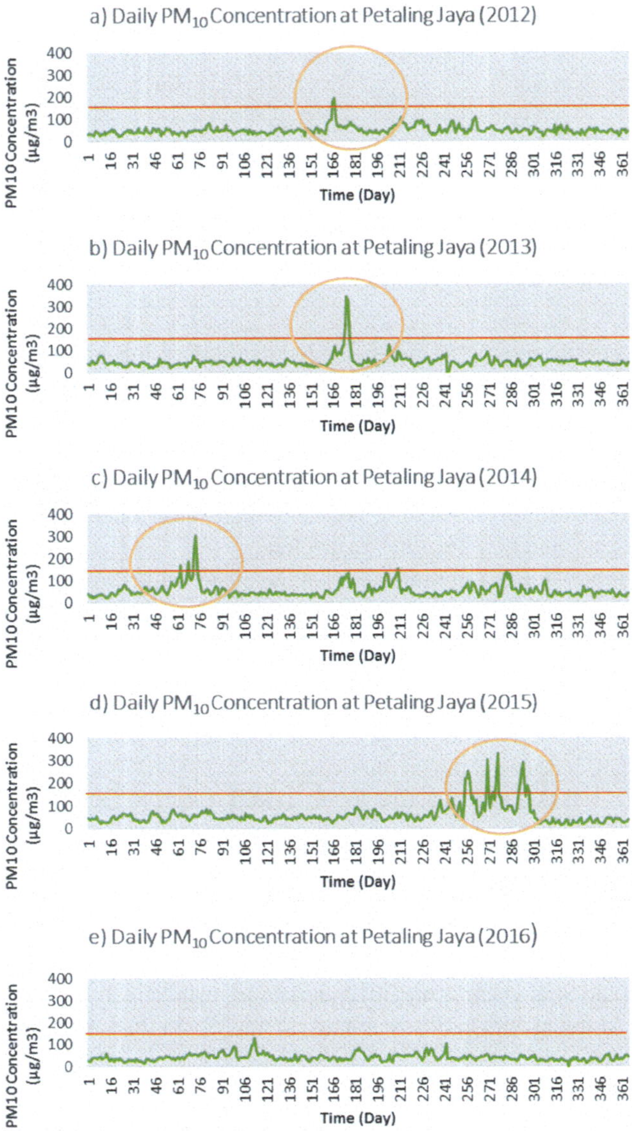

Red solid line: MAAQG limit
Orange circle: High PM$_{10}$ concentration

Fig. 1 Time series plot for PM$_{10}$ Concentration at Petaling Jaya

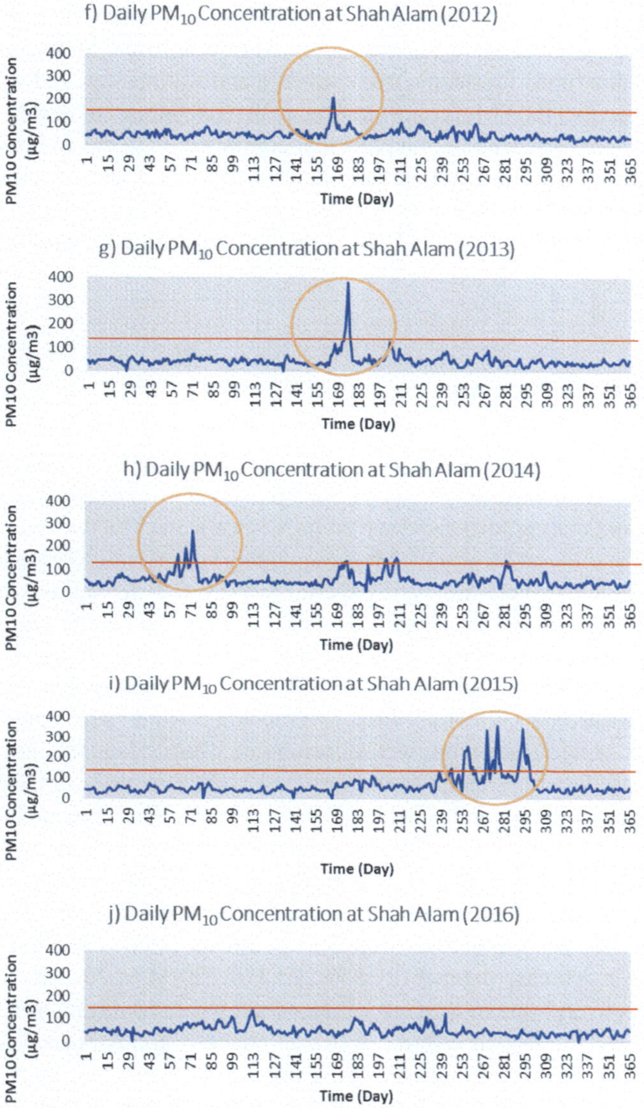

Red solid line: MAAQG limit
Orange circle: High PM_{10} concentration

Fig. 2 Time series plot for PM_{10} Concentration at Shah Alam

exceedances occurred with fluctuate during September and October. In 2015, haze episodes had affected at Malaysia during period of Southwest Monsoon where the massive land and forest fires came from Sumatra and Kalimantan, Indonesia [12]. It can be observed that the PM_{10} concentration in 2016 are almost at healthy air quality level. Malaysia experienced the La Nina that continued up to the end of year where the temperature of sea surface that near the central equator and eastern part of the Pacific Ocean were decreased. The high concentration of PM_{10} circled in the time series plot were used to model the PM_{10} concentration during high particulate event (HPE).

3.3 Wind Rose Diagram

The distributions of wind direction and speed over the entire period for Petaling Jaya and Shah Alam according to the two main monsoons in Malaysia are shown in Figs. 3 and 4 in Petaling Jaya and Shah Alam, respectively. The two main seasons in Malaysia are Southwest Monsoon (May–September) where and Northeast Monsoon (November–March) where Malaysia experienced dry and wet season, respectively. There are also two short transitional from April to May and October to November.

In Petaling Jaya, the wind rose shows wind blowing in all directions throughout the study period but dominant in the north-east (NE) quadrant with weak winds prevailing in the north-west (NW) direction. Low and moderate wind speeds predominate in Petaling Jaya, while strong winds were not common. During January until March, the dominant wind directions came from north-east (NE) directions (25.76%, 32.67% and 30.97% respectively) and considered as the highest wind directions among the months. The wind blow frequently from south-east (SE) directions with 3.33% in January, 4.08% in February, and 3.87% in March. During this time, Malaysia experience north-east (NE) monsoon where part of Peninsula Malaysia received heavy rainfall [13]. The wind directions were still dominant in NE direction in April to August but in a decrease manner (24.53%, 23.77%, 20.34%, 20.34%, 21.17% and 17.22% respectively) due to a change in monsoon where Malaysia generally experienced southwest monsoon where dry season phenomenon happened. During this period, the wind from the southwest bring the air pollutants that discharged from biomass burning and forest fires for agricultural purposes from neighboring country, Indonesia. Hence, the PM_{10} concentrations were peak during the south west monsoon for certain months [10]. However, started from September until December, the wind highly blew in west.

In Shah Alam, the wind directions were not consistent where there were no dominant directions of wind of all the months that can be identified. However, from the diagram obtained, there were dominant wind blew that came from NE in January and February (25.18% and 32.16% respectively). During this time, Malaysia experience north-east (NE) monsoon where part of Peninsula Malaysia received heavy rainfall [13]. From the Fig. 2, the lowest direction in January and February came from South (S) with percentage of 3.15% and 4.02%. In March, April and May, the dominant

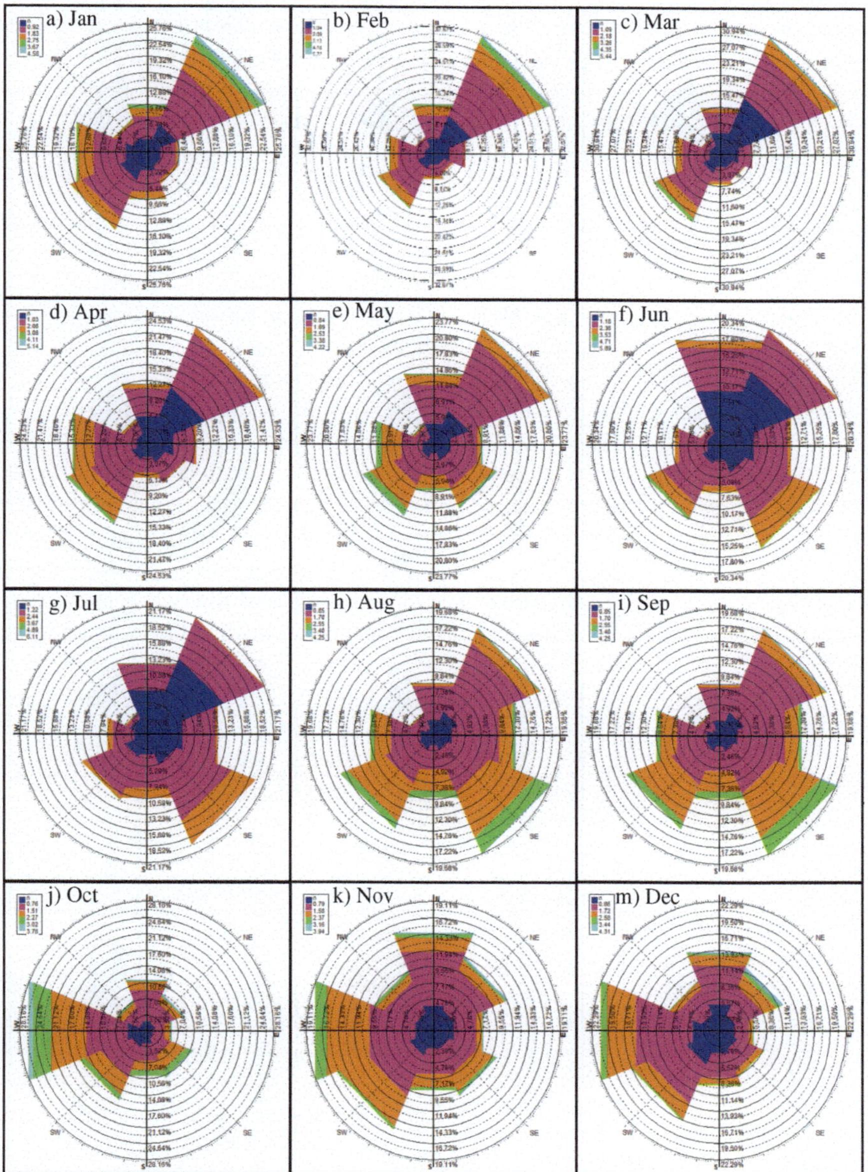

Average PM_{10} Concentration

Jan = 41.96 μg/m^3 Feb = 44.25 μg/m^3 March = 59.81 μg/m^3 April = 48.36 μg/m^3
May = 41.01 μg/m^3 June = 61.65 μg/m^3 July = 56.53 μg/m^3 Aug = 53.69 μg/m^3
Sep = 65.02 μg/m^3 Oct = 65.13 μg/m^3 Nov = 37.67 μg/m^3 Dec = 37.62 μg/m^3

Fig. 3 Monthly wind rose diagram in Petaling Jaya

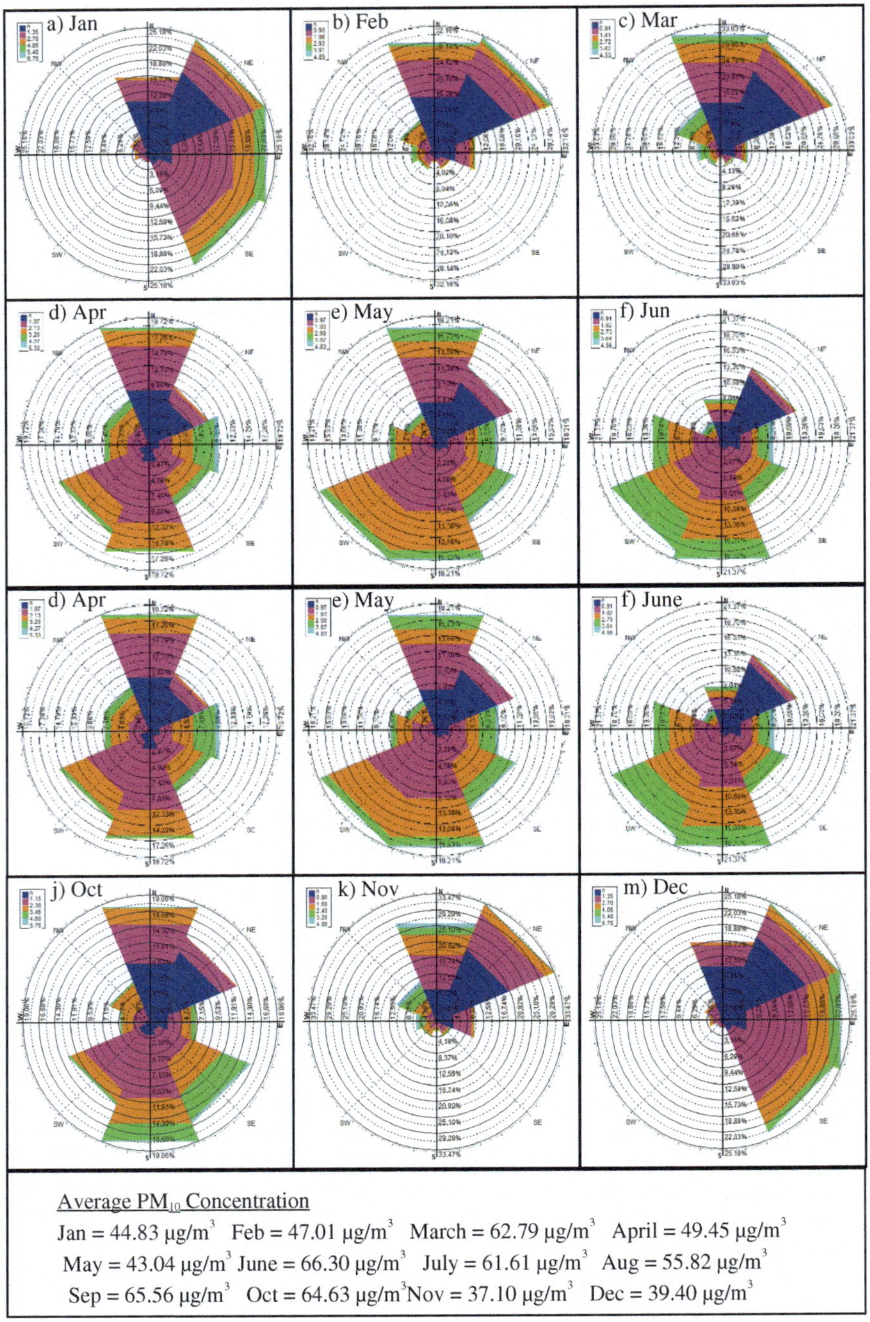

Average PM_{10} Concentration

Jan = 44.83 μg/m^3 Feb = 47.01 μg/m^3 March = 62.79 μg/m^3 April = 49.45 μg/m^3

May = 43.04 μg/m^3 June = 66.30 μg/m^3 July = 61.61 μg/m^3 Aug = 55.82 μg/m^3

Sep = 65.56 μg/m^3 Oct = 64.63 μg/m^{3}Nov = 37.10 μg/m^3 Dec = 39.40 μg/m^3

Fig. 4 Monthly wind rose diagram in Shah Alam

wind direction came from North with 33.03%, 19.72% and 18.21% respectively. The wind direction only came on south in June where the percentage recorded 18.7%. In July and August, the wind blew dominantly from South-west (36.60% and 19.79%, respectively). Subsequently, in September and October, the wind mostly came from north direction with value of 26.39% and 19.06%. Finally, the wind mostly blew in northeast directions in November and December where they recorded a percentage of 33.47% and 25.18%, respectively. For wind speed, it can be concluded that the wind dominantly blew a moderate wind speed in a range of 2.88 km/h until 9.72 km/h.

From both study locations, the wind mostly blew from northeast in January until February as Malaysia experienced northeast monsoon where rainfall happened. However, Shah Alam received stronger wind compared to the Petaling Jaya because the topography of city. Dotse et al. [14] stated that the low wind system coupled with the hilly lowlands, swampy plains and alluvial valleys topographic features thus could play a major in the transport pollutants to the country throughout the year.

4 Conclusion

PM_{10} is among the main air pollutants in Malaysia and seems to be crucial to the Malaysian Air Pollution Index (API) calculation. A dataset of PM_{10} concentration in Petaling Jaya and Shah Alam for 5 years (2012–2016) was obtained from DOE and was analysed temporally and spatially. A total of 3 parameters, consists of particulate matter (PM_{10}), and weather parameters i.e. wind speed and wind direction had been used in this research. The focus of this research was to examine the temporal and spatial characteristic of PM_{10} concentration during High Particulate Event (HPE). The concentration exceeded the Malaysian Ambient Air Quality Standard (MAAQS) at both areas in 2015 as the concentration at Petaling Jaya and Shah Alam where the concentrations are 60.13 $\mu g/m^3$ and 66.22 $\mu g/m^3$ respectively. The mean values for Petaling Jaya and Shah Alam were higher compared to median which indicated that there were extreme event occurred during that year. Results demonstrated that year 2015 recorded the highest PM_{10} concentration at all study location due to local emission coupled with trans-boundary haze from neighboring country. It suggested that Malaysia experienced a prolonged period of extreme haze due to the degradation of air quality in Sumatra and Kalimantan from August to September in 2015 attributable to enormous land as well as wild fires.

References

1. Chen J, Li C, Ristovski Z, Milic A, Gu Y, Islam MS, Guo H (2017) A review of bi-omass burning: Emissions and impacts on air quality, health and climate in China. Sci Total Environ 579:1000–1034. https://doi.org/10.1016/j.scitotenv.2016.11.025
2. Kusumaningtyas SDA, Aldrian E (2016) Impact of the June 2013 Riau province Sumatera

smoke haze event on regional air pollution. Environ Res Lett 11(7):75007. https://doi.org/10.1088/1748-9326/11/7/075007

3. Mead MI, Castruccio S, Latif MT, Nadzir MSM, Dominick D, Thota A, Crippa P (2018) Impact of the 2015 wildfires on Malaysian air quality and exposure: a comparative study of observed and modeled data. Environ Res Lett 13(4). https://doi.org/10.1088/1748-9326/aab325
4. Jewaratnam J, Aghamohammadi N, Sahani M, Jing C (2018) Impact of regional haze towards air quality in Malaysia: a review. Atmos Environ 177:28–24. https://doi.org/10.1016/j.atmosenv.2018.01.002
5. Khan MF, Shirasuna Y, Hirano K, Masunaga S (2010) Urban and suburban aerosol in Yokohama, Japan: a comprehensive chemical characterization. Environ Monit Assess 171(1–4):441–456. https://doi.org/10.1007/s10661-009-1290-1
6. Sahani M, Zainon NA, Mahiyuddin WRW, Latif MT, Hod R, Khan MF, Tahir NM, Chan CC (2014) A case-crossover analysis of forest fire haze events and mortality in Malaysia. Atmos Environ 96:257–265. https://doi.org/10.1016/j.atmosenv.2014.07.043
7. Asmat A, Norhayati S, Tarmizi M, Zakaria NH (2018) Seasonal particulate matter (PM_{10}) concentration in Klang Valley, Malaysia. Int J Eng Technol 7(3):162–167. https://doi.org/10.14419/ijet.v7i3.11.15953
8. Sukatis FF, Noor NN, Zakaria NA, Ul-Saufie AZ, Annas S (2019) Estimation of missing values in air pollution dataset by using various imputation methods. Int J Conserv Sci 10(4):791–804
9. Kozanis S, Christofides A, Mamassis N, Efstratiadis A, Koutsoyiannis D (2010) Hydrognomon–open source software for the analysis of hydrological data. Geophys Res Abstr 12:12419
10. Sansuddin N, Ramli NA, Yahaya AS, Yusof NFFM, Ghazali NA, Madhoun WA, Al. (2011) Statistical analysis of PM10 concentrations at different locations in Malaysia. Environ Monit Assess 180(1–4):573–588
11. Department of Environment (DOE) (2015) Chronology of haze episodes in Malaysia. http://www.doe.gov.my/portalv1/wp-content/uploads/2015/09/Chronology-of-Haze-Episodes-in-Malaysia.pdf. Accessed 23 Nov 2018
12. Mohamad ND, Ash'aari (2015) Preliminary assessment of air pollutant sources identification at selected monitoring stations in Klang Valley, Malaysia. Proc Environ Sci 30:121–126
13. Jamaludin S, Sayang MD, Wan ZWZ, Abdul AJ (2010) Trends in Peninsular Malaysia rainfall data during the southwest monsoon and northeast monsoon seasons: 1975–2004. Sains Malaysiana 39(4):533–542
14. Dotse SQ, Dagar L, Petra MI, De Silva LC (2016) Influence of Southeast Asian Haze episodes on high PM10 concentrations across Brunei Darussalam. Environ Pollut 219:337–352. https://doi.org/10.4209/aaqr.2019.06.0313

The Influence of Atmospheric Stability on PM_{10} Concentrations in Urban Areas

György Deak, Natalia Raischi, Lucian Lumînăroiu, Marius Raischi, Nicolae-Valentin Vladut, Raluca Prangate, and Norazian Mohamed Noor

Abstract The stability of air masses in the atmosphere influences not only the dispersion but even the accumulation of pollutants in certain areas. The most common methods for determining the stability in the atmosphere are the Richardson number, the Monin-Obukhov length and the Pasquill-Gifford method. For this study, the Monin-Obukhov length method was used to determine the stability of atmospheric air in Bucharest, Romania. The determined values highlighted the fact that the atmosphere is unstable during the day, of the 2 years analyzed (2017 and 2018), favoring the dispersion of pollutants in the atmosphere. Thus, the PM_{10} concentrations exceeded the limit value allowed by the Directive 2008/50/EC) in 44 days in 2017 and in 34 days in 2018, mostly in October–March.

Keywords Atmospheric stability · Monin-Obukhov length · PM_{10} · Pollutant concentrations · Urban areas

1 Introduction

The stability of air masses in the atmosphere influences the dispersion and the accumulation of pollutants in certain areas and it might be determined by several methods [1, 2]. According to Schnelle et al. [3], most of these methods are taking into consideration the air turbulences.

G. Deak · N. Raischi (✉) · L. Lumînăroiu · M. Raischi · R. Prangate
National Institute for Research and Development in Environmental Protection, 060031 Bucharest, Romania

N.-V. Vladut
National Institute for Research-Development of Machines & Installations for Agriculture, 013813 Bucharest, Romania

N. M. Noor
Sustainable Environment Research Group (SERG), Centre of Excellence Geopolymer and Green Technology (CEGeoGTech), Universiti Malaysia Perlis, 01000 Kangar, Perlis, Malaysia
e-mail: norazian@unimap.edu.my

N. Mohamed Noor et al. (eds.), *Proceedings of the 3rd International Conference on Green Environmental Engineering and Technology*, Lecture Notes in Civil Engineering 214,
https://doi.org/10.1007/978-981-16-7920-9_7

In urban areas there are a wide variety of air pollution sources due to gas emissions from industrial or transport sectors, as well as landfills and the burning of fossil and non-fossil fuels. To these are added the influence of other factors such as chemical transformations of pollutants, the weather impact and topography, as well as newly formed and previously deposited particles [4, 5].

According to the 2008/50/EC Directive, to protect human health, the annual limit value for PM_{10} was established at 40 $\mu g/m^3$, while the daily limit at 50 $\mu g/m^3$.

The main objective of this paper was to determine the stability of the atmosphere in Bucharest, during the day, from the seasons with decreasing temperatures (October–March), for a period of 2 years (2017 and 2018). Monin-Obukhov length was estimated using the meteorological parameters recorded in all four seasons of the 2 years of the study. Subsequently, the stability of the atmospheric air and its influence on PM_{10} concentrations were estimated with the data obtained by calculation.

2 Materials and Method

The estimations of the atmospheric stability were made for the central area of Bucharest city by taking into account the convective boundary layer (CBL) where the unstable conditions are found, which is specific during the day [6].

Subsequently, to estimate atmospheric stability during the day, the Monin-Obukhov length was used, which is one of the most common methods [7–10]. Monin and Obukhov (1954) proposed a similarity theory that introduces two scaling parameters, the friction velocity (u_*) and the Monin-Obukhov length ($L_{M\text{-}O}$) for determining turbulence. Atmospheric stability, according to the Monin-Obukhov method, is directly proportional to the friction velocity and inversely proportional to the sensible heat flux on the ground surface [2]. For interpretation of estimates, $L_{M\text{-}O}$ is positive when the atmosphere is stable and negative in the case of an unstable atmosphere. The Eq. (1) underlies the calculation of Monin-Obukhov length:

$$L_{MO} = \frac{-\left(\frac{u_*^3}{k}\right)}{\left(\frac{gH}{C_p \rho T}\right)} \Leftrightarrow L_{MO} = -\frac{\rho C_p T u_*^3}{kgH} \tag{1}$$

$$u_* = \frac{ku}{\ln\left(\frac{z-d}{z_o}\right) - \psi_m\left(\frac{z-d}{L_{MO}}\right) + \psi_m\left(\frac{z_o}{L_{MO}}\right)} \tag{2}$$

where u_* represents the friction velocity in Eq. (2), k is the von Karman constant, g is the gravitational acceleration, Cp is the specific heat at constant pressure, H is the sensible heat flux, T is the temperature, z is the height, u is the wind speed and d is 0.5 m.

The similarity function for momentum can be expressed according to Eqs. (3) and (4):

$$\psi_m\left\{\frac{z-d}{L_{MO}}\right\} = 2\ln\left(\frac{1+\mu}{2}\right) + \ln\left(\frac{1+\mu^2}{2}\right) - 2\tan^{-1}\mu + \frac{\pi}{2} \tag{3}$$

$$\psi_m\left\{\frac{z_0}{L_{MO}}\right\} = 2\ln\left(\frac{1+\mu_0}{2}\right) + \ln\left(\frac{1+\mu_0^2}{2}\right) - 2\tan^{-1}\mu_0 + \frac{\pi}{2} \tag{4}$$

The main materials underlying this study are the daily values of PM_{10} for Bucharest, which were obtained from the National Air Quality Monitoring Network (RNMCA) for the period January 2017–December 2018.

3 Results and Discussions

By calculating the Monin-Obukhov length, it was highlighted that the atmosphere is unstable during the day, in the seasons with lower temperatures (October–March), compared to those in which temperatures begin to rise (April–September), when the dispersion of pollutants in the atmosphere is favorized (Fig. 1).

Thus, the lowest value for 2017 was determined for January, −102.46 m, with a average temperature of −3.36 °C and a sensible heat flux of 38.94 W/m^2. The highest value was determined for July, −22.98 m, with a average temperature of 25.94 °C and a sensible heat flux of 148.04 W/m^2. In 2018, the lowest value was also determined for January, −99.60 m with a average temperature of 2.89 °C and a sensible heat flux of 41.64 W/m^2. The highest value was determined for July, −9.50 m, with a average temperature of 24.73 °C and a sensible heat flux of 143.46 W/m^2.

For a stable atmosphere, the value of Monin-Obukhov length is positive. It occurs mainly at night. In conditions of the unstable atmosphere, specific during the day, the value of Monin-Obukhov length is negative. At the same time, the sensible heat flux has negative values when the atmosphere is stable and positive values when it is unstable [11].

The average daily values of PM_{10} during the two cold seasons vary between 26.9 μg/m^3 and 44.03 μg/m^3 in 2017 and between 22.23 μg/m^3 and 42.17 μg/m^3 in 2018 (Fig. 2a). In 2017, there were 44 days in which the value of PM_{10} concentrations

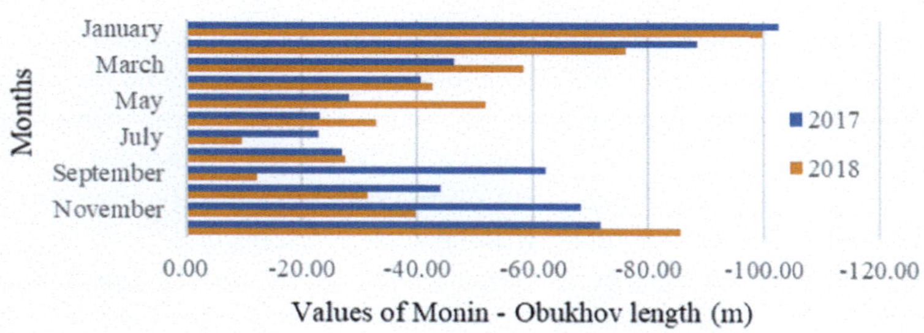

Fig. 1 Determined values of Monin-Obukhov length

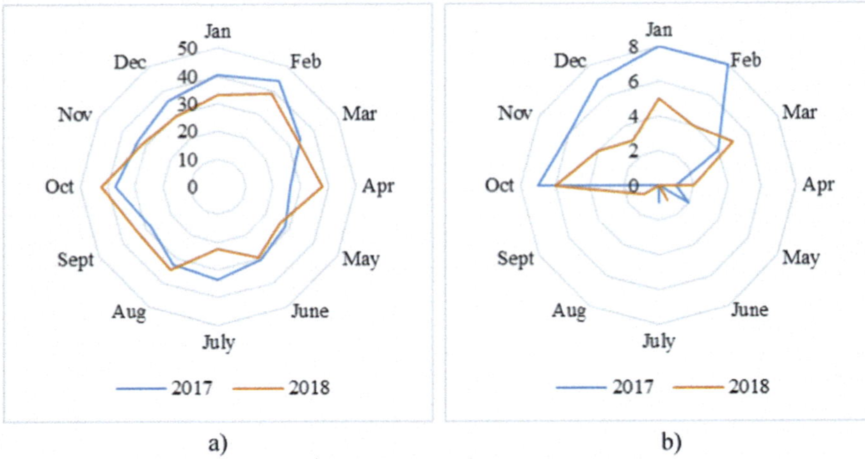

Fig. 2 **a** PM_{10} average concentration. **b** Number of days with exceedances of PM_{10} limit value

was over the limit and 90% of these have been recorded during the cold season. In 2018, there were 34 days in which the value of PM_{10} concentrations was over the limit and 79,4% have been recorded in October and March (Fig. 2b).

However, the maximum daily concentration for 2017 was recorded in January, with a value of 100.01 $\mu g/m^3$, exceeding the average value by 250%. In 2018, the maximum daily concentration was 90.36 $\mu g/m^3$, recorded in October, exceeding the average value by 215%.

4 Conclusions

This paper presents the results of the study regarding the influence of atmospheric stability on PM_{10} concentrations during the seasons with lower temperatures (October–March), compared to those in which temperatures begin to rise (April–September), on diurnal period for 2 consecutive years, using the Monin-Obukhov length method.

In the cold seasons the Monin-Obukhov length has lower values than in the warmer seasons, which highlights the importance of atmospheric stability on exceeding the limit value of PM_{10} concentrations during the analyzed period. The negative values of the Monin-Obukhov length show that on diurnal period the atmosphere is unstable. Exceedances of daily concentrations of over 200%, were mainly recorded during the cold seasons. These values are also influenced by the occurrence of the thermal inversion. Due to the increase in temperature with height, this phenomenon creates a barrier that limits the pollutants dispersion in the upper layers of the atmosphere, retaining them on the ground surface and producing increases in concentrations.

Acknowledgements This paper was developed through Nucleu Progamme—Contract number 39N/2019 carried out with the Romanian Ministry of Research and Innovation support, within the project "*Evaluation of air quality at the surface and in the underground spaces in Bucharest*".

References

1. Arya SP (1999) Air pollution meteorology and dispersion, vol 310. Oxford University Press, New York
2. Briggs GA (1988) Analysis of Diffusion Field Experiments. In: Venkatram A, Wyngaard JC (eds) Lectures on air pollution modeling. American Meteorological Society, Boston, MA. https://doi.org/10.1007/978-1-935704-16-4_3
3. Schnelle KB, Dey PR (2000) Atmospheric dispersion modeling compliance guide/; Karl B. Schnelle, Jr., Partha R. Dey. https://totearicenposa.blogspot.com/2020/01/pdf-atmospheric-dispersion-modeling.html
4. Holban E, Diacu E, Matei M, Ghita G, Raischi M, Daescua A, Gheorghea IP, Iliea M, Szepc R, Daescua V, Dumitrua D, Marinescua F, Tociua C, Popescua I, Gh. Popescu CR (2017) Assessment of atmospheric pollution in a cement factory area situated in the eastern part of Romania. J Environ Prot Ecol 18(3):819–830. https://doi.org/10.13140/RG.2.2.15090.61121
5. Deak G, Raischi N, Matei M, Boboc M, Cornateanu G, Raischi, Matei S, Yusuf SY (2020) Meteorological parameters and air pollution in urban environments in the context of sustainable development. IOP Conf Ser Earth Environ Sci 616(1):012003. IOP Publishing. https://doi.org/10.1088/1755-1315/616/1/012003
6. Cuculeanu G (2009) Gaussian approach of the atmospheric pollutant diffusion. A patra conferinţă internaţională Efectele crizei globale asupra economiilor în dezvoltare, Bucureşti. http://store.ectap.ro/articole/664_ro.pdf
7. Mohan M, Siddiqui TA (1998) Analysis of various schemes for the estimation of atmospheric stability classification. Atmos Environ 32(21):3775–3781. https://doi.org/10.1016/S1352-2310(98)00109-5
8. Zoras S, Triantafyllou AG, Deligiorgi D (2006) Atmospheric stability and PM10 concentrations at far distance from elevated point sources in complex terrain: Worst-case episode study. J Environ Manage 80(4):295–302. https://doi.org/10.1016/j.jenvman.2005.09.010
9. Yassin MF (2013) Experimental study on contamination of building exhaust emissions in urban environment under changes of stack locations and atmospheric stability. Energy build 62:68–77. https://doi.org/10.1016/j.enbuild.2012.10.061
10. Perez AI, Garcia AM, Torre B, Sanchez LM (2009) Ann Geophys 27:339–349
11. Cimorelli AJ, Perry SG, Venkatram A, Weil JC, Paine RJ, Wilson RB, Brode RW (2005) AERMOD: A dispersion model for industrial source applications. Part I: General model formulation and boundary layer characterization. J Appl Meteor 44(5):682–693. https://doi.org/10.1175/JAM2227.1

Environmental Management and Protection

A Comparative Study on Generation and Composition of Food Waste in Desa Pandan Kuala Lumpur During Covid-19 Outbreak

Irnis Azura Zakarya, Nur Adilah Rashidy, Tengku Nuraiti Tengku Izhar, Muhammad Haizar Ngaa, and Lucian Laslo

Abstract Food waste is a type of solid waste that is heavily influenced by consumers. The composition (%) and the total weight generation (kg) of food waste were determined during the Covid-19 outbreak. The data was obtained by direct weighing of the food waste collected where it was generated throughout the day. Findings have shown that 63.3% of the composition of uncooked food waste types identified is generated by internal organs, vegetables, fruits, and eggshells, while 36.7% of the identified composition of cooked food waste types is generated by rice and noodles, bones, vegetables, fruits, and others in Desa Pandan. During six days of collection, the total food waste generated by 30 households, 10 restaurants, and 3 schools was 146.5 kg (0.20 kg/capita/day), 231.7 kg (0.026 kg/capita/day), and 155.4 kg (0.010 kg/capita/day) respectively. A food waste awareness survey was conducted online with 100 respondents using Google Forms. Findings show that the Covid-19 pandemic did influence people's attitudes and practices regarding food purchase, management, and consumption habits, all of which have a substantial impact on reducing food waste generation at the household level.

I. A. Zakarya (✉) · N. A. Rashidy · T. N. T. Izhar
Faculty of Civil Engineering Technology, Universiti Malaysia Perlis, 02600 Arau, Perlis, Malaysia
e-mail: irnis@unimap.edu.my

N. A. Rashidy
e-mail: s171130602@studentmail.unimap.edu.my

T. N. T. Izhar
e-mail: nuraiti@unimap.edu.my

Sustainable Environment Research Group (SERG), Center of Excellence Geopolymer and Green Technology, Universiti Malaysia Perlis, 01000 Kangar, Perlis, Malaysia

M. H. Ngaa
Equine Commercial Center, Tekno RE Consulting Sdn Bhd, S-07-11, Trio PermaiJalan Equine Bandar Putra Permai, 43300 Seri Kembangan, Selangor, Malaysia
e-mail: haizar@tekno-re.com

L. Laslo
National Institute for Research and Development in Environmental Protection Bucharest (INCDPM), Splaiul Independentei Street, 6th District, 294060031 Bucharest, Romania

N. Mohamed Noor et al. (eds.), *Proceedings of the 3rd International Conference on Green Environmental Engineering and Technology*, Lecture Notes in Civil Engineering 214,
https://doi.org/10.1007/978-981-16-7920-9_8

Keywords Food waste · Composition · Generation · Awareness · Management · Covid-19

1 Introduction

As stated by the SWCorp Malaysia (Solid Waste Management and Public Cleansing Cleansing Corporation), Malaysia's daily domestic waste reaches 37,500 tons/day. The remaining 44.5% of the residue is food waste with approximately 16,687.5 tons, of which 12,682.5 tons is the unavoidable food waste (76%), while 4005 tons of food waste can be avoided (24%) that can be consumed by 2,970,000 people for three meals. SWCorp also reports that each person is estimated to produce 1.17 kg of unavoidable food per day with an average weight of one serving per person being 0.45 kg/meal. The growing purchasing power in line with the booming economy has driven people to buy more than they need and a healthier lifestyle, which is expected food waste to expand to 30% by 2020 [1].

Food waste can be further classified info food loss, unavoidable food waste and avoidable food waste. Food loss and food waste represent a misuse of resources that are used to produce it. Food loss refers to the decreased in quantity or quality of food that is lost during the preparation and production of the food supply chain. As for food waste it is part of food loss, unavoidable food waste refers to the inedible parts of food that include fruit core and peels. Avoidable food waste is edible food lost generated at any level within the food chain, which includes production, processing, distribution, and consumption [2].

The announcement of the Movement Control Order (MCO) happened all of a sudden after Covid-19 had been declared a pandemic on March 11, 2020, by the World Health Organisation [3]. According to SWCorp Malaysia, during the month of March, including the period before MCO was enforced on March 18, food waste generated daily in Kuala Lumpur averaged 2.1 tonnes and in April, it went down to 1.7 tonnes a day, reflecting a 0.3-tonne reduction in food wasted daily [4]. The MCO period, nevertheless, had a silver lining as there was a marked decrease in the quantity of food waste that ended up in landfills as people are urged to stay at home and go out only to meet the most urgent needs, such as buying food, during lockdowns [5]. Therefore, Covid-19 lockdowns may have an effect on citizens' daily lives, including impacts on food behaviors, food waste habits, and household consumption. Higher frequency of activity at home due to stay-at-home orders has possibly increased the percentage of household waste being produced through over-purchasing, especially the food waste [6]. Overall, this study determined the total weight of food waste generated (kg), the total composition (%) of uncooked and cooked food waste and the consumer awareness on food waste management during the Covid-19 pandemic in Desa Pandan, Kuala Lumpur.

2 Materials and Method

2.1 Food Waste Sampling

The study area covered in Desa Pandan, Kuala Lumpur at coordinates 3.149478° North and 101.737575° east as shown in Fig. 1. Previously a small town, Desa Pandan grew and developed enormously. It occupies an area of approximately 0.54 square kilometers. The study was conducted in 3 different sectors, which are residential, commercial, and institutional, in Desa Pandan.

As shown in Table 1, 30 households in 5-story apartment buildings on Jalan 5/76a, 5 restaurants on Jalan 1/76, 5 restaurants on Jalan 1/76d, and 3 schools from Maluri were chosen to study food waste composition in Desa Pandan. The study area was chosen because the residents are from my hometown. The restaurants were chosen

Fig. 1 Study area in Desa Pandan (Google Maps Application)

Table 1 Description of selected sampling site in Desa Pandan, Kuala Lumpur

Sectors	Sampling site	Location
Residential	30 number of households	Jalan 5/76a
Commercials	5 restaurants	Jalan 1/76a
	5 restaurants	Jalan 1/76d
Institution	SMK Aminuddin Baki	Jalan Kampung Pandan, Maluri
	SMK Datok Lokman	
	SMK Taman Maluri	

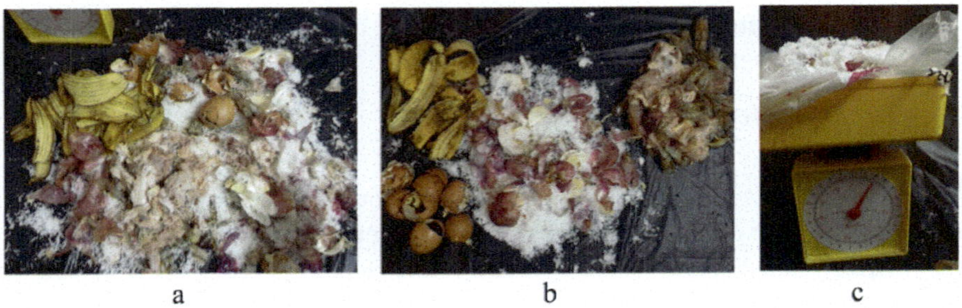

Fig. 2 Food waste; a Separating, b Categorize, c Weighing

because they are in close proximity to one another, which is in a row of buildings. While for the school area, they are in one location near to each other too to make it easier for the food waste audits.

Food waste was collected with each sector consumed six times of sample collection. Food waste from each industry had been collected on different timelines, beginning on February 15th and ending March 26th, 2021. The garbage bags with a 50 L for households, 120 L for restaurants and institutions were provided for each study area during the sampling period. Food waste composition analysis is the process of physically separating, categorizing waste, and weighing as shown in Fig. 2a–c. This sorting process determine the total amount of food waste and organize the various types of foods discarded and distinguish between uncook and cooked food waste in Desa Pandan.

Uncook food waste is categorized as uncooked food waste and is generated from food preparation, handling, storage, processing, and disposal of unwanted parts of raw food, while cooked food waste is generated during and after the cooking process, serving, and unfinished food. Lastly, a statistical graph was used to analyze the results of the study where the data was analyzed based on food waste composition types and food waste generation of each sector in Desa Pandan, which presented in the quantity of food waste generated per capita per day (kg/capita/day).

2.2 Food Waste Awareness Survey

The awareness levels of consumers with respect to food waste during Covid-19 outbreak and knowing how knowledgeable they are about the whole food waste phenomenon were carried out to address the very critical issue of the food waste. 100 respondents participated in this field survey through generated online questionnaires using Google Forms. The questions consist of 27 question items. The first eight questions of Section A were used to collect demographic data. The following 19 questions of Section B consisted of questions on thoughts and behavior on food purchase, management, consumption as well as food waste. All 19 food waste questions were originally developed by referring to several previous studies on food waste

[7]. Participants were requested to respond using a 4-point Likert scale ranging from "1 = Strongly disagree" to "4 = Strongly agree".

3 Result and Discussion

3.1 Food Waste Composition

Figure 3 shows the food composition of uncook food waste and cooked food waste generation for each of the sectors selected at Desa Pandan for this study. The results obtained for all sectors from the data collection were 63.3% category as the uncooked food waste and generated from food preparation, handlings, storage, processing, and disposal of unwanted parts raw food material. On the other hand, 36.7% of cooked food waste is generated during, after the cooking process, serving, and unfinished food. For 30 households, 58% was from uncook food waste and 42% was cooked food waste, 10 restaurants were 72% uncook food waste and 28% cooked food waste, and 3 schools were 60% uncook food waste and 40% cooked food waste respectively.

Figure 4a illustrates the composition of uncook food waste were classified into meat and fish (internal organs), vegetables, fruits, and eggshells. The data analysis shows that every sector produced different major contributors to uncook food waste. Households and restaurants made the highest composition of the vegetable component, which are (20%) and (28%) respectively, and schools had meat and fish (internal organs) (27%). The results show that vegetables mostly have a higher percentage of uncooked food waste. This is because in cooking, only the edible part of vegetables are being used. The process of vegetable preparation generates waste such as peels, skins and roots.

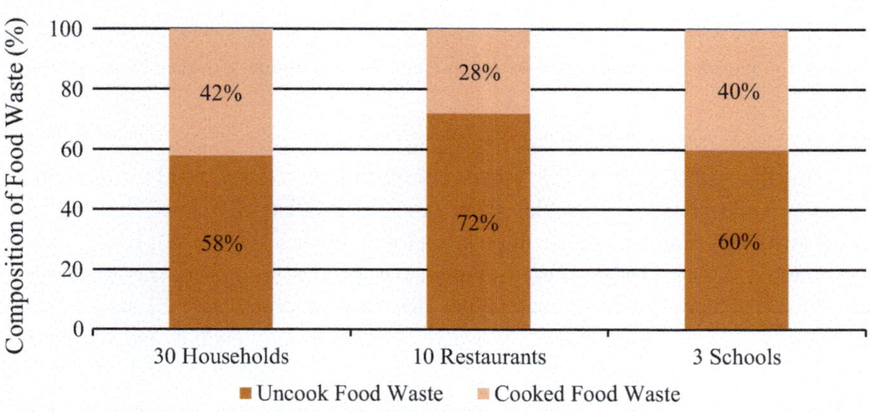

Fig. 3 Total composition of uncook food waste and cooked food waste (%)

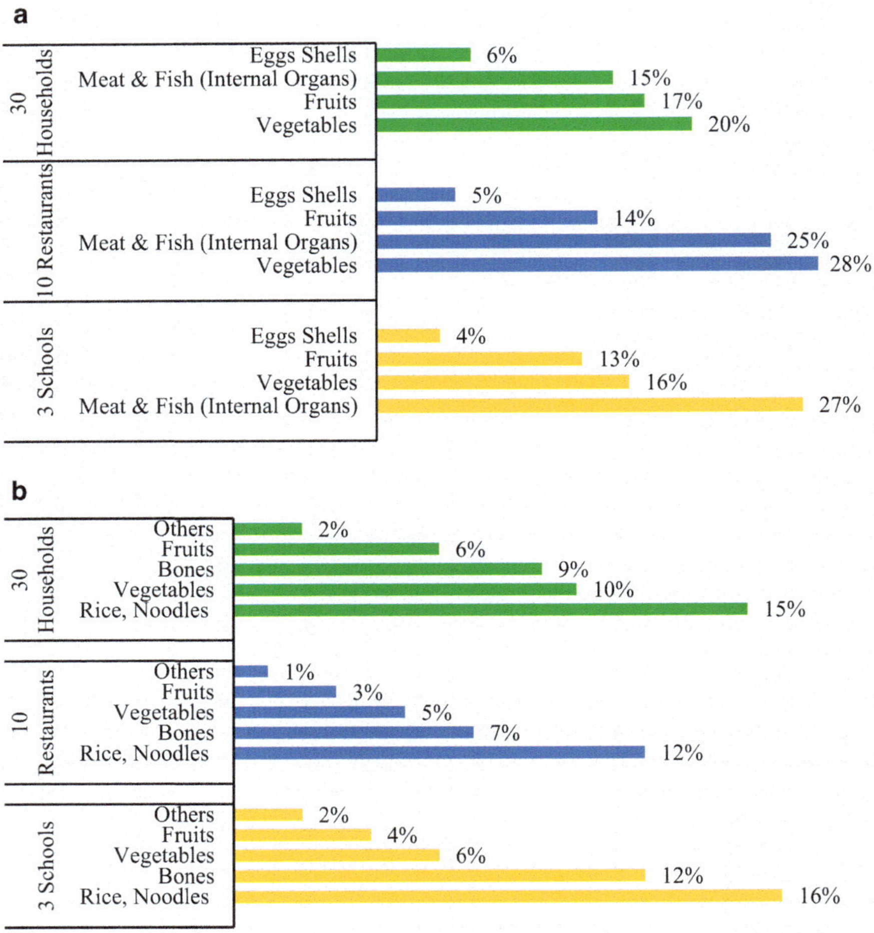

Fig. 4 **a** Composition of uncook food waste types (%). **b** Composition of cooked food waste types (%)

On the other hand, refer to Fig. 4b, the composition of cooked food waste includes rice and noodles, bones, vegetables, fruits, and others at Desa Pandan. The data analysis shows the rice and noodles composition for 30 households (15%), 10 restaurants (12%), and 3 schools (16%) differed significantly than the second contributor for each sector for these cooked food waste types. Rice and noodles, as well as bones, are the main contributed to cooked food waste generation, and each of the sectors has similar types of the major contributor to cooked food waste. Rice and noodles are the primary food source in Southeast Asia, contributing to rice's generation caused by unfinished eating and excess cooking and over-cooking [8]. Additionally, bones are mainly from fish and chicken during food preparation, as well as from unfinished eating.

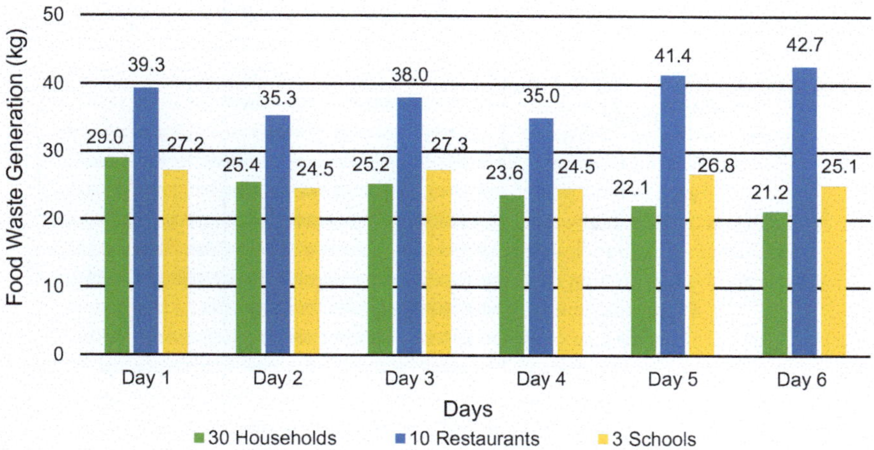

Fig. 5 Food waste generation of 30 households, 10 restaurants and 3 schools

3.2 *Food Waste Generation*

Figure 5 shows the total food waste generated during six days of collection at 30 households, 10 restaurants and 3 schools were 146.5 kg (0.20 kg/capita/day), 231.7 kg (0.026 kg/p/day) and 155.4 kg (0.010 kg/p/day) respectively.

As the date of collection was during the week of Chinese New Year celebrations, there was a significant increase in food waste for the first two days, which were 29.0 kg (Day 1) and 25.4 kg (Day 2) total. Strict regulations imposed by national health authorities forced people to stay at home and plan their meals carefully, resulting in a shift in household food waste behaviour [5]. The food waste generation from restaurants could be attributed to the official government announcement to open certain business sectors and commercial activities as Kuala Lumpur enters the first week of Conditional Movement Control Order (CMCO). Despite this, the food waste recorded the highest amounts of 41.4 kg (Day 5) and 42.7 kg (Day 6), indicating an increase in food waste generation. People are beginning to dine in restaurants instead of relying on delivery or takeout [2]. For the school canteen, they provided meals in packages, and the food was pre-ordered ahead of time to reduce the number of pick-ups required, and the school only allowed students of SPM candidates to attend. This resulted in the least amount of food waste generated, which was 155.4 kg (0.010 kg/capita/day).

3.3 *Food Waste Awareness*

This section is aimed at determining how the Covid-19 pandemic has affected people's attitudes and behaviors toward food waste. Figure 6 findings reveal that

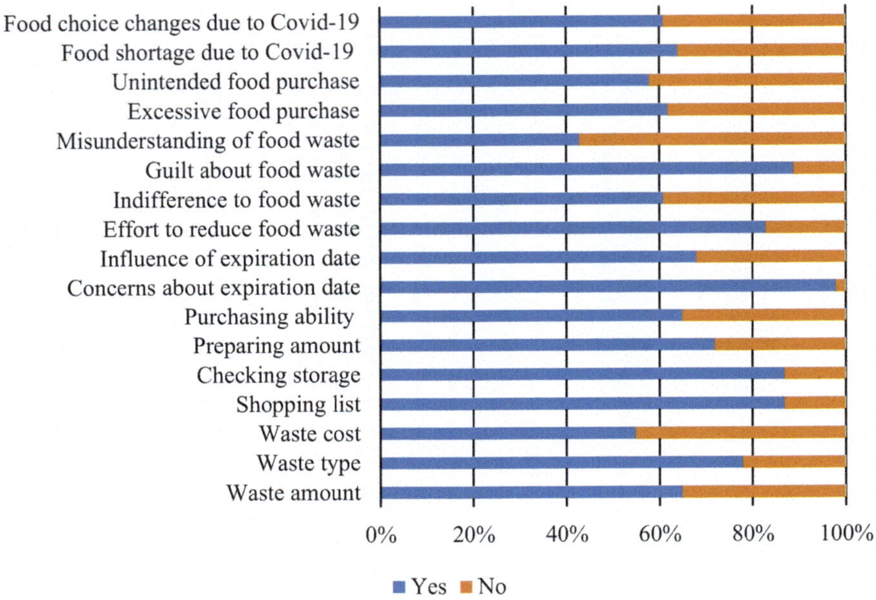

Fig. 6 Food waste awareness of the respondents during Covid-19 outbreak

people in pandemic-affected areas such as in Desa Pandan, pay more attention to the (78%) types, (55%) amount, and (55%) cost of daily household food waste. Most pandemic-affected areas, such as Kuala Lumpur's metropolitan areas, have a larger population, higher food costs, less household storage space, and a relatively more risky food supply [5].

Food preparation denotes how participants reduce food waste when purchasing food, preparing food, and cooking. 87% of the respondents list the food to buy before a shopping trip and claim that they check their food storage first before buying. 72% of respondents can exactly cook and prepare the food they need while 65% can precisely buy the amount of food their household needs. The findings revealed that the amount of eatable food thrown away is the result of several interconnected factors, each with varying degrees of importance for households. Socio-cultural and socio-demographic features, as well as eating habits and food preparation practices, may have an impact on the amount of food discarded [4].

The highest level of agreement among respondents was related to guilt on throwing away food with 89% agreeing with the statement. When it comes to food waste, customers are concerned about food poisoning, and by thrown away already unconsumed food can reduce the risks of food-borne illness [9]. Finally, the respondent is asked about suggesting whether the participants take note of the dates imposed on food productions and are influenced by them. 98% of the respondents are concerned as they will pay attention to food date labels such as "use by" and "best before" dates when buying food but show a significant drop in the influence of expiration date, which is 68% as the respondent will make different decisions depending on the date label. Consumers can use this information to plan their food stock accordingly. Unfortunately, variation in date labeling, terms, and uses has resulted in significant food waste. This scenario may hinder efforts to reduce food waste [10].

4 Conclusion

In conclusion, the total weight of the food waste generation (kg) and total composition (%) of uncooked food waste and cooked food waste was determined during the Covid-19 outbreak. Total food waste generated during six days of collection at 30 households, 10 restaurants and 3 schools were 146.5 kg (0.20 kg/capita/day), 231.7 kg (0.026 kg/capita/day) and 155.4 kg (0.010 kg/capita/day) respectively. Furthermore, findings show that food waste awareness, as well as behaviour on food purchase, management, and consumption, have significant impacts on the reduction of food waste generation at the household level, where the Covid-19 pandemic influenced people's attitudes and practises toward household food waste in Desa Pandan, Kuala Lumpur. Overall, food waste is a type of solid waste with a classification that is heavily influenced by consumers. As a result, if food waste management is not managed efficiently, several issues will arise, including environmental pollution and human health issues.

References

1. Fegalo K, Ismail TH (2018) Household purchase and generation of food waste in Malaysia (Sri Serdang and Taman Connaught Cheras Kuala Lumpur). Adv Recycl Waste Manag 02. https://doi.org/10.4172/2475-7675.1000139
2. Amicarelli V, Bux C (2020) Food waste in Italian households during the Covid-19 pandemic: a self-reporting approach. Food Secur. https://doi.org/10.1007/s12571-020-01121-z
3. Portugal T, Freitas S, Cunha LM, Rocha AMCN (2020) Evaluation of determinants of food waste in family households in the greater porto area based on self- reported consumption practices. Sustainability (Switzerland) 12(1–12):2. https://doi.org/10.3390/su12218781
4. Jarjusey F (2017) Consumers' awareness and knowledge about food waste in Selangor, Malaysia. Int J Bus Econ Aff 2:91–97. https://doi.org/ https://doi.org/10.24088/ijbea-2017-22002
5. Ismail MH, Ghazi TIM, Hamzah MH, Manaf LA, Tahir RM, Nasir AM, Omar AE (2020) Impact of movement control order (MCO) due to coronavirus disease (Covid-19) on food waste generation: a case study in Klang valley, Malaysia. Sustainability (Switzerland) 12:1–17. https://doi.org/ https://doi.org/10.3390/su12218848
6. Roberts M, Downing P (2020) Food waste and Covid-19-survey 2: lockdown easing. Wrap 1–25
7. Qian K, Javadi F, Hiramatsu M (2020) Influence of the Covid-19 pandemic on household food waste behavior in Japan. Sustainability (Switzerland) 12:1–14. https://doi.org/ https://doi.org/10.3390/su12239942
8. Amirudin N, Gim THT (2019) Impact of perceived food accessibility on household food waste behaviors: a case of the Klang valley, Malaysia. Resour Conserv Recycl 151:104335. https://doi.org/10.1016/j.resconrec.2019.05.011

9. Qi D, Roe BE (2016) Household food waste: Multivariate regression and principal components analyses of awareness and attitudes among u.s. consumers. PLoS ONE 11:1–19. https://doi.org/10.1371/journal.pone.0159250
10. Zainal D, Hassan KA (2019) Factors Influencing Household Food Waste Behaviour in Malaysia. Int J Res Bus Econ Manag 3:56–71

A Study on the Environmental Impact During Distribution and Disposal Stages for the 3-Ply Face Masks by Using Life Cycle Assessment (LCA)

Chow Suet Mun Christine, Tengku Nuraiti Tengku Izhar, Irnis Azura Zakarya, Sara Yasina Yusuf, Ayu Wazira Azhari, and Madalina Boboc

Abstract The demand of face masks had increased tremendously due to pandemic outbreak of COVID-19, leading to the increment production rate of face masks in Malaysia. Waste is also produced at the same time, resulting impacts towards the environment. Due to the land scarcity issue in Malaysia, the end of life treatment for the waste is taken into consideration. The study tools used in this study is life cycle assessment (LCA) to identify the significant potential environmental impact produced during the life cycle stages for distribution and disposal through GaBi Education Software. The disposal stage between landfill and incineration of the 3-Ply face masks is studied to determine the end of life treatment for it. The impact assessment method selected in this study is CML 2001-Jan 2016 with the environmental indicator of Global Warming Potential (GWP), Acidification Potential (AP), Eutrophication Potential (EP) and Ozone Depletion Potential (ODP). GWP results

C. S. M. Christine
Faculty of Civil Engineering Technology, Universiti Malaysia Perlis, Kompleks Pusat Pengajian Jejawi 3, 02600 Arau, Perlis, Malaysia
e-mail: s171130574@studentmail.unimap.edu.my

T. N. Tengku Izhar (✉) · I. A. Zakarya · S. Y. Yusuf
Sustainable Environment Research Group (SERG), Centre of Excellence Geoploymer and Green Technology (CeGeoGTech), Universiti Malaysia Perlis, 02600 Arau, Perlis, Malaysia
e-mail: nuraiti@unimap.edu.my

I. A. Zakarya
e-mail: irnis@unimap.edu.my

S. Y. Yusuf
e-mail: sarayasina@unimap.edu.my

A. W. Azhari
Water Research and Environmental Sustainability Growth, Centre of Excellence (WAREG), Universiti Malaysia Perlis, 02600 Arau, Perlis, Malaysia
e-mail: ayuwazira@unimap.edu.my

M. Boboc
National Institute for Research and Development in Environmental Protection Bucharest (INCDPM), Splaiul Independentei Street, 6th District, 294060031 Bucharest, Romania

N. Mohamed Noor et al. (eds.), *Proceedings of the 3rd International Conference on Green Environmental Engineering and Technology*, Lecture Notes in Civil Engineering 214,
https://doi.org/10.1007/978-981-16-7920-9_9

in producing highest impact to the environment during both distribution and disposal stages. The impact of GWP also relates to the climate change. Modern incineration is recommended to overcome the issue of land scarcity in Malaysia as the amount of waste by 3-Ply face masks are increasing due to the pandemic COVID-19, reducing the impacts towards the environment.

Keywords Environmental impact · Distribution · Disposal · LCA · 3-Ply face masks

1 Introduction

On March 2020, World Health Organization (WHO) had announced the coronavirus disease (COVID-19) as the pandemic disease [1]. The virus can be transmitted easily through saliva droplets when the infected person coughs or sneezes. WHO had encouraged the public to wear face masks and practice social distancing to reduce the risk from being infected. The target material for the study is 3-Ply face masks, focusing in the life cycle stages for distribution and disposal in Malaysia.

The usage of face masks speeds up tremendously causing the production rate of the face masks to increase in order to meet the demand of public. Based on the statistics value, up to 80% of the public started to wear face masks in public area in Malaysia since April 2020 [2]. As the usage increases, the waste produce also increases, causing plastic pollution to the environment due to the presence of microparticles in the fabrics [3]. 3-ply face mask is made up of polymeric materials, taking around 450 years to degrade completely [4]. Face masks are usually treated as general waste, ending up in landfills. However, due to the land scarcity issues in Malaysia, it is looking into the possibility of incineration process to address the issue [5].

The aims of this study are to identify the significant potential impacts produced during the life cycle stages for distribution and disposal with the input of 1000 kg 3-Ply face masks reflecting on real- life situation and also to analyze the end of life treatment with the uniform input of 1000 kg 3-Ply face masks waste to incineration and landfill.

Life cycle stages of the 3-Ply face masks focused in this study is distribution and disposal stages, where the disposal stage includes incineration and landfill. Only road freight and 1 way travelling direction is involved in the study. Life cycle assessment (LCA) based on the standardization of ISO 14040 and ISO 14044 is used as the study tool to determine the potential impact towards the environment. The source of data is retrieved from the database in GaBi education software, using CML 2001-Jan 2016 as the impact assessment method. GWP, AP, EP and ODP are the environmental indicator focused in the study. The results were analyzed, normalized and interpreted (Table 1).

Table 1 Scope of study area

Life cycle stages	Description
Distribution	• Manufacture to shop • Shop to user • Waste from user to incinerator • Waste from user to landfill • Waste to others
Disposal	• Incineration • Landfill

2 Materials and Method

2.1 Material

The target material in this study is 3-Ply face masks. 3-ply face mask is made up of 3 layers of fabrics, mainly non-woven bonded fabrics, having a better bacteria filtration [6]. The fabrics of the layers attributes in the stretching of the celluloses of O–H and C–H showing that the fabrics are produced from polymeric materials [3]. It also undergoes the process of spunbond and melt- blown where the fibers bond can be bond closely to each other producing the filaments that have ultra-fine submicron filaments.

2.2 Method

The objectives of the study were first figured out, which is to determine the significant impacts produced during life cycle stages for distribution and disposal stage for the 3-Ply face masks and the end of life treatment between incineration and landfill for the disposal stage. The data used were from secondary source and also the database system in GaBi Education Software. Ganzheitliche Bilanzierung is known as GaBi where is allows to create a life cycle assessment inventory through modeling the life cycle of the product and calculate the different balance throughout the system.

The method used to generate the results of impact assessment is CML 2001-Jan 2016 in GaBi Education Software. The software consists of an extensive database set where the inputs of data were extracted from it based on the unit process studied. GaBi is used to determine the potential impacts produced based on the process flow in the system. Midpoint results were used in the study focusing on the environmental indicator such as GWP, AP, EP and ODP. The results generated were then normalized, analyzed and interpreted, making conclusion and recommendations for the study.

For the emission value from each environmental indicator based on 1000 kg waste of 3-Ply face masks, the formula used is as below, derived from the ratio and proportional method.

Table 2 Waste mass distribution to its end of life treatment based on real-life situation [7]

End-life treatment	Ratio	Amount (kg)
Landfill	0.532	532
Incineration	0.168	168
Others	0.3	300
Total	1	1000

$$\text{Emission value (1000 kg)} = \frac{\text{Emission value} \times 1000\,\text{kg}}{\text{Mass (kg)}} \tag{1}$$

The emission value is the results generated from the software based on its respective mass while the mass indicates the mass calculated based on the reference ratio mass in Table 2. Normalization is used in this study to allow the interpretation of the results characterized, which is relative to the environmental impacts categories studied [8]. The formula of normalization used is as shown below:

$$\text{Normalized of the impact category, } N = \frac{\text{Emission of indicator (kg eq)}}{\text{Nornref}\left(\frac{\text{kg eq}}{\text{capita yr}}\right)} \tag{2}$$

3 Result and Discussion

Figure 1 shows the process flow of the life cycle stages created using GaBi education software. Only distribution and disposal stage were focused in the study. As for disposal stage, only incineration and landfill were involved. The basis of waste involved is 1000 kg, with extracting database from GaBi education software to generate out the result. The impact assessment method used in CML 2001-Jan 2016 (Table 3).

3.1 Significant Potential Impact Produced During Distribution and Disposal Stage of 3-Ply Face Masks

Face masks were distributed to their respective location and ended -up as general waste in the disposal area. Each mass input for the waste to its disposal area is based on their ratio reference respectively [7]. Based on the study by Johnke [10] and also Xin et al. [11], the main emission produced during incineration are carbon dioxide, methane and nitrous oxide which mainly leads to GWP. As for landfilling, the study by Tonini et al. [12] stated that the main emission during landfilling is from the emission of landfill gas and leachate.

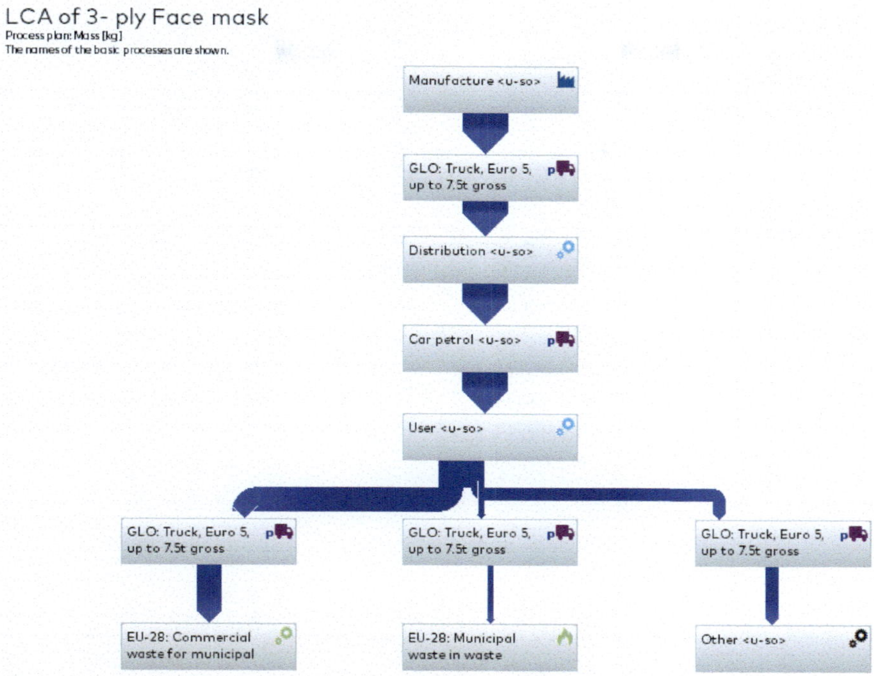

Fig. 1 Process flow diagram of LCA of 3-Ply face masks using GaBi education software

Table 3 Global normalization reference [9]

Impact category	Unit	Normref
Climate change	kg CO_2 eq	8700
Acidification	kg SO_2 eq	59
Eutrophication	kg PO_4^{3-} eq	19
Ozone depletion	kg CFC 11	0.103

Table 4 show the results that among the life cycles stages of the 3-Ply face masks studied, GWP produced the highest emission among the indicator studied while ODP produced the lowest emission. During distribution stage, the emission value for GWP is 130.57 kg CO_2 eq, while during the disposal stage, the GWP emission value is 718.68 kg CO_2 eq.

GWP is caused by the presence of GHGs which are made up of high percentage of carbon dioxide and methane gas [13]. The low emission value of ODP is due to the absent and low concentration of the compound of CFC and HCFC that will damage the ozone layer which is caused by the chlorine molecules react with ozone layer and converting them to oxygen [14].

Figure 2 shows the emission results during the distribution stage based on different travel distance. Truck A travels the route from transporting the products from manufacturer to seller. Vehicle A is from seller to users using lighter vehicle such as van.

Table 4 Overview on the potential impact produced during distribution stage reflecting on real-life situation

Environmental indicator	Life cycle stages		
	Distribution	Disposal	
		Incineration (168 kg)	Landfill (532 kg)
GWP (kg CO_2 eq)	130.57	560	158
AP (kg SO_2 eq)	0.2345	0.118	0.0629
EP (kg PO_4^{3-} eq)	0.06725	0.483	0.0131
ODP (kg CFC 11)	0	8.54E−14	6.77E−14
Total value (kg emission eq)	–	560.601	158.076
Total emission (kg emission eq)	130.8718	718.677	

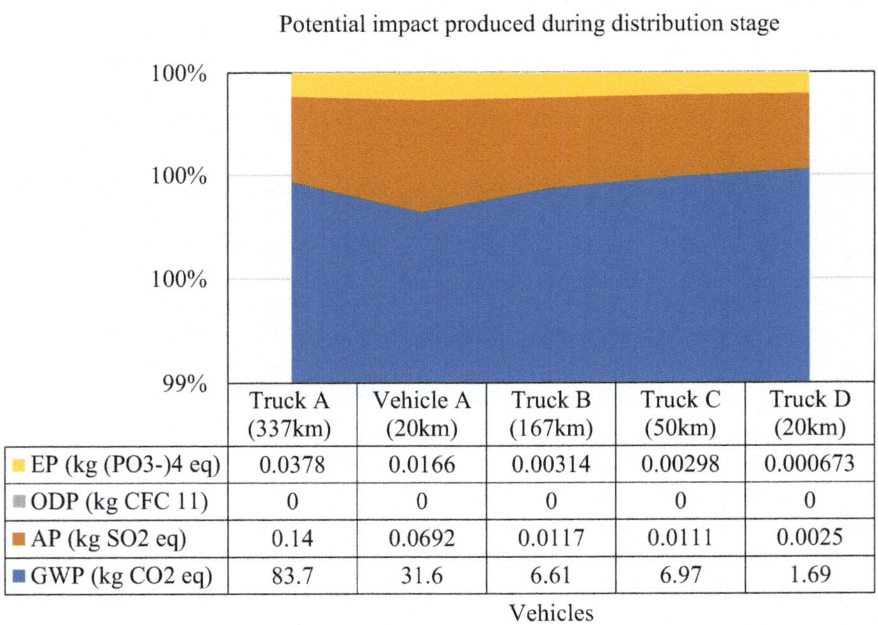

Fig. 2 Potential impact produced during distribution stage

Truck B and C are the transportation of waste to incinerator and landfill site respectively while truck D presented the other method of face masks disposal. The emission on each vehicle cannot be compared due to the different travel route distance and also different vehicles involved, thus only an overview can be discussed.

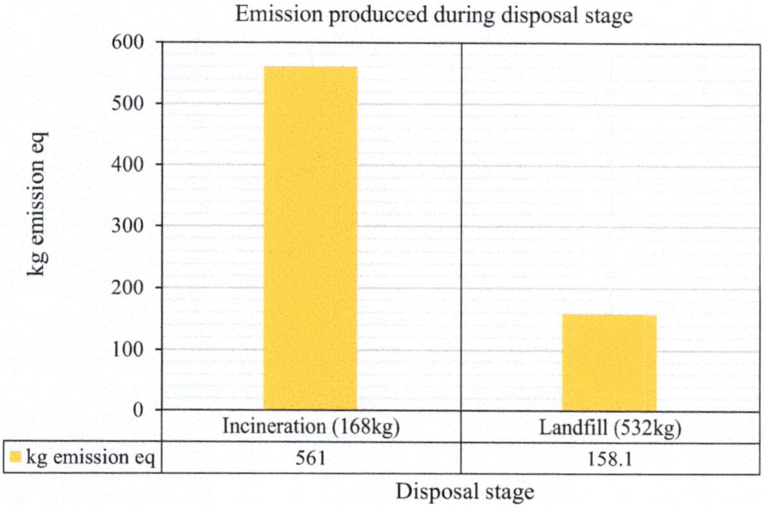

Fig. 3 Potential impact produced during disposal stage

The results for the emission value during disposal stage are shown in Fig. 3. Although the waste ratio sent off for incineration is less than landfill, however, the emission value towards the environment is higher than landfill. The high emission produced during incineration can be clearly determined on the impact towards the environment compare to landfilling.

3.2 *Emission Value from Incineration and Landfill Based on 1000 kg Waste of 3-Ply Face Masks*

1000 kg waste of 3-Ply face masks are sent off to the incinerator and landfill each respectively for their end of life treatment. The emission value is calculated based on the formula derived from ratio and proportion method, Eq. (1). The results show that GWP produced the highest emission among the indicator studied during both end of life treatment. The lowest emission produced is ODP, due to the absent and low concentration of compound such as CHCs and HCFC that will damage the ozone layer [14].

Table 5 above shows the overview on the emission between landfill and incineration based on 1000 kg waste. The total emission value for incineration is much higher than landfill with the value of 3336.907 kg emission eq to 297.1328 kg emission eq. From Fig. 4, the significant emission value produced by GWP is potentially higher among the indicator studied.

The emission value from GWP is high during incineration, mainly caused by the production of carbon dioxide and methane gas during the incineration of plastic

Table 5 Emission during disposal stage based on 1000 kg waste

Environmental indicator	Disposal stage	
	Landfill (1000 kg)	Incineration (1000 kg)
GWP (kg CO_2 eq)	296.99	3333.33
AP (kg SO_2 eq)	0.1182	0.7024
EP (kg PO_4^{3-} eq)	0.0246	2.875
ODP (kg CFC 11)	1.2726E−13	5.0833E−13
Total emission (kg emission eq)	297.1328	3336.907

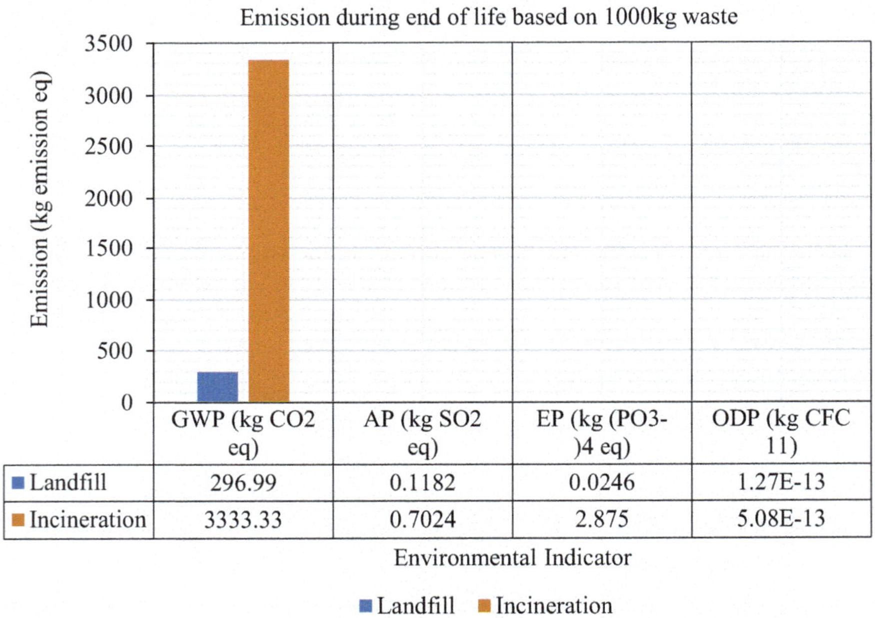

Fig. 4 Emission during the end of life treatment based on 1000 kg waste

waste [15]. Referring on the study by Kumar, centralized incineration produced high amount of carbon dioxide, which leads to GWP [16]. Contribution of greenhouse gas (GHGs) in landfill site that leads to GWP is due to the degradation of waste, combustion of diesel fuel by the vehicles transportation on site during the maintenance of the landfill [17].

Table 6 shows the normalized emission results during the life cycle stages of the 3-Ply face masks which is calculate using the formula (2) mention above. During the distribution stage, it results in the highest emission value at 0.015 for the climate change while during the disposal stage; the highest emission also comes from climate change for both landfill and incineration with 0.018 and 0.064. Among the 4 impact

Table 6 Normalized emission value during the life cycle stages for 3-Ply face masks

Impact category	Life cycle stages			Total
	Distribution	Disposal		
		Landfill	Incineration	
Climate change	0.015	0.018	0.064	0.097
Acidification	3.97E−3	1.066E−3	2E−3	7.036E−3
Eutrophication	3.53E−3	6.89E-4	0.025	0.029
Ozone depletion	0	6.57E-13	8.29E−13	1.486E−12
Total	0.0225	0.02	0.091	0.133

categories studied, climate change shows the highest emission which results in related to the emission of carbon dioxide and methane that leads to GWP.

From the study by Xin et al. [11], large amount of GHG is produced during distribution and disposal stage of waste, leading to climate change impact towards the environment. Figure 5 above shows the graph in the normalization of the impact categories. The graph significantly shows that the highest emission for the impact categories is climate change while the least emission of the impact categories is ozone

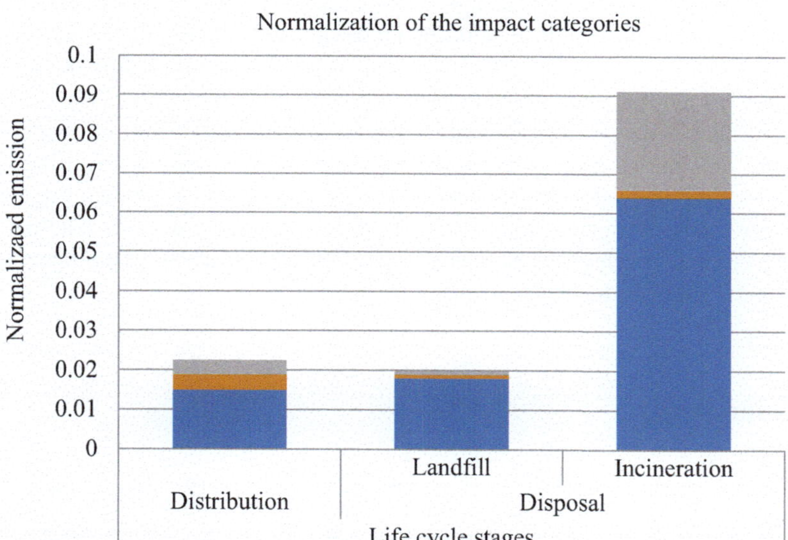

	Distribution	Landfill	Incineration
Ozone depletion	0	6.57E-13	8.29E-13
Eutrophication	3.53E-03	6.89E-04	0.025
Acidification	3.97E-03	1.07E-03	2.00E-03
Climate change	0.015	0.018	0.064

Fig. 5 Normalization of the impact categories

depletion. The impact of ozone depletion is not significant compare to the impact of climate change.

In order to achieve precise and efficient implementation to reduce pollutant emissions to the environment, it is critical to identify the link between environmental consequences and that specific life cycle stage [18]. The consideration to reduce pollutant emissions can be concentrated especially in the process of raw material extraction to the production stage, to primarily decrease emissions towards global warming (greenhouse gases), acidic substances, and particulate matters, which will accumulate in the atmosphere to cause significant effects.

4 Conclusion

From the results data generated above, the significant potential impact produced during the life cycle stages of 3-Ply face masks among distribution and disposal is GWP, while the least impact produced is ODP. High GWP emission value during the distribution stage is mainly caused by the burning of diesel and fossil fuel for the operation of the vehicles, while during the disposal stage is due to the incineration of plastic waste.

Under the situation of having 1000 kg of waste ending up in each end of life treatment of incineration and landfill, incineration results in a higher value emission data of GWP. Incineration release harmful gases such as carbon dioxide and methane gas resulting high emission value of GWP.

Based on the current pandemic situation in Malaysia, the waste produce by the 3-Ply face masks used is relatively high and huge in amount. With the issues of land scarcity happening at the same time, it is important to implement suggested end of life treatment such as incineration to reduce the waste towards landfill. Modern incinerator technology can be used to reduce the impact produced during incineration of waste towards the environment.

References

1. World Health Organization (n.d.) Coronavirus. Accessed 12 June 2021, from https://www.who.int/health-topics/coronavirus#tab=tab_1.
2. Statista (n.d.) Share of people who wore face masks in public places during COVID-19 outbreak in Malaysia from February 2020 to August 2021. Accessed 12 June 2021, from https://www.statista.com/statistics/1110960/malaysia-wearing-masks-during-covid-19-outbreak/
3. Aragaw TA (2020) Surgical face masks as a potential source for microplastic pollution in the COVID-19 scenario. Mar Pollut Bull. https://doi.org/10.1016/j.marpolbul.2020.111517
4. Waste of Ocean org. (2020) Plastic masks take 450 years to decompose in nature. Accessed 13 June 2021, from https://www.wastefreeoceans.org/post/plastic-masks-take-450-years-to-decompose-in-nature.

5. Abd Kadir SAS, Yin CY, Rosli SM, Chen X, El-Harbawi M (2013) Incineration of municipal solid waste in Malaysia: salient issues, policies and waste-to-energy initiatives. Renew Sustain Energy Rev 24:181–186. https://doi.org/10.1016/j.rser.2013.03.041
6. Henneberry B (2020) How surgical masks are made. Thomas Publ. Co.
7. Samsudin MDM, Mat Don M (2013) Municipal solid waste management in Malaysia: current practices, challenges and prospects. Jurnal Teknologi 62(1). https://doi.org/10.11113/jt.v62.1293
8. Crenna E, Secchi M, Benini L, Sala S (2019) Global environmental impacts: data sources and methodological choices for calculating normalization factors for LCA. Int J Life Cycle Assess 24:1851–1877. https://doi.org/10.1007/S11367-019-01604-Y
9. Heidi K, Stranddorf, Hoffmann L, Schmidt A, FORCE Technology (2005) Impact categories, normalisation and weighting in LCA. https://www2.mst.dk/udgiv/publications/2005/87-7614-574-3/pdf/87-7614-575-1.pdf
10. Johnke B (1996) Emissions from waste incineration. Intergov Panel Clim Chang https://www.ipcc-nggip.iges.or.jp/public/gp/bgp/5_3_Waste_Incineration.pdf
11. Xin C, Zhang T, Tsai SB, Zhai YM, Wang J (2020) An empirical study on greenhouse gas emission calculations under different municipal solid waste management strategies. Appl Sci 10(5). https://doi.org/10.3390/app10051673
12. Tonini D, Manfredi S, Bakas I, Kai-Sørensen Brogaard L, Damgaard A (2018) Life cycle assessment of landfilling. In: Cossu R, Stegmann R (eds) Solid waste landfilling: Conc, pp 955–972
13. United States Environmental Protection Agency (n.d.) Overview of Greenhouse Gases. Accessed 13 June 2021, from https://www.epa.gov/ghgemissions/overview-greenhouse-gases.
14. Darment (n.d.) Environmental impact indicators for refrigerants : ODP, GWP, TEWI. Accessed 13 June 2021, from https://darment.eu/refrigerant-environmental-impacts-indicators-odp-gwp-tewi/
15. Assamoi B, Lawryshyn Y (2012) The environmental comparison of landfilling vs. incineration of MSW accounting for waste diversion. Waste Manag 32(5):1019–1030. https://doi.org/10.1016/j.wasman.2011.10.023
16. Kumar H, Azad A, Gupta A, Sharma J, Bherwani H, Labhsetwar NK, Kumar R (2020) COVID-19 Creating another problem? sustainable solution for PPE disposal through LCA approach. Environ Dev Sustain 23:9481–9432. https://doi.org/10.1007/s10668-020-01033-0
17. Manfredi S, Tonini D, Christensen TH, Scharff H (2009) Landfilling of waste: accounting of greenhouse gases and global warming contributions. Waste Manag. Res 27(8):825–836. https://doi.org/10.1177/0734242X09348529
18. Tengku Izhar TN, Yap VM (2020) Life cycle analysis of plastic packaging. Conf Ser Earth Environ Sci 616:012036. https://doi.org/10.1088/1755-1315/616/1/012036

Comparative Evaluation of the Thermal Efficiency Between Rehabilitated and Non-rehabilitated Blocks by Using Thermography

Natalia Enache, György Deák, Lucian Laslo, Anda Rotaru, Monica Matei, Mădălina Boboc, and Nur Liza Rahim

Abstract Monitoring and inspection of apartment building facades with the help of thermography aims to identify and evaluate construction defects and their effects on thermal efficiency. The analysis of facade cracks is of great importance in the investigation of construction issues, especially in the rehabilitation processes of buildings. In this paper we studied the thermal rehabilitation efficiency of apartment buildings in Bucharest. The used approach was to compare the facades of insulated with non-insulated ones in two periods of time to identify defects and detailed analysis of thermographs in areas with heat losses. The study focuses on the temperature gradient, temperature variation and distribution, but also on ΔT value which is a thermographic analysis tool that shows the temperature differences between two components, being a criterion to identify and evaluate defects.

Keywords Hydrodynamic modelling · Boundary conditions · Model calibration · Model validation · Danube river · Bystroe channel

1 Introduction

Monitoring the condition of the blocks with a thermal camera helps to identify the thermal energy losses that can be caused by the degradation of the thermal insulation, by an unsatisfactory insulation/workmanship from a qualitative point of view or by the total lack of insulation [1, 2]. Heat losses lead to higher thermal energy consumption which implicitly lead to problems with the ventilation system, condensation inside homes and an inadequate indoor climate [3].

N. Enache (✉) · G. Deák · L. Laslo · A. Rotaru · M. Matei · M. Boboc
National Institute for Research and Development in Environmental Protection Bucharest (INCDPM), Splaiul Independentei Street, 6th District, 294060031 Bucharest, Romania

N. L. Rahim
Sustainable Environment Research Group (SERG), Centre of Excellence Geopolymer and Green Technology (CEGeoGTech), Universiti Malaysia Perlis, 01000 Kangar, Perlis, Malaysia
e-mail: nurliza@unimap.edu.my

N. Mohamed Noor et al. (eds.), *Proceedings of the 3rd International Conference on Green Environmental Engineering and Technology*, Lecture Notes in Civil Engineering 214,
https://doi.org/10.1007/978-981-16-7920-9_10

In order to identify and evaluate the potential defects on the surfaces of the buildings in the thermographic image, emphasis is directed on the ΔT value. The identification of defects depends primarily on the heat flow, being influenced by the nature of the defect and the direction of heat flow [4]. ΔT up to 2 °C indicates abnormal thermal behavior, while ΔT at 4 °C provides strong evidence of it [5].

In order to compare the anomalies in the different heat fluxes, the thermographic contrast values of the anomalies (ΔT) can be corrected in relation to the average temperature of the region where the anomaly is located by adjusted ΔT, resulting ΔTa [6]. By using the ΔT criteria (temperature difference) in thermography, the severity of defects on unrehabilitated surfaces was evaluated, having as a control surface a similar surface, considered as normal temperature, without heat loss. Also, these ΔT criteria highlight the fluctuations of the temperature anomaly and is dependent on the atmospheric temperature [7].

The purpose of the study is to monitor the energy efficiency of apartment buildings included in a rehabilitation program of local authorities located in the neighborhoods of Crangasi and Drumul Taberei in Bucharest, by analyzing the behavior and temperature distribution on building surfaces and by detecting structural defects compared to apartment buildings that have not undergone thermal insulation.

2 Materials and Methods

2.1 Material

In this case study were selected 7 rehabilitated blocks (R) that were analyzed quantitatively and qualitatively by calculating ΔT, adjusted ΔT (ΔTa) and compared with other 7 unrehabilitated blocks (UR), but similar in architecture, height regime and construction materials. The use of ΔTa is justified for comparison of anomalies in different thermal flow conditions, such as anomalies in facades with different orientations or anomalies at different heights in the same facade.

In the first thermography campaign of the selected buildings from the winter season in February, both overview thermographs and detailed thermographs were performed, where heat losses were identified on the facade. In the second campaign that took place in the summer season, in June, only detailed thermographs were performed to highlight the dynamics of heat loss in areas with defects under the influence of air temperatures.

2.2 Methods

The FLIR E60 infrared camera with 320 × 240 IR Resolution, thermal sensitivity of 0.05 °C and temperature range of −20 °C to + 650 °C was used in the process of evaluating the thermal efficiency for the rehabilitated blocks. The thermal images were processed by the softwares specialized for the use of infrared camera: FLIR Tools and FLIR ResearchIR. With the help of FLIR Tools, the quantitative analysis of the ΔT values between the defective area and the no-defect area was performed. ΔT was obtained by $\Delta T = Ta - Tr$, where Ta is the temperature of the anomaly and Tr is the temperature of the defect-free region. Also, ΔT_{region} represents the temperature difference between two surfaces, one unrehabilitated and the other rehabilitated, but having the same dimensions and being similar in positioning on the building's facade. A ΔT adjustment method (ΔTa) is applied by $\Delta Ta = 100 * (\Delta T/Tm)$, where Tm represents the average temperature of the region where the anomaly is located. The influence of the external temperature on the heat flow on the surface was followed by determining the difference between the external ambient temperature and the maximum temperature on the building facade through $\Delta T_{air} = T_{air} - Sp_{max}$, where T_{air} is the outside temperature measured at the time of thermography and Sp_{max} represents the maximum temperature point on the facade of the building.

3 Result and Discussion

From the qualitative analysis of Fig. 1a, b, in which are represented the facades of rehabilitated buildings R1 and R2, as well as the unrehabilitated facades of buildings UR1.1 and UR2.1, we can see the significant temperature difference in the color's palette of thermographs and distribution of their non-uniformity caused by heat loss within the unrehabilitated facades. The conditions of the main parameters of thermography are: atmospheric temperature of 3.2 °C, distance of 50 m, relative humidity of 79.5% and emissivity set at 0.86.

Table 1 shows the values calculated for ΔT, ΔT_{region}, ΔTa and ΔT_{air}. ΔT was achieved by the difference between the maximum temperature point on the facade of the anomaly and a defect-free point. The highest ΔT calculated values were obtained

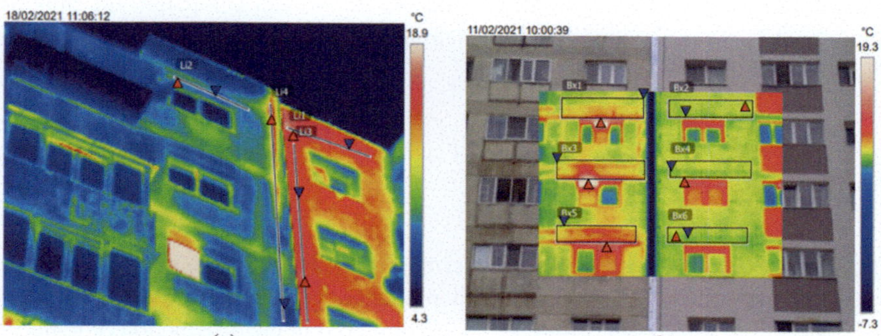

(a) (b)

Fig. 1 Comparison of the R1 surface (left) with the UR1.1 surface (right) within the same building (**a**) and comparison of the R2 (right) with the R2.1 surface (left) within the same building (**b**)

Table 1 Difference (ΔT) based on the comparison between similar components and atmospheric temperature

Building no	*Delta*T°_{air}C		$\Delta T^{\circ}_{region}$C		ΔT°C		ΔTa	
	First campaign	Second campaign	First campaign	Second campaign	First campaign	Second campaign	First campaign	Second campaign
R1	15.4	3	7.2	1.5	7.8	1.4	60	4.8
UR1.1	14.4	1.6						
R2	2.2	2.2	6.9	0.9	7.0	1.3	68.62	4.5
UR2.1	9.1	1.9						
R3	2	3.3	4.2	2.3	4.7	2.5	106	9.1
UR3.1	1.1	2.7						
R4	6.4	3.2	1.1	1.6	3.1	2.1	155	7.9
UR4.1	7,2	4.1						
R5	2,9	5.3	3.4	6.1	4.9	3.5	74.8	12.5
UR5.1	1.6	0.6						
R6	2	1.9	4.1	2.9	5.8	3.2	246.8	11.5
UR6.1	5.3	2.4						
R7	1.4	0.9	6.4	8	3.9	8.2	195	27.8
UR7.1	5.8	7						

for the UR1.1 building with a temperature range of 7.8 °C in the first campaign and UR7.1 building with a temperature of 8.2 °C in the second campaign. The lowest values were obtained for RU4.1 building with 3.1 °C in winter season and RU2.1 with a temperature of 1.3 °C in the summer season. ΔT_{region} calculation was made to have a clearer view of the amplitude of the anomaly on the surface.

The highest values were obtained for UR1.1 building, with temperatures of 7.2 °C, and the lowest values were for RU4.1 building with temperature 1.1 °C. The results for ΔT_{air} led to a maximum value of 15.4 for R1, at an outside temperature of 3.2 °C, and a minimum value of 1.1 °C for RU3.1 in the first campaign under the influence of an outdoor temperature of 7.4 °C. Also, in the second campaign, the highest value for ΔT_{air} was obtained for R5 which indicates a strong efficiency of thermal rehabilitation in these case.

4 Conclusion

Through the qualitative analysis of thermographs and the quantitative analysis consisting in the statistical processing of temperature values and their differences by ΔT calculation, significant differences were observed in the distribution of temperatures on surfaces, placement of defects on facades and severity of anomalies identified in unrehabilitated buildings. From the analysis of the ΔT criterion for surfaces with similar components and similar properties obtained between surfaces, but also between facades and ambient temperatures, a strong discrepancy of values obtained for rehabilitated and non-rehabilitated facades is indicated, with high efficiency for thermally insulated facades. It has been observed that insulation influences the decrease of temperatures on the exterior of facades, but has no long-term effect in maintaining a uniform distribution of surface temperatures, over time, appearing various defects caused by climatic conditions and external stressors. This situation is encountered in the case of buildings which, although insulated, have heat losses between the supporting plates or pillars and do not maintain the uniformity of the temperature distribution on the surface.

References

1. Mortarotti G, Morganti M, Cecere C (2017) Thermal analysis and energy-efficient solutions to preserve listed building Façades: the INA-casa building heritage. Buildings 7:56. https://doi.org/10.3390/buildings7030056
2. Ioja C, Patroescu M, Rozylowicz L, Nita M, Ioja A, Patroescu-Klotz I (2012) Evaluarea integrată a stării mediului în spaţiile rezidenţiale. https://doi.org/10.13140/2.1.3408.0327.
3. Hart JM (1991) A practical guide to infra-red thermography for building surveys, building research establishment. Watford
4. Sobota T, Taler J (2018) Determination of heat losses through building partitions. In: MATEC web of conferences, vol 240. EDP Sciences, p 05030
5. Bauer E, Pavón E, Barreira E, Kraus De Castro E (2016) Analysis of building facade defects using infrared thermography: laboratory studies. J Build Eng 6:93–104. ISSN 2352-7102. https://doi.org/10.1016/j.jobe.2016.02.012
6. Aidar LA, Bauer E (2019) Correction of thermographic contrast values in relation to the average temperature of the region for comparative analysis of anomalies exposed to different thermal flows. In: 15th international workshop on AITA (2019). Florence, Italy
7. Candoré JC, Bodnar J, Szeflinski A, Ibos L, Datcu S, Candau Y, Matteï S, Frichet JC (2009) Helps with the thermal diagnosis of the building: detection of defects of insulation by stimulated infra-red thermography. In: 9th international conference on quantitative infrared thermography. Krakow, Poland

Content Validity Index: Competencies of Quantity Surveyors and Enabling Technologies in Industry 4.0

Siti Nur Aishah Mohd Noor, Siti Uzairiah Mohd Tobi, and Mohamad Syazli Fathi

Abstract Competency is important for every profession where it is needed to ensure the efficiency of a person while carrying out their task. Furthermore, with the emergence of Industry 4.0 recently, there is a need for an improved competencies model for Quantity Surveyor. The objective of this study is to develop an improved competency model for Quantity Surveyor complete with Industry 4.0 enabling technologies that currently emerge nowadays. To achieve the above objective, a pilot study should be conducted on the target respondents, namely Quantity Surveyor. However, before the pilot study can be conducted, pre-testing on the research questions is done through the Content Validity Index (CVI) that involves the expert validators consist of academicians and industry experts. This paper intended to describe a systematic approach to quantify content validity in the form of content validity index best practice. The output of the CVI identified seven (7) parameters for non-technical competencies, ten (10) parameters for technical competencies, and twelve (12) parameters for enabling technological that are appropriate to be used in the questionnaire questions. Eventually, the output for this research will develop a comprehensive competency model for Quantity Surveyors where it is embedded with Industry 4.0 technology enablers.

Keywords Competency · Quantity surveyor · Questionnaire validation · Content validity index

S. N. A. M. Noor · S. U. M. Tobi (✉) · M. S. Fathi
Razak Faculty of Technology and Informatics, Universiti Teknologi Malaysia, Jalan Sultan Yahya Petra, 54100 Kuala Lumpur, Malaysia
e-mail: uzairiah.kl@utm.my

M. S. Fathi
e-mail: syazli@utm.my

Department of Built Environment & Technology, Faculty of Architecture, Planning and Surveying, Universiti Teknologi MARA Perak Branch, 32610, Seri Iskandar, Perak, Malaysia

N. Mohamed Noor et al. (eds.), *Proceedings of the 3rd International Conference on Green Environmental Engineering and Technology*, Lecture Notes in Civil Engineering 214,
https://doi.org/10.1007/978-981-16-7920-9_11

1 Introduction

Instrument content validity must be measured in order to assure construct validity of the instruments. The extent to which a measuring tool represents the measured concept is known as content validity, and it is used to establish the validity of measurement instrument. Due to the important of content validity in ensuring overall validity, it should be done in a systematic way based on research and best practice [1] mentioned that content validity refers to the degree that the instrument covers the content that it is supposed to measure. Meanwhile, [2] refers it as to the adequacy of the sampling of the content that should be measured. Therefore, [3] concluded that content validity measures the comprehensiveness and representativeness of the content of a scale. In this research, all parameters for each section of this questionnaire were obtained through a literature review, document review, and policy review on the competency of Quantity Surveyors.

2 Materials and Method

2.1 Method

Accordingly, Fig. 1 illustrate the methodology involve in the content validation as one of the pre-testing in the data collection.

In order to ease the understanding the research methodology, each of the step are elaborate as depicted in Table 1 below.

In this research, six panels of expert were selected to validate the content of Quantity Surveyor competencies. Content validity is usually carried out by seven or more experts [5]. However, [6] mentioned that five to ten panels are sufficient, whereas more than ten panels were likely unnecessary [7] highlighted criteria for selecting expert for content validity should be based on their job title, knowledge, experience in related fields and their availability to participate. Therefore, by adapting this criteria, several criteria in selecting the expert for this research were as follows:

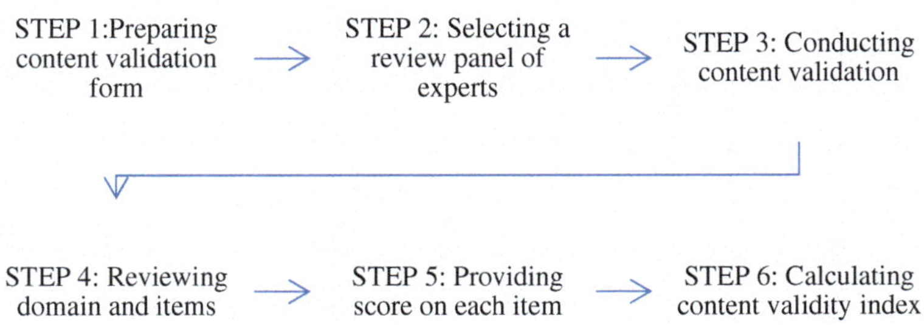

Fig. 1 Research methodology steps

Table 1 Research Methodology Elaboration

Step	Process	Description
1	Preparing content validation form	Content validation form is prepared to provide a platform for the expert panel to review and comments on each questions in the questionnaire. It is also to helps the panel of experts to have clear expectations and understanding about the task [4]
2	Selecting a review panel of experts	The selection of potential experts to review and critiques the assessment tool (questionnaires) where it is based on their expertise in Quantity Surveying fields as illustrated in Table 2
3	Conducting content validation	The content validation process was conducted through face-to-face and non-face-to-face during this pandemic phase. Ideally, face-to-face session is to facilitate and expedite the content validation process
4	Reviewing domain and items	The expert panel are requested to critically review the domain in each section before providing score of each item. Allocation of comment sections are to encourage the experts to provide suggestions to improve the relevance of the targeted domain
5	Providing score on each item	The expert is required to scale the item and submit their responses to the researcher once they have completely scored the items
6	Calculating content validity index	The calculation for the CVI indices are as illustrated in Table 3

Source Author Illustration adapted from [4]

1. Degree Holder in Quantity Surveying and
2. Professional Quantity Surveyor registered with Board of Quantity Surveyor (BQSM) or Royal Institute of Surveyor Malaysia (BQSM) or
3. Experienced Quantity Surveyor with over ten years of experience or
4. Academician who holds Doctor of Philosophy in Quantity Surveying fields.

Table 2 summaries the criteria for each of expert panels in this content validity process.

2.2 Data Analysis

Each of the items was calculated using Content Validity Indexing (CVI) and the final average of the I-CVI scores produces a scale level content validity score (S-CVI). Table 3 below illustrate the calculation for CVI indices.

Table 2 Criteria distribution of experts

Expert panels criteria	Expert 1	Expert 2	Expert 3	Expert 4	Expert 5	Expert 6
Degree holder in quantity surveying	*	*	*	*	*	*
Professional quantity surveyor registered with BQSM or RISM	*			*	*	*
Experienced quantity surveyor with over ten years of experience				*	*	*
Academician who holds Doctor of Philosophy (PhD) in quantity surveying fields	*	*	*	*		

Table 3 The definition and formula of I-CVI, and S-CVI [4]

The CVI indices	Definition	Formula
I-CVI (item-level content validity index)	The percentage of content experts rating an item's significance as 3 or 4	I-CVI = (agreed item)//(Number of expert)
S-CVI/Ave (scale-level content validity index based on average method)	The average score of all I-CVI over the number of item or everage of experts relevamce judgements	S-CVI/Ave = (sum of I-CVI scores)/(number of item)

3 Result and Discussion

The researcher analysed the results of the content validity of the scale as illustrated in Table 4. Accordingly, the item in the parameter that had CVI over than 0.79 are considered as appropriate and were remained. Meanwhile item score between 0.70 and 0.79 need some revision based on the panel of experts' recommendation and other items which score below 0.70 were discarded [7]. In addition, according to [8, 9], the acceptable CVI values involving six experts should be at least 0.83 [10]. However, items for 3.5- Big Data and Predictive Analytics with CVI value of 0.66 and 4 uniform agreements among expert panels; need to be revised accordingly to expert panels suggestions. The experts had agreed that the components are important for Industry 4.0 enabling technologies measurement, where some improvement is needed to ensure it can help to achieve the research objectives.

Based on the result showed, the research content and face validity of the instruments was established; the questionnaire reviewed by the panel of experts indicated that it was appropriate for the research purposes. On top of that, the content validity of the instrument was also evaluated. Therefore, based on the feedback received from the panel of experts, the questionnaire was revised and modified to ensure the degree of relevancy of each item in the questionnaire were adequate. Subsequently, this

Table 4 Result of content validity index

Section	Parameters	Expert agreement	I-CVI	UA	Rating
Non-technical	1.1 Communication Skills (CS)	6	1.00	1	Appropriate
	1.2 Critical Thinking and Problem Solving (CTPS)	6	1.00	1	Appropriate
	1.3 Teamwork Skills (TS)	6	1.00	1	Appropriate
	1.4 Lifelong Learning and Information Management Skills (LL)	6	1.00	1	Appropriate
	1.5 Entrepreneur Skills (ES)	5	0.83	0	Appropriate
	1.6 Professional Ethics and Moral (EM)	6	1.00	1	Appropriate
	1.7 Leadership Skills (LS)	6	1.00	1	Appropriate
	S-CVI/Ave		0.97		
Technical	2.1 Procurement	6	1.00	1	Appropriate
	2.2 Tendering	6	1.00	1	Appropriate
	2.3 Contract Administration	6	1.00	1	Appropriate
	2.4 Quantification & Measurement	6	1.00	1	Appropriate
	2.5 Bills of Quantities	6	1.00	1	Appropriate
	2.6 Estimating & Cost Planning	6	1.00	1	Appropriate
	2.7 Construction Knowledge	6	1.00	1	Appropriate
	2.8 Economic & Finance	6	1.00	1	Appropriate
	2.9 Information Technology & Software Literacy	5	0.83	0	Appropriate
	2.10 Law & Regulations	6	1.00	1	Appropriate
	2.11 Research & Development	3	0.50	0	Eliminate
	S-CVI/Ave		0.93		
Enabling Technologies	3.1 Building Information Modelling	6	1.00	1	Appropriate
	3.2 Internet of Things	5	0.83	0	Appropriate
	3.3 Blockchain	5	0.83	0	Appropriate
	3.4 Augmented Reality/Virtual Reality	5	0.83	0	Appropriate
	3.5 Big Data and Predictive Analytics	4	0.66	0	Need to revise
	3.6 3D Scanning & Photogrammetry/UAV Application	5	0.83	0	Appropriate

(continued)

Table 4 (continued)

Section	Parameters	Expert agreement	I-CVI	UA	Rating
	3.7 Cloud and Realtime Collaboration/Monitor Sensor	5	0.83	0	Appropriate
	3.8 Artificial Intelligence	5	0.83	0	Appropriate
	3.9 Advance Building Materials	5	0.83	0	Appropriate
	3.10 Smart Construction Site	5	0.83	0	Appropriate
	3.11 Smart Factory/Prefabricated/Modular	5	0.83	0	Appropriate
	3.12 3D Printing & Additive Manufacturing	5	0.83	0	Appropriate
	S-CVI/Ave		0.83		

I-CVI: Item Content Validity Index
S-CVI/Ave: Average score of I-CVI
UA: Universal Agreement

instrument was judged to have a good face validity and good content-related validity based on the S-CVI/Ave value of more than 0.90.

4 Conclusion

In conclusion, a total of 7 non-technical competencies, 11 technical competencies and 12 enabling technologies were identified from literature and were validated by panel of experts. Respectively, based on the CVI calculation, 7 non-technical, 10 technical competencies and 12 enabling technologies were selected from the outcome of Content Validation Survey. The established competencies and enabling technologies were collectively agreed upon and confirmed to be applicable for Quantity Surveyors competencies. Through this finding, this research has proceeded into main data collection stage in order to achieve the main objectives of this research.

References

1. Bush CT (1985) Nursing research. Reston Publishing Company, Virginia
2. Polit DF, Hungler BP (1991) Nursing research principles and methods, 4th edn. JB Lippincott Company, Philadelphia
3. Yaghmaie (2003) Content validity and its estimation. J Med Educ 3(1):25–27. https://doi.org/10.22037/jme.v3i1.870
4. Yusoff MSB (2019) ABC of content validation and content validity index calculation. Educ Med J 11(2):49–54. https://doi.org/10.21315/eimj2019.11.2.6

5. Parsian N, Dunning T (2009) Developing and validating a Questionnaires to measure spiritually: a physchometric Process. Global J Health Sci 1(1):2–11. https://doi.org/10.5539/gjhs.v1n1p2
6. Saidin MT, Mohammad IS, Zain FMY, Lop NS (2019) Establishing the content validity index of post occupancy(POE) of green building in Malaysia. Malaysian J Sustain Environ 7(1):75–100. https://doi.org/10.24191/myse.v7i1.8911
7. Yun J, Ulrich DA (2002) Estimating measurement validity: a tutorial. Hum Kinet J 19(1):32–47. https://doi.org/10.1123/apaq.19.1.32
8. Rodrigues IB, Adachi JD, Beatti K, Macdermid JC (2017) Development and validation of a new tool to measure the facilitators, barriers and preferences to exercise in people with osteoporosis. BMC Musculoskelet Disord 18(1):540. https://doi.org/10.1186/s12891-017-1914-5
9. Polit DF, Beck CT (2006) The content validity index: are you sure you know whats being reported? Critique and recommendations. Res Nurs Health 29(5):89–97. https://doi.org/10.1002/nur.20147
10. Polit DF, Beck CT, Owen SV (2007) Is the CVI and acceptable indicator of content validity? appraisal and recommendations. Res Nurs Health 30(4):59–67. https://doi.org/10.1002/nur.20199

Cost Analysis of Different Scenarios for Extracting Groundwater Contamination

Marwan Kheimi, Sultan K. Salamah, and Mohd Ikhmal Haqeem Hassan

Abstract Groundwater is an essential product for arid and semi-arid with dry climate region. The essence of groundwater is portrayed in areas where is low precipitation amount is occurred. Therefore, protecting the recharged aquifer from contamination plays a huge role to provide clean and easy to use pumped water for agricultural, industrial, commercial, and public use. In this paper, a conceptional of different scenarios is used to minimize the cost of extraction and transported plume pollutants from groundwater. The Interactive Groundwater model (IGW) has been implemented in Volusia County, north of Orlando area by utilizing parameters of hydraulic conductivity, storativity (specific yield) and layers depth from the SSURGO soil database. Also, the results of the model were compared with USGS groundwater levels from nearby well. The flow rates are grounded on the hydraulic conductivities (K), which have an initial average value of 11.3 m^3/day, and it is randomly varied throughout the domain using spectral distribution function, the thickness (b), and the hydraulic gradient (h) comprises the boundary conditions for the modeled area. Three different scenarios have been adopted based on the EPA technical report to seek cost efficient and realistic scenario of removing pollutants. The model has depicted that the best-case scenario to implement is three injection wells and two pumping wells to get the best results needed.

Keywords Groundwater modeling · Groundwater contamination · Plume transport

M. Kheimi (✉)
Department of Civil Engineering, Faculty of Engineering-Rabigh Branch, King Abdulaziz University, Jeddah 21589, Saudi Arabia
e-mail: mmkheimi@kau.edu.sa

S. K. Salamah
Department of Civil Engineering, Faculty of Engineering, Taibah University, Madinah, Saudi Arabia
e-mail: szain@taibahu.edu.sa

M. I. H. Hassan
Faculty of Civil Engineering and Built Environment, Universiti Tun Hussein Onn Malaysia, 86400 Parit Raja, Batu Pahat, Johor, Malaysia

N. Mohamed Noor et al. (eds.), *Proceedings of the 3rd International Conference on Green Environmental Engineering and Technology*, Lecture Notes in Civil Engineering 214,
https://doi.org/10.1007/978-981-16-7920-9_12

1 Introduction

Groundwater is one of the essential sources for irrigation and public use for drinking and other purposes. Unfortunately, groundwater is susceptible to pollutants [1, 2]. Groundwater contamination occurs with the interference of human-induced products such as gasoline, oil, road salts, and chemicals when they get into the groundwater and cause it to become unsafe and harmful. Therefore, it becomes more challenging to be used as a natural source of water [2]. During the past decade, groundwater scientists and engineers have advised several methods to contain and remediate soil groundwater contamination. Pumping and treatment is one of the remediation methods used to contain and remove the area of contamination [3]. So, this study aims to evaluate the time and cost for planning the extraction of contamination by using practical concepts adopted from the Emergency and Protection Agency (EPA) [4]. However, in terms of practical use, collection of three-dimensional characterization of the soil properties is essential to enable us to model the best practice method and allow us to have a reasonable prediction of the treatment plan within a specific budget.

The main objectives of this study are prevent the contamination from reaching the nearby lake, minimize the cost of contamination extraction project, suggest realistic scenarios for contamination removal, and model the scenarios using the Interactive Groundwater model (IGW).

2 Materials and Method

2.1 Description of the Model

In this study, the authors used an Interactive Groundwater Model (IGW) on the study area. The IGW is using rapid complex calculations and multi-model analysis to generate an accurate graphical representation of the input data results for groundwater modeling. The model uses 3DFLOW, UNCERT engines for incremental and first-order flow modeling and uses 3DTRANS for transport modeling. In addition, IGW solves sparse matrix systems to acquire iterative solutions using 3DTRANS. To present the data visually in 1D, 2D, and 3D, the IGW uses the GSLIB, OLECTRA CHART, GIS MAP-OBJECTS, and VTK engines to construct the graphical representation of the results. Using these engines in the base design of the IGW model allows for sophisticated interactive graphical solutions for the groundwater aquifer framework. In addition, IGW can assist in monitoring wells management by computing and graphically presenting concentrations and heads in time series manors [5, 6].

2.2 Data Collection

In order to run the model to analyze the movement of contamination in groundwater and how the pumping wells could extract the contamination, first step is to find the value of several parameters, mainly on soil characteristics, for our area of analysis. This included hydraulic conductivity, storativity, depth as well as water table level. Therefore, a shapefile for our area of interest was created and the associate data was downloaded from the SSURGO database available on the Web Soil Survey website of the United States Department of Agriculture [7]. Figure 1 below shows the area of interest.

Several shapefiles and Microsoft Access files were downloaded from the [8]. The main shapefile used in the study consists of 17 soil polygons. Our analysis further, using map unit keys (mukeys) from the SSURGO database (Fig. 3 below), showed that 7 of these polygons had the same mukeys and covered surface water areas. Moreover, two other pairs of the polygons had the same mukeys. This meant that, in total, we had eight different soil polygons in our area of interest. In order to process the downloaded data, we needed to understand the structure of each soil polygon. Figure 2 below, illustrates the structure of each soil polygon from SSURGO database. Accordingly, each polygon can consist of multiple soil components; each component itself may consist of multiple horizons with different soil characteristics.

Given the modeling limitations in the software package used, it is important to find a single value for hydraulic conductivity, storativity, and depth for each of the soil polygons. Therefore, the hydraulic conductivity was calculated for each soil polygon by summarizing the available data for its components and horizons and calculating an average value. The depth of each polygon was calculated by first adding the depth of soil horizons for each component and then calculating the average depth of the polygon's components. For storativity, since an unconfined aquifer was considered for this analysis, finding the specific yield of each soil polygon was needed. From

Fig. 1 Study area (Modified from [7])

Fig. 2 (On the left side) Soil polygons in the area of analysis with their mukeys and (on the right side) structure of each soil polygon in SSURGO database (modified from [7])

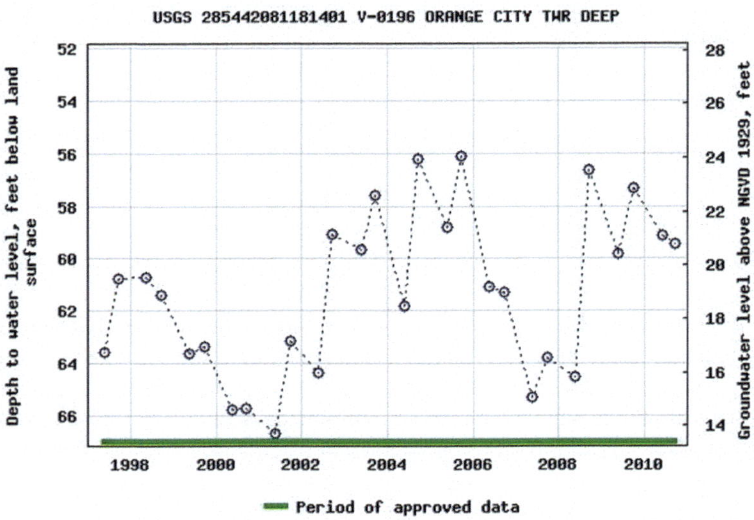

Fig. 3 Water table levels at the USGS gauging station [11]

SSURGO database, there was some data available on particle size for each soil component. Then, some data on the specific yield of various materials was acquired from a document by US Geological Survey (USGS), [9]. By summarizing the data, each polygon's soil components and making assumptions where data were missing, storativity or specific yield of each of our soil polygons was calculated. For porosity, since there was no data available on satiated water content for the area of analysis from SSURGO database, the typical porosity values for materials in each soil component was used. These typical porosity values were acquired from [10]. Then the porosity of each soil polygon was estimated. Table 1 below summarizes the analysis results on the required soil characteristics for each polygon in the study area.

In order to run our model, we also needed to provide water table level in our area of analysis. Therefore, we searched for the closest groundwater gauging station to our area of interest to estimate the water table level. The relative groundwater gauging station to the analysis area is on the right-hand side of Konomac Lake. We then

Table 1 Soil characteristics of each polygon in the area of analysis

Mukey	K-saturated (micrometers per second)	Depth (cm)	Storativity or specific yield (percent)	Porosity
1,544,109	172.6	131.4	16.5	0.4
1,544,117	162.4	163.1	26	0.37
1,544,121	60.3	299.6	20.4	0.39
1,544,130	125.7	209.8	26	0.37
1,544,132	133.9	207.3	26	0.37
1,544,139	201.6	199.6	26	0.37
1,544,166	103.5	185.2	26	0.37
1,544,171	84.3	135.7	16.5	0.4
1,544,185	Water	Water	Water	water

extracted the data on water table level at this site from USGS website, presented in Fig. 3. The data showed that, on average, the water table level was about 60 ft below the surface at this site. Since we intended to run the model with the assumption of an unconfined aquifer near the land surface, we decided to assume here for the water level table. Therefore, assuming that the water table level in our area of analysis was at one meter below the land surface.

2.3 The Model Domain in IGW

The area of the modeled domain is nearly 700 thousand square kilometers (700 m × 1000 m), surrounded by two lakes in the left and right of the domain. The model is comprised of three different layers as follow (with containment residing in the unconfined aquifer) (Fig. 4):

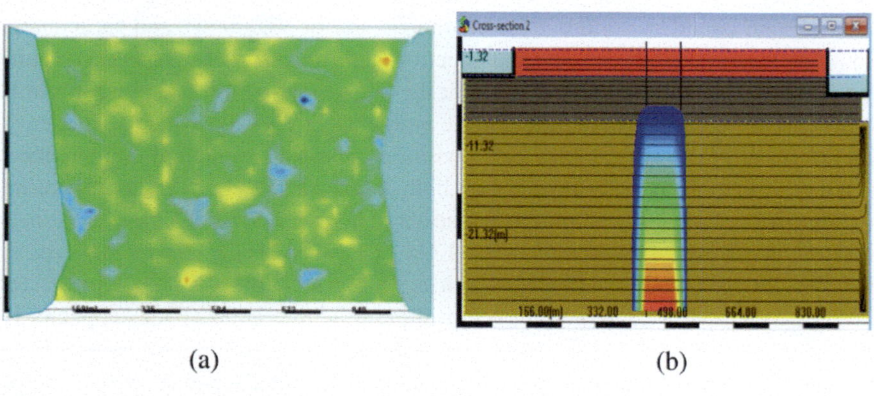

Fig. 4 **a** Area of analysis in IGW, **b** side view of layers below ground surface level

1. Unconfined layer with a depth of 3 m.
2. Confining layer of 5 m below the ground surface level.
3. Unconfined layer of 40 m in depth.

2.4 *Boundary Conditions*

The boundary conditions are shown Fig. 4: for a lateral boundary. The river boundary is the head difference that we assumed a head of 1 m below the ground surface for the lake on the left side in Fig. 4 while 3 m below the ground surface in the lake on the right, to allow the flow to move from left to right. In the north and south regions, we assigned no-flow boundary conditions as they are impermeable boundaries. The flow rates are based on the hydraulic conductivities (K), which have an initial average value of 11.3 m^3/day, and it is randomly varied throughout the domain using spectral distribution function, the thickness (b), and the hydraulic gradient (h).

2.5 *Proposal for Remediation*

Implementing three different scenarios acquired from [4] with a consideration that the below scenarios must be done without allowing the contamination to reach the nearby lake. First scenario is applying one well. Second scenario is applying pumping and injection wells (Doublet method). Finally, the third scenario is applying five wells (3 injection wells and two pumping wells).

3 Result and Discussion

Each scenario has been analyzed to see the effect of time and pumping rate with the same domain boundary conditions, containment size of 58,000 m^2, and parameters. However, the number of pumping and injection wells may vary based on the scenario implemented for the same conditions. For the first scenario with one pumping well with 2500 m^3/day, contamination can be extracted in 3 years, which seems easy and practical. However, we found that with one pumping wells and one injection well with 2500 m^3/day pumping rates for both, it takes nearly 3.7 years to extract the contamination. Lastly, in scenario three, the extraction process took 0.31 years, which is 89.7% faster than scenario one, and 91.6% faster than scenario two. However, it may also depend on other parameters such as pump's location and the pumping rate. Assuming that the price of a water removal pump with a 2500 m^3/day capacity is approximately $15,000 and the labor cost is around $20 per hour, we were able to estimate the cost associated with each scenario. Accordingly, the scenarios would respectively cost around $540,000; $678,000; and $129,000. Therefore, scenario three here seems to be the most cost-effective solution to the problem, with about 76.1% and 80.9% cost reduction from scenarios one and two [12].

4 Conclusion

Groundwater contamination is a significant threat to the provision of sustainable water resources for human and ecological uses. There have been creative methods devised to remove contamination from groundwater, including the use of extraction wells. In this study, we analyzed three different scenarios with different numbers of extraction wells for the removal of a plume of contamination from the area in Volusia County, north of Orlando. This area is located between two lakes, so the study analyzed scenarios that could remove the contamination before reaching the nearby lakes. We acquired data from SSURGO database [7] and used them in the IGW software package to analyze the performance of each scenario. Based on the results of our analysis, we concluded that a scenario where we use five wells (3 injection wells and two pumping wells) would provide the most cost-effective solution to the problem of contamination removal from groundwater in our area of study.

References

1. Hui Qian JC (2020) Assessing groundwater pollution and potential remediation processes in a multi-layer aquifer system. Environ Pollut 263, Part A. https://doi.org/10.1016/j.envpol.2020.114669
2. Peiyue Li DK (2021) Sources and consequences of groundwater contamination. Arch Environ Contam Toxicol 80:1–10. https://doi.org/10.1007/s00244-020-00805-z
3. Akujieze CN (2021) Remedies for soil and groundwater pollution: an overview. African Sci 7
4. EPA (1996) Pump-and-treat ground-water remediation: a guide for decision makers and practitioners. EPA, Washington DC
5. Li S-GQ (2006) A real-time, interactive steering environ-ment for integrated ground water modeling. Ground Water 44:758–763. https://doi.org/10.1111/j.1745-6584.2006.00225.x
6. Shu-Guang Li, QL (2004) Interactive Ground Water (IGW): an innovative digital laboratory for groundwater education and research. Comput Appl Eng Educ 11(4). https://doi.org/10.1002/cae.10052
7. Survey, W. S. (2016). United States Department of Agriculture, SSURGO database. Accessed 22 Sept 2016 from http://blogs2.esri.com/
8. Web Soil Survey (2019) (USDA-NRCS). Accessed 22 June 2016, from https://websoilsurvey.sc.egov.usda.gov/
9. Johnson A (1963) Compilation of specific yield for various materials. U.S. Geological Survey. Accessed 09 Feb 2021 from https://pubs.usgs.gov/of/1963/0059/report.pdf
10. Freeze AR, Cherry JA (1979) Groundwater. Prentice Hall, Englewood Cliffs, p 604
11. USGS (2015) SJRWMD Hydrological Data, v-0196. Accessed 22 Sept 2016, from http://webapub.sjrwmd.com/agws10/hdsnew/data.aspx?tp=3&stn=05601057_NGVD1929_POR&des=V0196%20Orange%20City%20Fire%20Tower%20(WL)%20FA$285439.08$811815.03$Lake%20Woodruff%20Unit$Volusia$Well$Water%20Level%20(WL)
12. Homeguide (2021) (Liaison Ventures, Inc.) Accessed 22 July 2021, from https://homeguide.com/costs/well-pump-cost

Knowledge, Attitudes, Awareness and Practices on Household Hazardous Waste Disposal Among Undergraduate Students in Selangor, Malaysia

Nurhidayah Hamzah, Nur Syazwina Marzuki, Fauzi Baharudin, Nur Liza Rahim, Nor Amani Filzah Mohd Kamil, Nor Azliza Akbar, and Nur Shazlinda Mohd Zin

Abstract The exponential growth of population in Malaysia and the consequently growing number of residences have aggravated the problem of household waste. Like many other towns of Selangor, a small town located in every district of Selangor also faces serious problems in terms of household hazardous waste disposal. The households of every district of Selangor generate and discharge a huge amount of untreated hazardous waste daily. In this study, an attempt has been made to analyze the scenario of household hazardous waste disposal in Selangor by means of a Knowledge, Attitude, Awareness and Practices (KAP) survey. Interview and personal observation were used to collect the data. The data collected were tabulated and processed to supplement the analysis through them. This study was using Microsoft Excel to

N. Hamzah (✉) · N. S. Marzuki · F. Baharudin
School of Civil Engineering, College of Engineering, Universiti Teknologi MARA, 40450 Shah Alam, Selangor, Malaysia
e-mail: nurhidayah0527@uitm.edu.my

F. Baharudin
e-mail: fauzi1956@uitm.edu.my

N. L. Rahim
Faculty of Civil Engineering Technology, Universiti Malaysia Perlis, Kompleks Pusat Pengajian Jejawi 3, 02600 Arau, Perlis, Malaysia
e-mail: nurliza@unimap.edu.my

N. A. F. Mohd Kamil · N. S. Mohd Zin
Faculty of Civil Engineering and Built Environment, Universiti Tun Hussien Onn Malaysia, Batu Pahat, Johor, Malaysia
e-mail: noramani@uthm.edu.my

N. S. Mohd Zin
e-mail: nursha@uthm.edu.my

N. A. Akbar
School of Civil Engineering, College of Engineering, CawanganPulau Pinang, Universiti Teknologi MARA, Kampus Permatang Pauh, 13500 Permatang Pauh, Pulau Pinang, Malaysia
e-mail: norazliza049@uitm.edu.my

N. Mohamed Noor et al. (eds.), *Proceedings of the 3rd International Conference on Green Environmental Engineering and Technology*, Lecture Notes in Civil Engineering 214,
https://doi.org/10.1007/978-981-16-7920-9_13

obtain percentage and correlation. The findings show that the average level of knowledge among students was 89.8%. While the average awareness level was 93.7%. The average attitude level for them is 91.8%. Lastly was 44% for practice level. Generally, this paper validates how knowledge influences attitude that subsequently determines behavior particularly in household hazardous waste management as intervened by appropriate environmental education. This study also indicates towards many positive aspects along with the negative ones. The positive aspects are mostly related to the awareness and efforts made by the students, whereas the negative aspects are mostly related to the improper methods of waste disposal adopted by the students and the existing scenario of the disposed waste. It is suggested that the household hazardous disposal practice can be improved by introduce more extensive program in university collaborated with related agencies.

Keywords Household Hazardous Waste (HHW) · Household Hazardous Waste Management (HHWM)

1 Introduction

Waste is defined by the original consumer as unused residues, disposed residues and items or artefacts which are no more usable. These items are by-products of human activities such as preparing, sorting, shipping, reassembling, disassembling, refurbishment, construction restoration and extraction. Fast urbanization, growing economy, industrialization and a rise in people's well-being have contributed to rising growth of disposal of waste [1]. The increase in atmospheric waste generation and disposal without recycling is seen as a risk to the planet and health impacts [2].

Household hazardous waste is classified as a fraction of household waste that contains corrosive, destructive, flammable, poisonous, inflammable, or reactive materials and is difficult to dispose of or endangers human health and the environment due to its biochemical nature. Dangerous waste from households is the unwanted portions of those products which contain hazardous ingredients. Household waste is a large portion of municipal solid waste, 4% or more of which can potentially affect both the environment and human health. In general, hazardous product divided into five categories: automobile, sanitation and polishing, paint and associated solvents, pesticides, and various things (examples include batteries, fingernail polish remover, certain cosmetics, and shoe polish). These materials are considered dangerous because they can be poisonous, combustible, corrosive and/or cause violent reactions to the chemical [3].

A good environmental controls good health and increases the quality of life for humans. Knowledge and awareness about waste management for households is very important [4]. Proper waste disposing is important for environmental conservation. Lack of awareness, erratic and unplanned waste dumping are the key causes of inappropriate waste disposal. Poor awareness of waste management is the main human health problem. People need to have information on disposal of household waste. It is important to be conscious between people about dealing with waste. Inadequate and inappropriate information of household waste handling can have severe health effects as well as the environmental impacts become more significant.

Moreover, Jatau [4] explained that a good understanding about household waste disposal will avoid infectious diseases and ensure safe environment. The attitude towards waste management is affected by their level of understanding. Inadequate collection and insufficient treatment of waste facilitates the transmission of bacteria, induces diseases such as cholera and diarrhea, and provides breeding grounds for disease vectors such as mosquitoes (malaria, dengue fever), flies (diarrhea) and rodents [5]. People ought to get good practice in their homes to dispose of household waste. Moreover, inadequate waste disposal activities contribute to environmental degradation by rising the burden of illness and disease among people. Therefore, a survey was conducted in Selangor because of previous study stated that huge household waste was collected as Petaling Jaya in 2010 was found to have 1.70 kg per household and 0.34 kg per person per day [6]. The population estimate living in this area (Bandar Baru Sungai Buloh, Shah Alam, Selangor, Malaysia) is 8000 people with a total population of approximately 2000. Bandar Baru Sungai Buloh is one of the old housing estates in Selangor is currently facing problems in the management of their domestic waste. Thus, this study aims to determine the level of understanding on the hazardous household waste management amongst undergraduate students by determine the percentage of knowledge, attitude, awareness, and practices. It is expected from the results obtained; more efficient program/policy can be performed by universities to achieve sustainable environment.

2 Materials and Method

Setting: The research has been conducted the undergraduate students in Malaysia that live in Selangor.

Research Design: A descriptive cross-sectional study has been carried out the undergraduate students in Malaysia that live in Selangor.

Population: The study population has female and male of students age between 20 to 30 years.

Sampling: Data was randomly collected of 100 samples. The research is primarily subjective and incorporates data obtained from the student questionnaire survey.

Research Instrument: Data was collected by pre-designed adopted questionnaire [7, 8], interview technique on different variables household hazardous waste, knowledge, attitude and practices.

Data Gathering Procedure: Convenience sample technique has been used to collect data on demographic variables, knowledge, attitude, and practice among undergraduate students.

(i) Section 1—Background of Respondents: Respondents are asked to provide details such as name, age, student's residence, and current semester. It is necessary to understand the student's background to get a general view of the respondent.

(ii) Section 2—Knowledge on HHW: This section aims to access the student's knowledge regarding household hazardous waste. Such as, do the respondents know how to classify the hazardous household waste. There are 5 questions in this section which require student to answer agree or disagree. Agree means students has knowledge on that question while disagree shows otherwise.

(iii) Section 3—Awareness on HHW: In this section, there are 3 questions. First question is designed by using the Likert scale. For this, the waste disposal was divided into four levels: Excellent awareness (very agree), Good awareness (agree), Satisfactory awareness (disagree) and Poor awareness (very disagree). Another two question are designed by using single choice answer (Yes or No).

(iv) Section 4—Attitude on HHW: In this section, there are 5 questions. These questions are designed by using the Likert scale. For Attitude on HHW analysis, assessing type of respondents' attitude towards household hazardous waste management was divided into three: positive attitude (very agree and agree), not sure (neutral) and negative attitude (disagree and very disagree).

(v) Section 5—Practices on HHW: In this section, there are 8 questions. These questions are designed by using the single choice and multiple choice.

Analysis of Data: Data analysis was done using Microsoft Excel. Data was analyses to obtain percentage of each section and the correlation analysis. The percentage of each section was calculated by determined the average value of the questions.

Study Timeline: Study has been conducted in 10-month duration from March 2020 to January 2021.

3 Result and Discussion

3.1 Profile of the Respondents

The study has done for undergraduate students in Malaysia that live in Selangor. The study indicates that majority of respondents were among the female group with 58% which equal to 58 female respondents while the rest 42 are among male respondents. Therefore, the overall results from this study can be concluded as more women preference rather than men. This can relate with some studies have shown that women are more in line with social comparisons than men [9]. Rather, it seems that men and women vary in their form of interdependence. For example, in view of the primary duty of women for washing, food preparation, family wellbeing, laundry and domestic maintenance, women and men can view household waste and its disposal differently. They may have different knowledge of waste or garbage. They will also handle waste differently and set different goals for disposal.

From this survey, the largest range of age that contributed to the percentage was 20–21 years old with a percentage of 51%. The second-largest range of age was

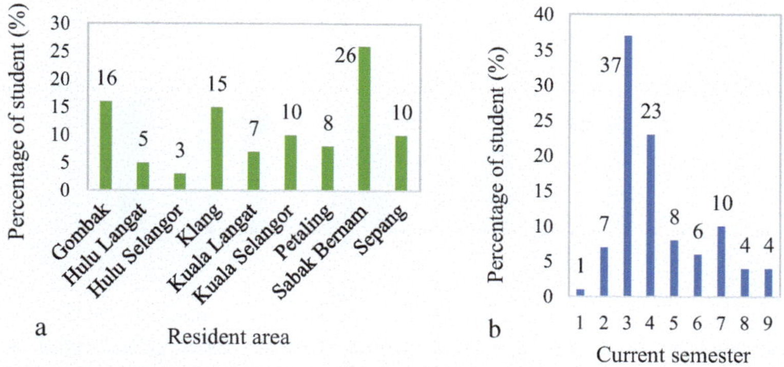

Fig. 1 Demographic Characteristic of Respondents which shows **a** current semester of the respondents and **b** resident area of the respondents

22–23 years old with a percentage of 43%. Then, the data followed by age 24 years old above with the percentage of 6%.

According to Fig. 1a the highest respondents were coming from semester 3 which consists of 37 respondents. While the lowest respondents were coming from semester 1 which consists of 1 or 1% of the respondents. Another semester for semester 2, 4, 5, 6 and 7 were 7%, 23%, 8%, 6% and 10% respectively. Semesters 8 and 9 were equal which were 4%. This study shows that most of respondent was come from semester 3 which enroll the environmental studies in their curriculum plan. For respondents' resident (Fig. 1b), the highest percentage was 26% which at Sabak Bernam. For the lowest percentage was 3% which at Hulu Selangor. For Kuala Selangor and Sepang got same percentage which was 10%. For Gombak, Hulu Langat, Hulu Selangor, Klang, Kuala Langat and Petaling were 16%, 5%, 3%, 15%, 7% and 8% respectively.

3.2 Knowledge on Household Hazardous Waste (HHW)

Figure 2 reveals the level of knowledge of respondents. In knowledge on HHW analysis, calculating the extent of knowledge of respondents about household hazardous waste disposal was split into two: knowledge (agree) and less knowledge (disagree). More than 80% respondents were agreed and less than 20% were disagree with the question given. Therefore, the analysis showed that an average of 89.8% of the respondents had knowledge of household hazardous waste disposal. This value is consistent with study made by Barloa et al. [10] and Shahzadi et al. [7] which shows satisfactory knowledge of 73.4% and 72% respectively on solid waste management.

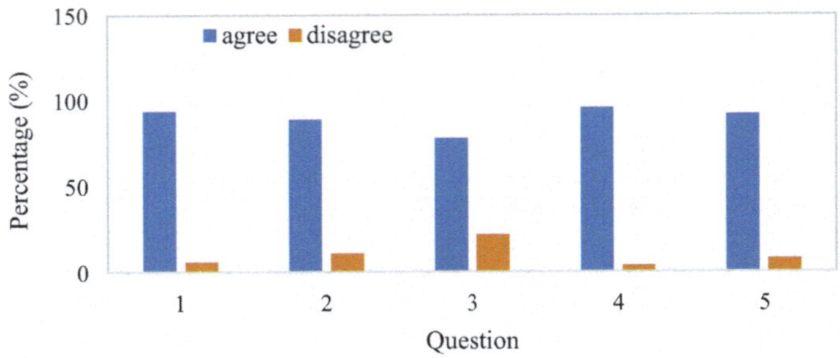

Fig. 2 Respondent's knowledge on HHW disposal which asked about (Q1) Chemical waste is considered as HHW, (Q2) Electronic and Electrical waste are considered as HHW, (Q3) Sharp waste are considered as HHW, (Q4) Education of family members is effective on quality and quantity of HHW, (Q5) The type and area of residential building are effective on quality and quantity HHW

3.3 Awareness on Household Hazardous Waste

This study estimated the level respondents' awareness towards HHW as shown in Fig. 3a which the average result of awareness level in this study showed that 45% respondents were excellent, 48.67% were good, 3.67% were satisfactory and 2.66% were poor in awareness. The study found a total of 93.7% of the respondents that had good and excellent awareness towards HHW disposal. This value is aligning with similar study conducted in South Africa which indicated 80.0% of the community had knowledge of household waste disposal and were aware of the detrimental effects of excessive waste disposal. Most of the population (83.0%) is conscious that further waste generation could damage the environment [7]. As shown in Fig. 3b an 83% of respondents had an environment topic in curriculum, so they have knowledge on HHW disposal. This clearly shown by 75% of respondents were knew how to dispose household waste. According to Maddox et al. [11] and Wang et al. [12],the awareness of students about environmental problems and solutions can be increased through

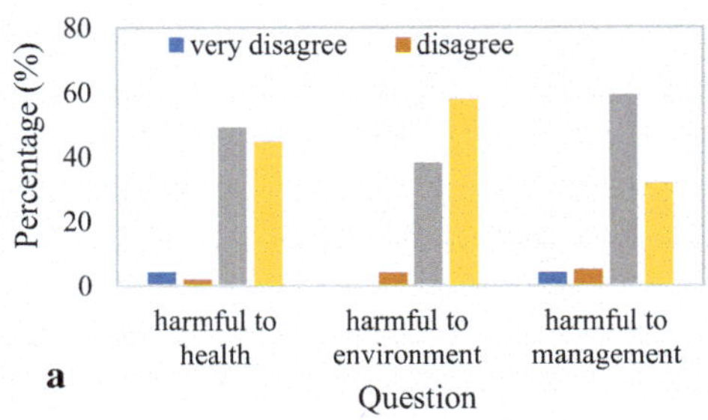

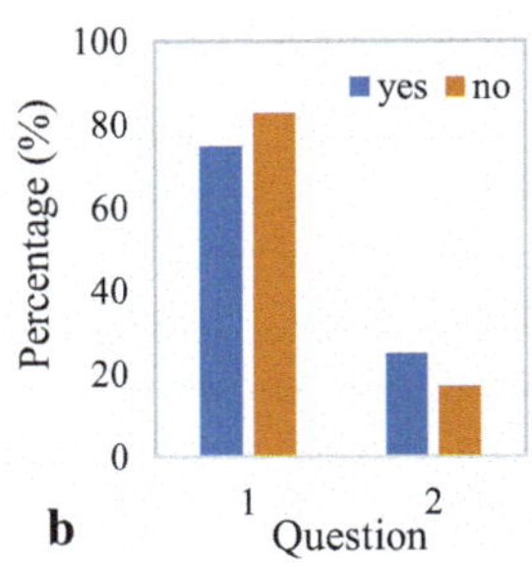

Fig. 3 Respondent's Awareness on HHW disposal that shows **a** awareness on health, environment, and management **b** question 1 asked if respondents know how to dispose HHW and question 2 asked if respondents have an environmental subject in their curriculum

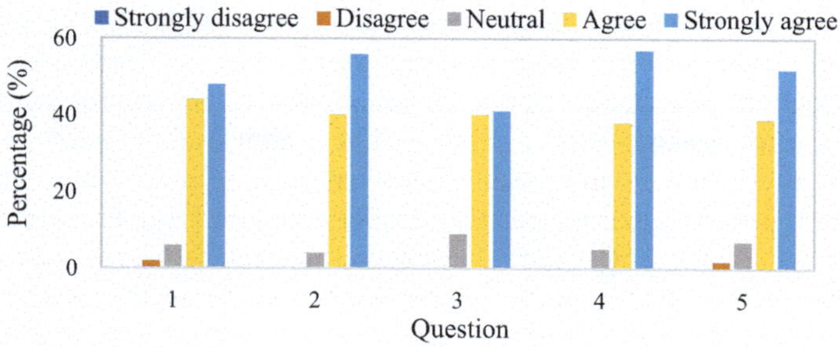

Fig. 4 Respondent's Attitude on HHW disposal which each question asked about the importance of (Q1) considering hazardous household waste material and their properties, (Q2) health and environment effects of hazardous household waste, (Q3) adverse effect of hazardous waste management on the proper management of normal household waste, (Q4) proper management (collection, processing, and disposal) of household hazardous waste, (Q5) media training in management of household hazardous waste

education. Thus, the particular skills and knowledge gained from environmental education would help in changing human behavior towards the environment [13]. Moreover, students with some knowledge and skills in environmental education are more motivated to take part in environmental protection activities and plans or known as ecological behavior [14], thus creating new ideas for solving environmental problems.

3.4 *Attitude on Household Hazardous Waste*

Based on Fig. 4, most of the respondents showed positive attitude towards household hazardous waste management with 91.8% of respondents were agree and very agree with the questions given. There was a strong positive correlation (0.86) between attitude and knowledge. According to Goytay and Thatte [15], value between 0.5 and 1 indicated strong positive correlation. This supported by Ghanbari et al. [16], which attitude was found to be positively correlated with family background knowledge. This study proves that the degree of continuity between environmental attitudes and behavior is influenced by a person's knowledge and understanding, public verbal engagement and sense of obligation [17, 18].

3.5 *Practice on Household Hazardous Waste*

For Practice on HHW analysis, a total of 61% respondents had worst practice on household hazardous waste disposal which relying on family members to dispose the HHW. This study was supported by the previous study which weak correlation was found between students' knowledge and environmental practices [19]. In general, people with a good level of knowledge and attitude did not show a satisfactory practice, which may be due to lack of a comprehensive and appropriate plan for household hazardous waste disposal. Furthermore, a high level of understanding and a strong perspective on environmental issues have not been shown to necessarily transform into positive environmentally friendly practices [20]. This supported by Tatlonghari and Jamias [21] which did not find any relational evidence between attitudes and practices, taking into account their observations on the clear association between degrees of awareness and practice.

Moreover only 44% respondents were separated the household waste while 56% were not separate the waste. This showed that half of the respondents were not good in practices on household hazardous waste. Most of the respondents knew that excessive disposal of waste would affect the lives of humans. In the study of the quality of knowledge and practice undertaken in Ogun State, Nigeria, Ayodeji [22] reported the same findings. It was noted that secondary school students were aware of the waste problems in their school compounds in the sampled areas, but the same students had ineffective waste management activities.

As shows in the Fig. 5a, none of the respondent separate battery and medical waste from their household waste and 53% of respondents did not separate their household waste at all. This study was supported by Ismail et al. [23] which showed that municipal waste management (MWM) practices were low in all classes of employees surveyed at the tertiary health institute in Dakshina, India, and generally, it appears that the prevalence of unsafe MWM practices in developing countries is troubling due to lack of proper training and allocation of medical waste to poorly trained personnel. About 72% of respondents were using the polythene to dispose the household waste as shown in Fig. 5b. This showed that lack of sustainable environmental method of hazardous household waste being disposed as polythene is known to contribute to the plastic pollution. According to Budhiarta et al. [24], the plastic waste generated in Kuala Lumpur, Malaysia is 21% which is considered high. The result also consistent

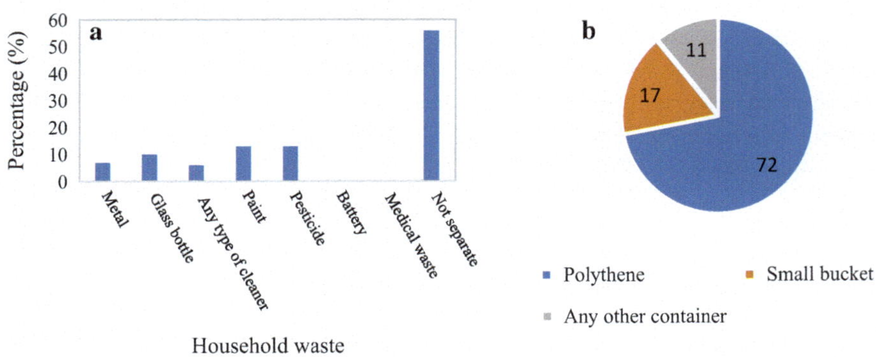

Fig. 5 Respondent's Practice on HHW disposal which **a** shows the separation items in the household and **b** the type of container respondent used to dispose the HHW

with the review made by Chen et al. [25] that the inconsistent application of policy initiatives by state governments, lack of public awareness and household recycling practices were found to be the major factors to plastic pollution in Malaysia.

4 Conclusion

The results showed that the students that live in Selangor have responded to 89.8% of knowledge level with an excellent 93.7% of attitude, 91.8% of awareness, but only 44% were good in their practice of household hazardous waste management. Most of the students are aware of the necessity of proper waste disposal. Therefore, it can be suggested that the study undertaken undergraduate students at Selangor revealed many attention-grabbing facts relating to the perception of the residents on household waste disposal. It helped in analyzing the present scenario of household waste disposal in student's resident. Environmental education is recommended, by held a program with emphasis on issues regarding household hazardous waste management and recycling, it should also be included in the basic curriculum or certain course works of college students that all students must be participated, to expand their knowledge on how to improve practices on household hazardous waste management. Relevant seminars and programs on environmental protection and waste management should be organized by Kementerian Pelajaran Malaysia (KPM) and all the higher institutions in Malaysia to encourage students and the public to become environmentally responsible citizens.

Acknowledgements Authors would like to thank College of Engineering, Universiti Teknologi Mara, Shah Alam for the encouragement and funding this project.

References

1. Brown DP (2015) Garbage: how population, landmass, and development interact with culture in the production of waste. Resour Conserv Recycl 98:41–54. https://doi.org/10.1016/j.resconrec.2015.02.012
2. Domingo JL, Marquès M, Mari M, Schuhmacher M (2020) Adverse health effects for populations living near waste incinerators with special attention to hazardous waste incinerators. a review of the scientific literature. Environ Res 187:109631. https://doi.org/10.1016/j.envres.2020.109631
3. Slack, R.; Letcher, T.M (2011). Chemicals in Waste: Household Hazardous Waste; Elsevier Inc., ISBN 9780123814753.
4. Jatau AA (2013) Knowledge, attitudes and practices associated with waste management in jos south metropolis, Plateau State. Mediterr J Soc Sci 4:119–127. https://doi.org/10.5901/mjss.2013.v4n5p119
5. Adogu POU, Uwakwe KA, Egenti NB, Okwuoha AP, Nkwocha IB (2015) Assessment of waste management practices among residents of owerri municipal imo state Nigeria. J Environ Prot (Irvine,. Calif) 06:446–456. https://doi.org/10.4236/jep.2015.65043

6. Mohd Yatim SR, Arshad MA (2010) Household solid waste characteristics and management in low cost apartment in Petaling Jaya. Selangor Heal Environ J 1:58–63
7. Shahzadi A, Hussain M, Afzal M, Gilani SA (2018) Determination the level of knowledge, attitude, and practices regarding household waste disposal among people in rural community of Lahore. Int J Soc Sci Manag 5:219–224. https://doi.org/10.3126/ijssm.v5i3.20614
8. Amouei A, Reza Hosseini S, Khafri S, Tirgar A, Aghalari Z, Faraji H, Barari R, Namvar Z (2016) Knowledge, attitude and practice of iranian urban residents regarding the management of household hazardous solid wastes in 2014. Arch Hyg Sci @Bullet J Homepage 55:1–8
9. Martínez-Borreguero G, Maestre-Jiménez J, Mateos-Núñez M, Naranjo-Correa FL (2019) Knowledge analysis of the prospective secondary school teacher on a key concept in sustainability: waste. Sustain 11:1173. https://doi.org/10.3390/su11041173
10. Barloa EP, Lapie LP, Paul C, Cruz PD (2016) Knowledge, attitudes, and practices on solid waste management among undergraduate students in a Philippine state university. Enviorn Earth Sci 6:146–153
11. Maddox P, Doran C, Williams ID, Kus M (2011) The role of intergenerational influence in waste education programmes: the THAW project. Waste Manag 31:2590–2600. https://doi.org/10.1016/j.wasman.2011.07.023
12. Wang H, Liu X, Wang N, Zhang K, Wang F, Zhang S, Wang R, Zheng P, Matsushita M (2020) Key factors influencing public awareness of household solid waste recycling in urban areas of China: a case study. Resour Conserv Recycl 158:104813. https://doi.org/10.1016/j.resconrec.2020.104813
13. Lawson DF, Stevenson KT, Peterson MN, Carrier SJ, Strnad R, Seekamp E (2018) Intergenerational learning: are children key in spurring climate action? Glob Environ Chang 53:204–208. https://doi.org/10.1016/j.gloenvcha.2018.10.002
14. Otto S, Pensini P (2017) Nature-based environmental education of children: environmental knowledge and connectedness to nature, together, are related to ecological behaviour. Glob Environ Chang. 47:88–94. https://doi.org/10.1016/j.gloenvcha.2017.09.009
15. Gogtay NJ, Thatte UM (2017) Principles of correlation analysis. J Assoc Physicians India 65:78–81
16. Ghanbari R, Mousazadeh M, Naghdali Z, Moussavi SP, Soheyli M, Rostami R (2018) Evaluation of knowledge, attitude and behavior of qazvin university of medical sciences students towards household. Hazard Waste Manag Keyword 6:6–10
17. Zhang H, Liu J, Wen ZG, Chen YX (2017) College students' municipal solid waste source separation behavior and its influential factors: a case study in Beijing. China J Clean Prod 164:444–454. https://doi.org/10.1016/j.jclepro.2017.06.224
18. Zhang B, Lai K, Hung; Wang, B., Wang, Z. (2019) From intention to action: how do personal attitudes, facilities accessibility, and government stimulus matter for household waste sorting? J Environ Manage 233:447–458. https://doi.org/10.1016/j.jenvman.2018.12.059
19. Ahmad J, Noor Md, S., Ismail, N. (2015) Investigating students' environmental knowledge, attitude, practice and communication. Asian Soc Sci 11:284–293. https://doi.org/10.5539/ass.v11n16p284
20. Echegaray F, Hansstein FV (2017) Assessing the intention-behavior gap in electronic waste recycling: the case of Brazil. J Clean Prod 142:180–190. https://doi.org/10.1016/j.jclepro.2016.05.064
21. Tatlonghari RV, Jamias SB (2010) Village-level knowledge, attitudes and practices on solid waste management in Sta. Rosa City, Laguna, Philippines 13:35–51
22. Ayodeji I (2012) Waste management awareness, knowledge and practices of secondary school teachers in Ogun state. Nigeria J Solid Waste Technol Manag 37:221–234
23. Ismail IM, Kulkarni AG, Kamble SV, Borker SA, Rekha R, Amruth M (2013) Knowledge, attitude and practice about bio-medical waste management among personnel of a tertiary health care institute in Dakshina Kannada. Karnataka Al Ameen J Med Sci 6:376–380

24. Budhiarta I, Siwar C, Basri H (2021) Current status of municipal solid waste generation in Malaysia. Int J. Adv Sci Eng Inf Technol 2:16–21. https://doi.org/10.2307/j.ctv11318vf.52
25. Chen HL, Nath TK, Chong S, Foo V, Gibbins C, Lechner AM (2021) The plastic waste problem in Malaysia: management, recycling and disposal of local and global plastic waste. SN Appl Sci 3:1–15. https://doi.org/10.1007/s42452-021-04234-y

Study of Slope Failure in North Sumatera: Case Study Medan—Berastagi Road

Ika Puji Hastuty, Fauziah Ahmad, Roesyanto, Moh Sofian Asmirza, Ahmad Perwira Mulia, and Ridwan Anas

Abstract North Sumatra is one of the areas where landslides often occur. The road along Medan-Berastagi is a landslide-prone area that has a rough topography with undulating hillside reliefs and slope angle between 60° and 90°. The literature on landslide susceptibility analysis in Sibolangit and Karo areas, where the two areas are located along Medan—Berastagi road, will be reviewed in this paper. This article presents summary review and classification of the main approaches that have been developed in Sibolangit and Karo as the parameters, methodologies, and tools used to obtain a landslide hazard map. The results of the review will be validated using landslide events that have occurred in 2020 by direct field observations.

Keywords Slope failure · Landslide · Risk · Hazard · Geographic Information System (GIS)

1 Introduction

The Medan-Berastagi road section plays an important role in the community's economy, but currently the capacity and geometric of the road and slopes cannot accommodate the volume of vehicles that pass through the road. Medan-Berastagi Road is traversed by 2 regencies, namely Deli Serdang Regency and Karo Regency.

I. P. Hastuty · Roesyanto · M. S. Asmirza · A. P. Mulia · R. Anas
Department of Civil Engineering, Universitas Sumatera Utara, Dr. Mansyur No. 58 Medan Baru, Sumatera Utara 20153, Indonesia
e-mail: ika.hastuty@usu.ac.id

A. P. Mulia
e-mail: a.perwira@usu.ac.id

R. Anas
e-mail: ridwan.anas@usu.ac.id

F. Ahmad (✉)
School of Civil Engineering, Universiti Sains Malaysia, Engineering Campus, 14300 Nibong Tebal, Pulau Pinang, Malaysia
e-mail: cefahmad@usm.my

N. Mohamed Noor et al. (eds.), *Proceedings of the 3rd International Conference on Green Environmental Engineering and Technology*, Lecture Notes in Civil Engineering 214,
https://doi.org/10.1007/978-981-16-7920-9_14

Sibolangit District is a highland area with an altitude of 400–700 m above sea level, it has a rough topography with undulating hillside reliefs with slopes ranging from 60° to 90°. Karo Regency is also located in the highlands of Bukit Barisan with the lowest elevation of ± 140 m above sea level (Mardingding) and the highest at ± 2,451 m above sea level (Mount Sinabung) [1, 2].

2 Method

The main purpose of this review is to analyze the research that has been carried out in North Sumatera, especially in Sobolangit and Karo; regarding the potential and mapping of landslide-prone risk using Geographic Information System (GIS). The review result will be validated predictively using landslide events that have occurred and review result of studies that have been carried out by other researchers.

3 Result and Discussion

3.1 Historical Area Along Medan—Berastagi Road

Research on landslide vulnerability analysis has previously been carried out in Sibolangit District using Geographic Information System (GIS) [3] this research was carried by observing population density and slope angle factors. Base on data from Badan Pusat Statistik [1], population density in Sibolangit is around 21.5−582.5 people/km^2. Final hazard map can be obtained by calculation of total score; the scoring/ weighting method referred to Puslittanak Bogor [4]:

$$\mathrm{LHI} = (0.4\mathrm{PD} + 0.2\mathrm{JD} + 0.2\mathrm{NW} + 0.1\mathrm{NE} + 0.1\mathrm{NC}) \tag{1}$$

where LHI = landslide hazard index, PD = Population Density, JD = Job Description, NW = Number of Women, NE = Number of Elderly, NC = Number of Children.

The parameters used in Karo Regency study were angle of slope, slope water management, soil conditions, seismicity, constituent rocks, vegetation, and rainfall where scoring and weighting referred to Peraturan Menteri Pekerjaan Umum No.22/PRT/M/2007 [5] which can be formulated as follows:

$$\mathrm{LHI} = (0.1\mathrm{CP} + 0.2\mathrm{SE} + 0.1\mathrm{PC} + 0.1\mathrm{SD} + 0.2\mathrm{BC} + 0.2\mathrm{PD} + 0.1\mathrm{ME}) \tag{2}$$

where LHI = landslide hazard index, CP = Cropping Patterns, SE = Slopes Excavation and Cutting, PC = Pond Construction, SD = Slope Drainage, BC = Building Construction, PD = Population Density, ME = Mitigation Efforts.

Based on Fig. 1, it can be seen that population density in areas with slopes above 45% affects the level of vulnerability to landslides in Sibolangit District and Karo Regency. The reason is that the land has changed its function due to the need for regional infrastructure. From Fig. 1a; it was concluded that three villages were classified as very high landslide prone areas with a population density of 423.81–582 people/km^2 and a slope of >45%. Study conducted by Ahmad [6] described >500 people categorized as very high density population, developed areas such as residential buildings with no less than 100 units; the number of people at risk is very high. The land is also not suitable for development because it is very risky to humans and the environment and is considered the most likely to experience slope failure. According to Althuwaynee [7], high population density is directly related to the high density of residential buildings where the population density is between 700 and 1500 person/acre, but this area is not the most risky areas to landslide. Slope instability generally occurs due to natural processes, but some of the results of uncontrolled human activities in exploiting nature can also be a factor causing slope instability which can lead to landslides.

From the Fig. 1b at Karo Regency, it was concluded that five districts were classified as high landslide prone areas. Uncontrolled human activity in exploiting nature has a big share in the occurrence of landslides. The construction of building, excavation, and slope cutting are the parameters that influence the occurrence of landslides the most. The reason is because it has an indicator weight by 20%, so that the high landslide hazard level is mostly located in areas that have a high level of construction and slope cutting [8].

3.2 Landslide Validation Along Medan—Berastagi Road

The validation carried out was predictive validation where the research results were adjusted to field conditions. Direct field observations at nine landslide locations along Medan—Berastagi road in 2020 can be seen in Fig. 2. The results of field observations show that the most prone areas to landslides are in accordance with mapping using GIS.

Based on data from Balai Besar Pelaksanaan Jalan Nasional II Medan [9], there were 7 landslides that occurred from 2017 to 2018, and 3 locations that are prone to landslides have been detected. The location can be seen in Table 1.

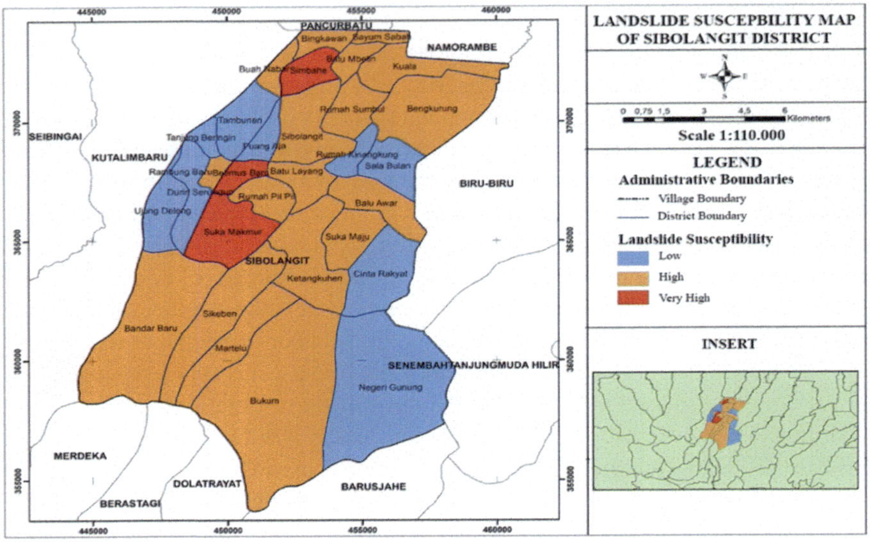

(a)

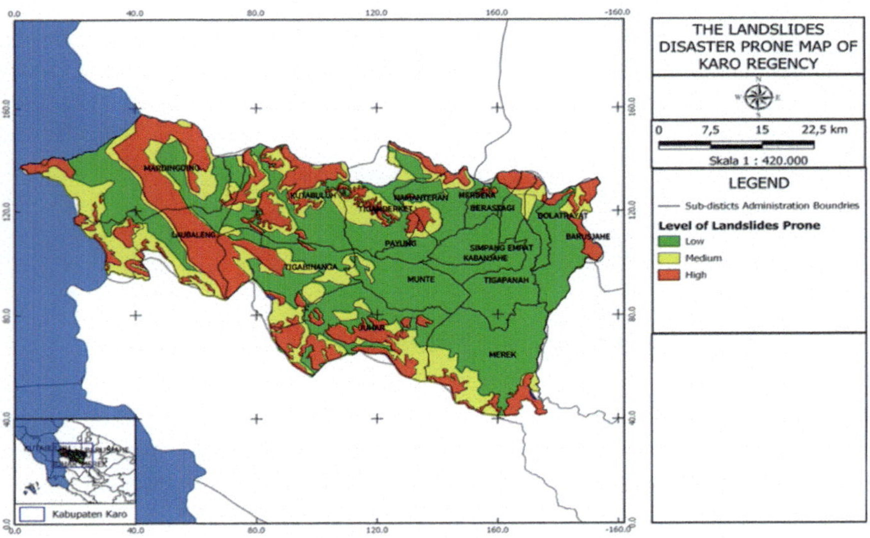

(b)

Fig. 1 Landslides susceptibility Map; **a** Sibolangit District **b** Karo Regency

Fig. 2 Landslides validation along Medan-Berastagi road

Table 1 Landslides data during 2017–2018 [9]

Road Section	Location (Km)	Remarks
Medan—Karo boundaries	36 + 700 L/s	Landslide on October 2018
Medan—Karo regency boundaries	37 + 550 R/s	Landslide prone on slopes
Medan—Karo regency boundaries	37 + 600 R/s	Landslide prone on slopes
Medan—Karo regency boundaries	37 + 700 R/s	Landslide prone on slopes
Medan—Karo regency boundaries	37 + 800 R/s	Landslide on September 2018
Medan—Karo regency boundaries	50 + 400 R/s	Landslide on December 2017
Medan—Karo regency boundaries	50 + 600 R/s	Landslide on November 2017
Deli Serdang boundaries—Ujung Aji	56 + 050 R/s, L/s	Landslide on October 2018
Deli Serdang boundaries—Ujung Aji	57 + 300 R/s	Landslide on November 2018
Deli Serdang boundaries—Ujung Aji	57 + 700 R/s	Landslide on November 2018

R/s = right side and L/s = left side

4 Conclusion

One of the causes of landslides in Siboangit District and Karo Regency is population density. Uncontrolled human activities in exploiting nature can be a triggering factor causing slope instability. Based on the landslide potential mapping that has been carried out in Sibolangit District and Karo Regency, it can be seen that the road along Medan-Berastagi has a very large potential for landslides and the main factors triggering landslide are not yet known clearly.

Acknowledgements This study is in collaboration between Universiti Sains Malaysia—Universitas Sumatera Utara (USM—USU) and supported by Universiti Sains Malaysia Research University Individual (RUI) Grant, No: 8014094

References

1. Badan Pusat Statistik (2018) Kecamatan Sibolangit dalam Angka
2. Badan Pusat Statistik (2020) Kabupaten Karo dalam Angka
3. Hastuty IP, Ahmad F, Anas MR, Hidayati YR (2020) Vulnerability analysis of landslide at Sibolangit using Geographic Information System (GIS) based on population density and slope gradient. IOP Conf Ser Mater Sci Eng 801(1):012008. https://doi.org/10.1088/1757-899X/801/1/012008
4. Puslittanak (2004) Laporan Akhir Potensi Bencana Kekeringan, Banjir dan Longsor di Kawasan Satuan Wilayah Sungai Citarum-Ciliwung, Jawa Barat Bagian Barat Berbasis Sistem Informasi Geografis. Bogor
5. Departemen Pekerjaan Umum (2007) Pedoman Penataan Ruang Kawasan Rawan Letusan Gunung Berapi dan Kawasan Rawan Gempa Bumi. Peraturan No 22
6. Ahmad F, Yahaya AS, Ali MM, Hairy SHA (2012) Qualitative risk assessment schemes using selected parameters for hillslope developments: a case study of Penang Island. Landslides 9(1):63–74. https://doi.org/10.1007/s10346-011-0275-7
7. Althuwaynee OF, Pradhan B (2017) Semi-quantitative landslide risk assessment using GIS-based exposure analysis in Kuala Lumpur City. Geomatics Natl Hazards Risk 8(2):706–732. https://doi.org/10.1080/19475705.2016.1255670
8. Hastuty IP, Ellysses Pratama O (2021) Geographical information system and satellite imagery for landslides analysis in Karo Regency based on human factors. IOP Conf Ser Mater Sci Eng 1122(1):012019. https://doi.org/10.1088/1757-899x/1122/1/012019
9. Balai Besar Pelaksanaan Jalan Nasional II Medan (2019) Lokasi Longsor dan Rawan Longsor PPK 18

The Barriers of Building Information Modelling (BIM) for Construction Safety

Nurul Hasanah Mohd Taat, Nor Haslinda Abas, and Muhammad Fikri Hasmori

Abstract Building Information Modelling (BIM) is technology moving in the world of engineering and architecture to the arena of construction companies in charge of construction operational that implement in the construction industry in Malaysia. However, most of the projects that implement BIM just focus on planning the building frame without considering safety planning. Hence, this paper is expected to fill the gap through the exploration of BIM-based on safety planning in Malaysia in previous literature. This paper presents a review of the barriers of BIM for safety planning in Malaysia's construction industry. All the barriers to adopting the BIM for construction safety has been found from previous research publications, such as lack of in-house expertise, lack of training/education on BIM, lack of awareness of BIM, lack of collaboration, client demand, and unsure of government commitment to BIM, high cost of software purchasing, lack of standardizing and resistance to change. All these barriers can be a guideline to the construction industry to solve this issue regarding BIM awareness in the construction industry. Proper safety planning in a project will surely benefit BIM in the construction industry. The research aim can be achieved and beneficial to all parties in the construction industry. This safety planning through BIM may minimize the accident at the project site, prevent hazard, and identify risk in the future.

Keywords BIM · Safety planning · BIM process · Barrier of BIM · Safety management

N. H. M. Taat · N. H. Abas (✉) · M. F. Hasmori
Faculty of Civil Engineering and Built Environment, Universiti Tun Hussein Onn Malaysia, 86400 Parit Raja, Batu Pahat, Johor, Malaysia
e-mail: nhaslin@uthm.edu.my

M. F. Hasmori
e-mail: mfikri@uthm.edu.my

N. Mohamed Noor et al. (eds.), *Proceedings of the 3rd International Conference on Green Environmental Engineering and Technology*, Lecture Notes in Civil Engineering 214,
https://doi.org/10.1007/978-981-16-7920-9_15

1 Introduction

Almost all new tender projects in Malaysia has started using BIM in their project as there are potential benefits if BIM is implemented in the project. Most of the clients in the private sector required BIM if the project cost is high. For the project that required a BIM, there is safety and health requirement that need to be fulfilled by contractors on how the contractor plan and manage their safety at the construction site. However, most of the projects that implement BIM focus on planning the building frame without considering safety planning, such as planning the best place to put the tower crane, workers cabin, and scaffolding in the BIM modeling. Thus, the site process, including the logistic aspect and safety, is ignored.

Apparently, in construction planning, safety planning should be seen to be part of them. Safety planning such as site layout planning ingress and egress, machineries storage and material storage, temporary site storage, the workers' rest area, falling protection equipment, and planning should be taken care of in the design stage to avoid an accident on site. This awareness of safety during the design stage can influence the risk of falling [1]. This is supported by Marefat et al. [2], as these safety rules, and geometric data can be used to generate information that, once implemented, can enhance safety education and training at the planning/design and construction stages. With proper implementation and adoption of BIM, the difficulties and complexities in managing a project would be reduced [3]. In addition, the performance and productivity will be increase throughout the project life cycle if BIM is implemented successfully with the proper sequence at the construction site [4].

To sum up, this research is to identify the awareness of safety planning in BIM-based projects and the implementation of safety planning during the design and construction stage. Hence, this paper is expected to fill the gap through the exploration of BIM-based on safety planning in Malaysia in previous literature.

2 The Concept of Building Information Modelling (BIM)

Building Information Modelling (BIM) is an intelligent 3D Model-based. Through BIM, it will help the engineering, architecture, and construction professional move towards more automated, more successful, and collaborative ways of working. BIM-based design techniques tended to reduce the difficulty and complexity of project management [3]. Adam et al. [5] stated that BIM is considered a strong driver in the construction industry for modernization, productivity, and new ideas.

There are two different terms of BIM, which are the process of building information modeling and the resulting model. BIM was conceived by Velasco [6] as a working methodology to improve the generation, management, transition, and visualization of knowledge throughout the entire life cycle of the building. According to Enshassi et al. [7], in the context of the visual, BIM technology can be used as support in terms of construction simulation and information statistics.

A construction project is the process of the life cycle that includes time, quality and cost. With proper of implementation and adoption of BIM, the difficulties and complexities in managing a project would be reduced [3]. In addition, the performance and productivity will be increase throughout the project life cycle if BIM is implemented successfully with the proper sequence at the construction site [4]. As state in Sardroud et al. [8], BIM, also known as project modeling, virtual building, virtual design, construction, and finally 4D modeling, is an extension of the computer-aided design (CAD) framework that offers more intelligence and interoperable details.

3 The Awareness of Building Information Modelling (BIM) in Safety Planning

This project performance and productivity could significantly increase with the awareness of BIM implementation and benefits to the project [9]. Awareness and potential benefit of this new technology not discovered by the majority of the construction player, including stakeholders as the implementation of BIM is not taken seriously in this construction. As the majority of construction players has 20 or 30 years works experience in this construction industry, the implementation is a bit challenging.

This BIM was recognized as a central point in technology for construction site safety related to planning and scheduling, enabling visualization of safety arrangements in construction projects, connecting the safety viewpoint more closely to construction planning, and provide more illustrative site plans for communication [10]. Previous research concluded that the major safety problems are associated with falls, moving on-site from task to other task and installation of the prefabricated unit [11] BIM can also be used for risk analysis and safety-related evaluation as visually plans first, then can automated risk identification in future. This can reduce the risk at construction projects as the project team can identify risk in the future and prepare the best way to minimize the risk. According to Marefat et al. [2], with the development of BIM, BIM technology, and the lifecycle realization in one holistic environment, the information of Health and Safety can be created directly. If this BIM technology can be linked successfully with the implementation and planning of health and safety, there are high chances to delivers significant benefits to the construction industry.

4 BIM Application in the Construction Industry

According to Benjaoran and Bhokha [12], BIM has been labeled as the tool to improve the construction industry due to its ability to run project information (n-dimension) throughout the process of the project lifecycle. It is also has been labeled as a set of digital tools that can manage the effectiveness of construction projects [13]. BIM is a set of technology processes and development that has transformed the infrastructure process in designed, constructed, analyzed and managed [14]. BIM also can improve and enhance the planning process and design of the construction framework [13].

BIM is a high technology and process to manage construction work. The communication through BIM-based visual 3D presentation can be easier to be present to the client and construction player. This BIM was recognized as a central point in technology for construction site safety related to planning and scheduling, enabling visualization of safety arrangements in construction projects, connecting the safety viewpoint more closely to construction planning, and provide more illustrative site plans for communication [10].

5 Barriers of Adopting BIM in Construction and Construction Safety

There is some of the studies have reported on the construction industry's reluctance to implement new technologies. According to Marefat et al. [2], the barrier to adopting BIM in the construction industry are summarized below (Note that summary of barriers is supported by relevant literature).

5.1 Lack of In-House Expertise

As BIM is the newest technology practice in Malaysia, it is crucial to have an expert in order to practice BIM. It is challenging to adopt a BIM if they have a deficiency of expertise to refer to. It is found by Marefat et al. [2] the most substantial barrier of adopting BIM for construction safety is the lack of in-house expertise. This is also supported by Gilligan and Kunz [15], that among the most common reasons for not adopting the BIM technology are lack of BIM professionals, lack of need, and lack of request by the client to execute the project planning using BIM [15]. Other than that, the qualification of BIM experts is also crucial to adopt BIM. The less of qualification and expert, it is hard to implement a BIM into a project. According to Chan [14], the lack of qualified in-house staff was the most significant BIM adoption barrier. Adam et al. [5] reported that construction players should have the necessary skill and knowledge of this technology.

5.2 Lack of Training/Education on BIM

The lack of training and education is one of the significant barriers of BIM. The lack of training and education in Malaysia is one of the problems that need to take seriously. Besides, the high cost of training is also one of the barriers that are taken into account why Malaysia has this barrier.

It is recommended to introduce BIM in the education phase, such as since in university to adopt a BIM. It is much easier to introduce and practice a BIM among the construction players with engineering background study. Training and education of BIM need to introduce since the early of education to make it easier for the student to explore more since university. According to Chan [14], to adopt BIM as a pedagogical platform, professional bodies should work together with educational institutions to review their curriculum and their course. This is support by Gilligan and Kunz [15] as training is needed to increase the expertise in BIM technology since early education in university. Ministry of higher education Malaysia has established various initiative programs for some universities, but the adoption is still low among the students [16]. Then, the lecturer's role is vital to spread awareness and introduce BIM to students.

Khemlani [17] reported that the lack of initiative and training, varied market readiness across geographic and reluctant to change the existing work-practice are the primary reasons on adopting the new technology. The lack of initiative, lack of training and knowledge, varied readiness across organizations and geographies, fractures existence in the construction industry, and the aversion of the industry to altering conventional working practices are the generic barrier to BIM adoption [18]. This is support by Abu Bakar et al., as one of the significant barriers is education and cost associated with training [19].

5.3 Lack of Awareness of BIM

The awareness of BIM in Malaysia is the significant barrier to BIM is fear of change [20]. Kassem at el. [21] were recognize the main obstacles to BIM is the shortage of experience within the workforce and lack of awareness by stakeholders. There were several challenges for safety planning to utilize BIM technologies which are the additional costs involved in developing/enhancing BIM models, lack of awareness of BIM use by construction players and technical concerns, such as the shortage of security modules and equipment available in the BIM software library. This is support by Gamil and Rahman [22], as one of the barriers is BIM knowledge among the project stakeholders.

The government has established various initiatives and the contractor should encourage their team to join the programs. The government's and client's role in this matter is crucial to shows the construction player that using this BIM, it will reduce the time consuming and save the cost of the project. This is supported by Chan

[14], that BIM benefit are minimized the time spent on project communication and documentation, BIM save the project cost, BIM improves the construction safety, and BIM will reduce the labor cost in the organizations in the long run. This BIM innovation adds benefits to deliver the project throughout the project lifecycle.

5.4 Lack of Collaboration, Client Demand and Unsure of Government Commitment to BIM

Support from the government is needed in order to adopt new technology, and this includes budget allocation to provide training to the employees and purchase these new technologies. According to Enshassi et al. [7], about 69% of contractors, 60% of client/owns and 60% of consultants had a low commitment in adopting this new technology to improve the safety performance in construction sites. Government enforcement is one of the barriers in adopting BIM technology [22]. Marefat et al. [2] indicated that non-BIM users gave the reasons there was inadequate client demand to use BIM and lack of trusted knowledge source to validate the potential benefit of BIM. Zakaria et al. [23] showed that the absence of support from the government in Malaysia is one of the barriers in adopting BIM.

Government should collaborate with the construction industry, education institutes, and professional bodies to establish a clear standard to adopt BIM and use BIM to construction players. In Singapore, Building Control Authority (BCA) has come up with five-year plan in order to enhance productivity and to move the industry towards the adoption of BIM Wong et al. [24]. This is a good step for the government to implement and adopting BIM to overcome the "fear of change" as the major barrier to BIM application [20].

5.5 High Cost of Software Purchasing

In order to implement new technology, an amount of money is needed to support and provide this new technology used to do campaigns, training, and buys the newsoftware. The software too expensive is one of the barriers to incorporating BIM in a firm [25]. It is stated that the cost of software is one of the principal barriers to BIM adoption in India [26]. This is supported as one of the major barriers the high cost of software purchasing [19]. This is a major problem as most of the contractor want to save the budget and use a little of the budget only for this new software.

This is support by Kumar and Mukherjee [14], stated by the respondent that BIM software is costly. According to Gamil and Rahman [22], the main factor is the high cost of BIM technology and the the limitation of finances Although there is a government initiative that provides free software in Malaysia, the allocation of budget is very tight, and the software needs a high specification of the computer to run smoothly. It is not an easy task to persuade and change the mindset of the private client to practice BIM in the project within a short period and by expecting them to pay more for the professional software, but this problem can be solved if this BIM software gets support from the government as government shows the benefit to the client, this BIM will shorten the project duration. By this, the client will substantially be adding more budget, especially to purchase this software or upgrade the computer to enable the teams to use the software.

Table 1 demonstrates the rank of barrier to adopting BIM for construction safety. The barrier is rank based on the most barrier that the previous research discovered in their research. The critical barrier that is the most barrier that previous research found is lack of in-house expertise discussed earlier in this literature review followed by the lack of knowledge and awareness as second strongest BIM Barrier and the third position was the lack of appropriate training educate the construction player. It shows that, the construction industry needs to organize more training to construction player so that the construction industry will produce more expertise and there is awareness among the construction players. The lack that has less critical discovered by previous literature are the lack of collaboration, high cost of software purchasing, lack of standardizing, and resistance of change.

All this barrier shows that in the adoption of BIM and the awareness of BIM is still low among the construction players. This also indicates that the awareness of BIM in safety is still lacking. The construction player is still not aware of the use of safety in BIM, the benefit of BIM in safety, and how safety can be use in this BIM project and modeling. To ensure the BIM is fully implemented in Malaysia and get the benefit of BIM, the Malaysian government, clients, consultants, and construction players must take part to overcome all these barriers. The BIM and safety in BIM should align together and spread the benefit if BIM and safety are proper planning. The awareness of safety performance among BIM qualified professionals, training for all construction players, government support, allocate some budget to adopt BIM in the Malaysian construction industry are what they can do to improve the implementation of BIM.

6 Conclusion

In conclusion, through this review of the barriers of BIM in construction safety, this research has shown several barriers in adopting BIM in the construction industry. The construction industry needs support from the government and private sector, such as provide more training with a free classes or cheap fee payments to construction players and those who are in the BIM project are required to attend this BIM course to educate them about the benefit of BIM, especially in safety planning. Construction organizations should consider the knowledge about BIM as a fundamental requirement for recruiting safety professionals. It is recommended to conduct the training courses on BIM technology, the concept and benefits of the application. As Malaysia

Table 1 Rank of barriers to adopt BIM

Barriers/Authors	Gilligan [27]	Gambatese and Hallowell [20]	Rajendran and Clarke [28]	Kassem et al. [21]	Abubakar et al. [29]	Chan [14]	Gamil and Rahman [22]	Marefat et al. [2]	Adam et al. [5]	Rank
Lack in house expertise	/		/	/		/	/	/	/	1
Lack of knowledge & awareness				/	/	/	/		/	2
Lack of training					/	/	/		/	3
Lack of collaboration	/					/	/			4
High cost of software purchasing						/	/	/		4
Lack of standardizing						/	/	/		4
Resistance to change		/						/	/	4

aims to adopt 80% of the BIM system by 2025, it is also recommended that government should start now at Selangor and Kuala Lumpur as both cities now have rapid development of construction projects to require all the construction players to have a certification of BIM. More research is needed to explore the safety application that can be used in this BIM modeling.

References

1. Fritjers ACP, Swuste PHJJ (2008) Safety assessment in design and preparation phase. Saf Sci 46(2):272–281. https://doi.org/10.1016/j.ssci.2007.06.032
2. Marefat A, Toosi H, Mahmoudi Hasankhanlo R (2019) A BIM approach for construction safety: applications, barriers and solutions. Eng Constr Archit Manag 26(9):1855–1877. https://doi.org/10.1108/ECAM-01-2017-0011
3. He Q, Wang G, Luo L, Shi Q, Xie J, Meng X (2017) Mapping the managerial areas of Building Information Modeling (BIM) using scientometric analysis. Int J Proj Manag 35(4):670–685. https://doi.org/10.1016/j.ijproman.2016.08.001
4. Bentley (2017) How AIA partner Bentley's BIM applications designed the world-class architecture of One Blackfriars
5. Adam V, Manu P, Mahamadu AM, Dziekonski K, Kissi E, Emuze F, Lee S (2021) Building information modelling (BIM) readiness of construction professionals: the context of the Seychelles construction industry. J Eng Des Technol. https://doi.org/10.1108/jedt-09-2020-0379
6. Velasco A (2013) Assessment of 4D BIM applications for project management functions. Dissertation, University of Cantabria
7. Enshassi A, Ayyash A, Choudhry RM (2016) BIM for construction safety improvement in Gaza strip: awareness, applications and barriers. Int J Constr Manag 16(3):249–265. https://doi.org/10.1080/15623599.2016.1167367
8. Sardroud J M, Mehdizadehtavasani M, Khorramabadi A, Ranjbardar A (2018) Barriers Analysis to Effective Implementation of BIM in the Construction Industry. In: Abstract of the 35th ISARC, Berlin, Germany. https://doi.org/10.22260/ISARC2018/0009
9. Al-Ashmori YY, Othman I, Rahmawati Y, Amran YHM, Sabah SHA, Rafindadi AD, Mikić M (2020) BIM benefits and its influence on the BIM implementation in Malaysia. Ain Shams Eng J 11(4):1013–1019. https://doi.org/10.1016/j.asej.2020.02.002
10. Kiviniemi M, Sulankivi K, Kähkönen K, Mäkelä T, Merivirta ML (2011) BIM-based safety management and communication for building construction. In VTT Tiedotteita - Valtion Teknillinen Tutkimuskeskus (Research Notes 2597)
11. Lappalainen J, Mäkelä T, Piispanen P, Rantanen E, Sauni S (2007) Characteristics of occupational accidents at shared workplaces. In: Abstract of NoFS 2007—Nordic Research Conference on Safety, Tampere, Finland, pp 13–15
12. Benjaoran V, Bhokha S (2010) An integrated safety management with construction management using 4D CAD model. Saf Sci 48:395–403. https://doi.org/10.1016/j.ssci.2009.09.09
13. Latiffi AA, Mohd S, Kasim N, Fathi MS (2013) Building Information Modeling (BIM) application in Malaysian construction industry. Int J Constr Eng Manag 2(4A):1–6. https://doi.org/10.5923/s.ijcem.201309.01
14. Chan CTW (2014) Barriers of implementing BIM in construction industry from the designers' perspective: a Hong Kong experience. J Sys Manag Sci 4(2)
15. Gilligan B, Kunz J (2007) VDC use in 2007: significant value, dramatic growth, and apparent business opportunity. Center for Integrated Facility Engineering, Report TR171. Stanford (CA): Stanford University. http://cife.stanford.edu/sites/default/files/TR171.pd. Accessed 18 Sep 2013

16. Mamter S, Mat Salleh N, Mamat ME (2014) Building Information Modeling (BIM) awareness among higher education Institution students. In: Abstract of 2nd International Conference Green Technology & Ecosystem for Global Sustainable Development, Putrajaya, Malaysia
17. Khemlani, L (2006). BIM symposium at the University of Minnesota, Building the Future (Article). http://www.aecbytes.com/buildingthefuture/2006/BIM_Symposium.html. Accessed 19 Jun 2021.
18. Gu N, London K (2010) Understanding and facilitating BIM adoption in the AEC industry. Autom Constr 19(8):988–999. https://doi.org/10.1016/j.autcon.2010.09.002
19. Abubakar M, Ibrahim YM, Kado D, Bala K (2014) Contractors' perception of the factors affecting Building Information Modelling (BIM) adoption in the Nigerian construction industry. In: Raymond Issa R (ed). Proceeding of Computing in civil and building engineering (2014). https://doi.org/10.1061/9780784413616.022
20. Gambatese JA, Hallowell M (2011) Enabling and measuring innovation in the construction industry. Constr Manag Econ 29:553–567. https://doi.org/10.1080/01446193.2011.570357
21. Kassem M, Brogden T, Dawood N (2012) BIM and 4D planning: a holistic study of the barriers and drivers to widespread adoption. J Constr Eng Proj Manag. 24:1–10. https://doi.org/10.6106/JCEPM.2012.2.4.001
22. Gamil Y, Rahman IAR (2019) Awareness and challenges of building information modelling (BIM) implementation in the Yemen construction industry. J Eng Des Technol 17(5):1077–1084. https://doi.org/10.1108/JEDT-03-2019-0063
23. Zakaria Z, Mohamed AN, Tarmizi HA, Marshall-Ponting A, Abd Hamid Z (2013) Exploring the adoption of Building Information Modelling (BIM) in the Malaysian construction industry: a qualitative approach. Int J Res Eng Technol 2(8):384–395
24. Wong AKD, Wong FK, Nadeem A (2009) Comparative roles of major stakeholders for the implementation of BIM in various countries. In: Abstract of International Conference on Changing Roles: New Roles, New Challenges, Noordwijk Aan Zee, The Netherlands (vol 5) 2009
25. Kumar JV, Mukherjee M (2009) Scope of building information modeling (BIM) in India. J Eng Sci Technol Rev 2(1):165–169. https://doi.org/10.25103/jestr.021.30
26. Nanajkar A, Gao Z (2014) BIM implementation practices at India's AEC firms. In: Abstract of ICCREM 2014: Smart Construction and Management in the Context of New Technology 2014
27. Gilligan B, Kunz J (2007) VDC use in 2007: significant value, dramatic growth, and apparent business opportunity. CIFE Technical Report #TR171. Stanford University Center for Integrated Facility Engineering
28. Rajendran S, Clarke B (2011) Building information modeling: safety benefits & opportunities. Professional Safety, pp 44–51
29. Abubakar M, Ibrahim YM, Kado D, Bala K (2014) Contractors perception of the factors affecting Building Information Modelling (BIM) adoption in the Nigerian construction industry The Sixth Annual Int. Conf. on Computing in Civil and Building Engineering (Florida, USA) pp 455–462

Environmental Sustainability and Development

Drought Assessment Using Standardized Precipitation Index (SPI) Case Study: Sulphur Springs Tampa FL

Mohammad A. Almadani

Abstract Climate change and extreme events with Uncontrolled Urbanization cause environmental degradation and long-term impacts on human life. The water as the most precious commodity on the global has been affected as a result from this environmental degradation. The availability of fresh water in term of quality and quantity is an essential matter to be monitored and controlled to avoid subsequences that could be threaten for human life. Drought is emerging as a sign of water resource deterioration which required to be monitored and diagnosed. Drought events can be diagnosed from changing in precipitation, temperature, groundwater, and water resources. The drought indices are very important as an effective early warning system to help the decision makers for an appropriate measure to reduce the impacts of drought. In this study we are trying to assess the regional drought and its effect on the Sulphur Springs Nitrate concentration and the causes of Sulphur Springs streamflow depletion. Standardized Precipitation Index SPI Standardized Discharge Index (SDI) are used to estimate the dry and wet conditions. These conditions can by estimated on time scales verity based on precipitation records and data. The calculation of SPI index is based on rainfall distribution and long-term period for a specific location. The precipitation records should be normally distributed for fitting in the Standardized Precipitation Index SPI. A location has been selected to apply the method, then to be validated for other arid and semi-arid locations. The availability of data and weather conditions play a significant role for this selection. Monthly precipitation records during the period Jan-1988 to 2014 over Sulphur Springs County in Tampa, Florid, has been obtained to apply the Standardized Precipitation Index (SPI). These data will be used to compute and verify the SPI and SDI results. SPI is calculated at one location in Tampa, Florida to monitor the drought phenomena in regional scale. The basic data which is used in this analysis is cumulative monthly precipitation to calculate the SPI for 3, 6, and 12 Monthly events and 3, 9, and 12 for the SDI computation. The SPI and SDI responds effectively to analyze drought conditions and monitor drought events. The precipitation has a stable trend along the period of

M. A. Almadani (✉)
Department of Civil and Environmental Engineering, Faculty of Engineeing, King Abdulaziz University, 21589, P.O.Box: 80200, Rabigh, Saudi Arabia
e-mail: malmadani@kau.edu.sa

N. Mohamed Noor et al. (eds.), *Proceedings of the 3rd International Conference on Green Environmental Engineering and Technology*, Lecture Notes in Civil Engineering 214,
https://doi.org/10.1007/978-981-16-7920-9_16

study while the Spring tend to have a decreasing pattern. The index shows that the spring in Tampa has suffered from drought between 2000 to 2004. These impacted water quality and human life in terms of quantity by clogging the springs and quality by increasing the pollutant concentration during the drought.

Keywords SPI · Standardized precipitation index · Drought · Precipitation · Dry

1 Introduction

Urbanization benefits are significant for population convenience and life efficiency as well as the integration at the level of global [1]. It is a transformation of people life and rapid changes including the surrounding environment [2].

On the other hand, uncontrolled urbanization has enormous adverse impacts on the communities and the natural resources [1]. At the same time, the uneven spatial and temporal distribution of water resources, the urbanization drives the fast increase in water demand, making water availability vulnerable and critical issue nowadays [3]. Drought and flood have been increasing since these conditions began to change, and these events become more threatening and harmful than before [4].

A metrological event can be a saver and defecating, recurring natural disasters, which occur due to deficient precipitation level. It can cause significant losses in the economy regardless of its short period [5]. However, we cannot stop drought event though we can monitor it, after that we can alleviate its impacts. Several springs has been got affected by drought according to several water districts reports. Sulphur spring is one of the most meteorologically affected with drought which caused an increasing in the Nitrate concentration. In Florida, there have been several investigations to study the effect of spring nitrate contamination which can lead to dying of the marine habitats, grass, and other underwater creatures. Therefore, the influence of big variability of rainfall detection of drought very important for the policy makers [6]. Their efforts to reprehend and restore those areas if needed. Recently, focusing on food-water security is very important issue around the Globe because the drought events are intimately associated with food-water security. Hence, research on drought prediction and drought impact, specifically drought monitoring, is essential to be prepared and to implement mitigation measures to reduce theses consequences [7]. In each district around Florida, many research have been conducting to investigate the impacts of drought events on many aspects, such as agriculture activities and food production, economic and social life, and the whole society [8].

Several methods have been used for drought diagnosis over the United States. Therefore, this research aims to study the drought pattern through spatial–temporal diagnose by applying the Standard Drought Index (SPI) method and anthropogenic impact on the nitrate concentration raise in Sulphur spring.

2 Threats

The water discharge is directly linked to the water quality, where the pollutants concentration tends to increase while the flow discharge decreases in the Springs. contamination of Florida Springs becomes the most threating issue for the water security in the region [9], which is a direct impact of drought events. The geology and land use practices affect the water quality and quantity within the recharging of springs basin [10]. Contaminants and constituents are conveyed and sent through runoff and infiltration from surface into the springs [11]. Stormwater runoff flows over impervious surfaces such as parking lots, paved streets, building, and roofs tops gathering and gushing excess pollutants and nutrients [12]. In the urbans area, the opportunity to filter these water and flows is nonexistent through the natural system by soil and plants [13]. Stormwater carries many pollutants during its journey to the waterbodies on the surface through Runoff, or the groundwater into water table and aquifers through seepage. These pollutants include fertilizers, grass and oil, pesticides, animal wastes, debris and remains, dirt and chemicals [14]. The nutrients and bacteria pollute underground storage and septic tanks through seepage. As a result, pollutants from ground water will be transported to the springs through cracks and rock fractures [15].

Seepage from surface by direct conduits such as sinkholes, recharges the groundwater that feeds springs through flow back to the surface [16]. As a result, activities and land uses within the spring recharge basin directly impact the health of spring systems. The recent guidelines for water and groundwater quality standards were developed for sustainability and drinking water protection [16]. Depending on the EPA standards and Florida Department of Environmental Protection, for nitrate, the regulations prescribe a maximum contamination level of 10 mg/l. Infants that drink water with nitrate levels above this threshold are more likely to develop methemoglobinemia (also known as "Blue Baby Syndrome"). Other human health concerns have been connected to increased nitrate consumption in other studies [9]. The domestic sewer system and septic tanks contributes to the groundwater contamination that feeds the springs. By 1986, contamination by coliform bacteria forced the permanent closure of the spring to public recreation. Salinity of the spring water is increasing, a symptom of reduced freshwater inflow due to plugged sinkholes and pumping of the aquifer for water supply [17].

3 Study Area

Sulphur Springs is located about five miles north of downtown Tampa (Fig. 1). It is one of the largest of springs that flow into the River of Hillsborough within the Tampa city, with a daily flow of 26 million gallons. since 1800s, The Sulphur spring has been very popular as a recreation site. In 1957, the City of Tampa was purchased the Spring for the recreation purposes, and to be a water resource in the future. Sulphur

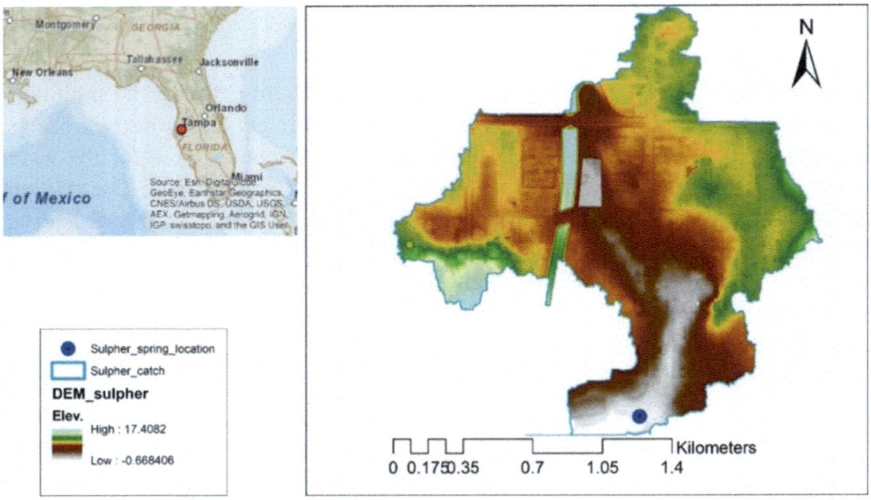

Fig. 1 Sulphur Springs location and elevation map

Spring's third of its flow is pumped to supplement reservoir of Tampa's water system. The recharge basin of the spring is extensively populated [18].

The climate in Sulphur Spring, Tampa is temperate and warm with significant rainfall during the year. 22.4 °C is the average temperature and about 1159 mm is the annual precipitation. The minimum rainfall is about 37 mm and falls in April. The most precipitation quantity is about 196 mm in August. Table 1 shows that the precipitation events and temperature in Tampa city, and the precipitation difference between the dry and wet month, which is 159 mm [19].

Land cover of Sulphur Springs has been rapidly changing through recent years (Fig. 2) [19]. As shown by Fig. 3 the urbanized area has been increased in general. Moreover, high intensity developed area appeared in 2011. On the other hand, there is obvious decrease in wetland areas and developed open space also, huge decrease in forest and green canopy, and disappearance of open water and cultivated areas [20].

Table 1 Passing ability of FA and BA incorporated in SCC by using the J-ring test

Month	1	2	3	4	5	6	7	8	9	10	11	12
	55	76	83	37	78	141	177	196	155	58	48	55
C	15.5	16.6	19.2	22.0	25.1	27.2	27.8	27.9	27.1	23.8	19.6	16.7
C (min)	10.0	11.1	13.7	16.3	19.6	22.5	23.4	23.5	22.6	18.7	14.1	11.2
C (max)	21.1	22.1	24.7	27.7	30.7	32.0	32.3	32.4	31.7	29.0	25.2	22.2
F	59.9	61.9	66.6	71.6	77.2	81.0	82.0	82.2	80.8	74.8	67.3	62.1
F (min)	50.0	52.0	56.7	61.3	67.3	72.5	74.1	74.3	72.7	65.7	57.4	52.2
F (max)	70.0	71.8	76.5	81.9	87.3	89.6	90.1	90.3	89.1	84.2	77.4	72.0

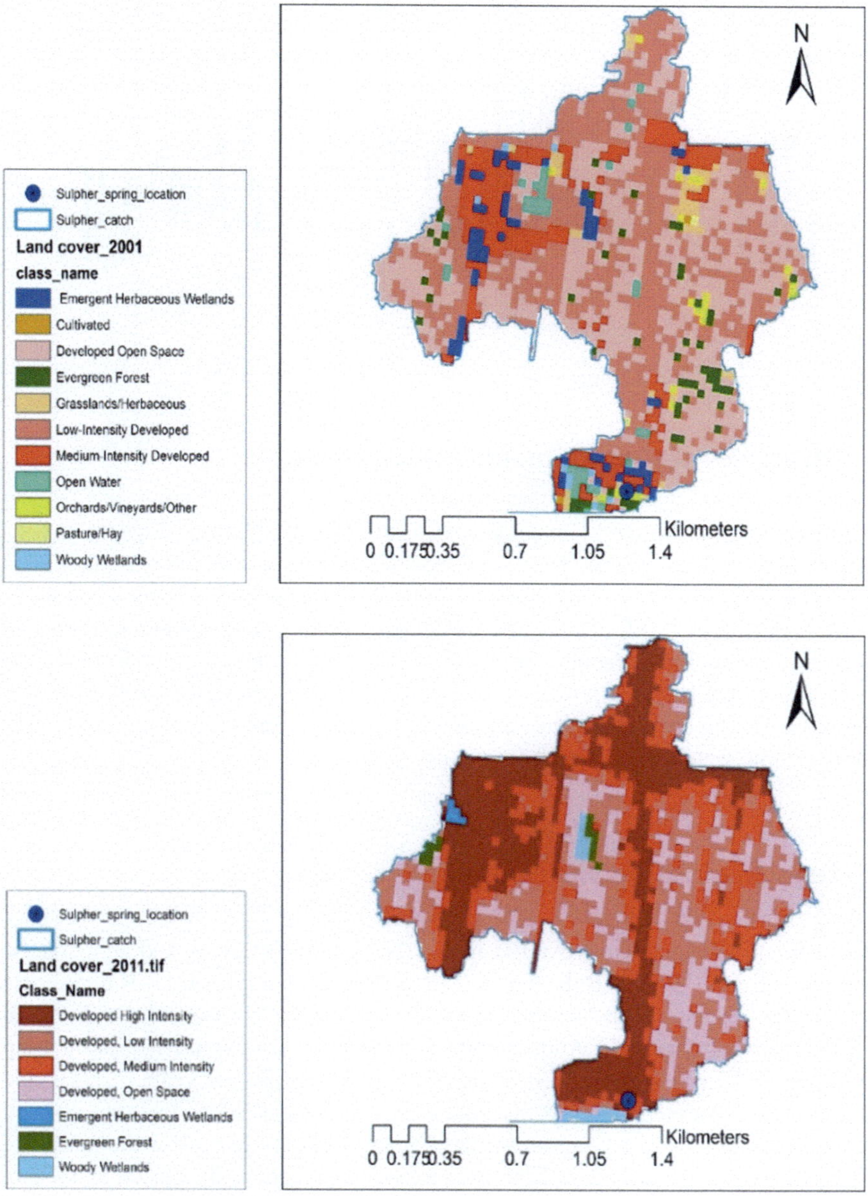

Fig. 2 Land use and land cover effect on the Sulphur Springs area 2001–2011

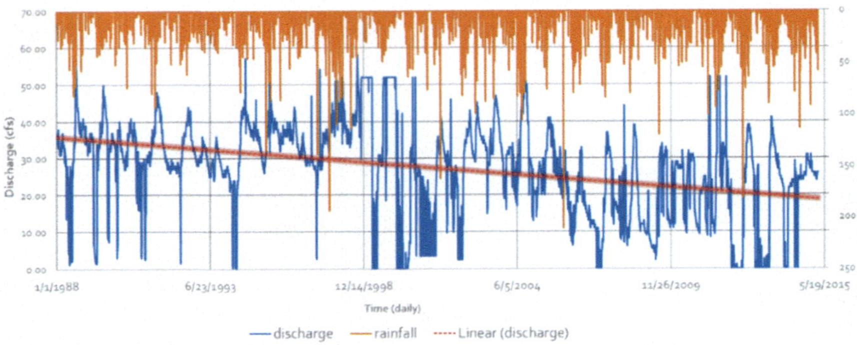

Fig. 3 Precipitation and discharge amounts for the period of 1988 to 2015

4 Strategies of Contaminants Management

Southwest Florida Water Management District (SWFWMD) developed a (MFL) rule to control spring runoff. Three recorders are being used by the US Geological Survey (USGS) in collaboration with the SWFWMD to measure continuous flow. In connection with the MFL project, SWFWMD will conduct biological monitoring of the spring run and manatee use. In addition, The City of Tampa and (SWFWMD) are working together on a project to reopen the clogged connection to the springs. The project could result in enhanced groundwater flow to feed the springs and lower salinity in the water discharged from springs [21].

5 Data Acquisition and Stations

The National Weather Service a division of the National Oceanic and atmospheric Administration's (NOAA) is the data that have been used included: 1—Rainfall data from 1987— current The National Weather Services, a division of the National Oceanic and Atmospheric Administration's (NOAA). 2–Digital Elevation Map DEM (30 × 30-m resolution). 3–Land cover and land use for years of 2001 and 201, retrieved from the United States Geological Survey (USGS), Institute of Land Cover. 4–Ground Water Usage from the USGS. 5–Water usage for the period of 1985 to 2010. 6–Water Quality and Stream discharge data for the period of 1988 to Feb 2015 from USGS.

The Standardized Precipitation Index (SPI) is calculated to assess drought's spatial and temporal extents and severity in the research area [22]. SPI is a drought detection and monitoring tool that is based on precipitation over different time intervals. The SPI has been shown in several studies to be an effective tool to detect and monitor the droughts events. The monthly records time series of the precipitation data for the study area can be used to figure the SPI for any picked month in the record for the previous months "I", where i = 1, 2, 3, 4,12... 24... 48 ... based on the interesting time scale. Henceforward, the SPI could be calculated for one to 48 months period. The drought categories are shown in a range of SPI values. These different values of SPI will be either positive value that consider wet season and negative values that consider dry seasons. SPI is computed for 3-, 6- and 12-month time period (hereafter 3 months series, 6 months series, and 12 months series) for a short-term length and long-time scale drought indexes detection. This study computes SPI at one station and one region case of study Sulphur spring from a monthly rainfall data acquisition and comparing it with the Nitrate concentration.

6 Standardized Precipitation Index (SPI)

Edwards and McKee recognized that in their previous works the developing Standard Precipitation Index SPI is very important to study the relative departures from normal condition [20]. SPI is the index computed based on specific period/periods of time (typically 1, 2, 3, 6, 12, 24 and 48 months) of total precipitation, to indicate how the precipitation in the selected period of time will be compared with the whole record such as 25, 50 or 100 years for the same location or station [23]. For example, if the index SPI in August 1998 is −2.00, then the rainfall is much less than normality comparing to the same period—Augusts' rainfall—in the total in the record. Similarly, if the SPI in August is + 1.00, then the rainfall for this August is above the average. The interpretation of the SPI values, it is essential considering that drought and wet events are related historically to the average of a location, specifically of a station, and it does not correlate to the total absolute precipitation values. To illustrate how the SPI index works, Aiguo Dai gave an example that is in Mexican Gulf, the very dry month may have −2.5 of SPI index, however the same precipitation level with the same magnitude in the driest northwestern State of Kansas gives + 3.0 of SPI index comparing to the same precipitation records [5].

6.1 SPI Calculation

The z-score is equivalent conceptually to the SPI, where the z-score expresses the score's distance of X from the average in the standard deviation units.

(Z-score = ((X – Average)) / (Standard Deviation))

Precipitation is generally positively skewed, thus a main pre-adjustment to this conventional z-score formulation reflects that. The gamma function converts the precipitation data into a more normal or Gaussian symmetrical distribution to account for this empirical reality. The SPI is calculated in the same way as the z-score formula and may be understood in the same way once the precipitation data has been converted [23].

Table 2 SPI of precipitation

SPI values	Precipitation periods more(percent)	Precipitation periods less (percent)	Nominal SPI class
3	0.13	99.87	Extremely wet
2.5	0.62	99.38	Extremely wet
2	2.28	97.72	Extremely wet
1.645	5.00	95.00	Very wet
1.5	6.68	93.32	Very wet
1.282	10.00	90.00	Wet
1	15.87	84.13	Wet
0.842	20.00	80.00	Slightly wet
0.524	30.00	70.00	Slightly wet
0.5	30.85	69.15	Normal
0.253	40.00	60.00	Normal
0	50.00	50.00	Normal
−0.25	60.00	40.00	Normal
−0.5	69.15	30.85	Normal
−0.52	70.00	30.00	Slightly dry
−0.84	80.00	20.00	Slightly dry
−1	84.10	15.87	Dry
−1.28	90.00	10.00	Dry
−1.5	93.30	6.68	Very dry
−1.64	95.00	5.00	Very dry
−2	97.72	2.28	Extremely dry
−2.5	99.38	0.62	Extremely dry
−3	99.87	0.13	Extremely dry

Table 2 represents the Probability Z and its correspondent SPI value and Nominal SPI class. Some of the methods to evaluate SPI are Excel Spread sheet, Matlap using Matlap Central (online) a function created by Taesam Lee (Appendix D), and program developed by the of National Mitigation Center using C+ + language [24], which is used in this study. In this study we are using the C+ + [24], a validated, well-well documented version from the National Drought Mitigation Center.

6.2 *SPI Results and Analysis*

Drought in Sulphur Springs may cause into an increasing of ground water pumping in the city of Tampa and the more frequent this happens the more negative impacts on the water resources, economy, social life, and on the environment [25]. During the year of 2000 to 2004 Tampa city has suffered from severe drought as it is shown in Figs. 2 and 3. As it is indicated in Fig. 1, the precipitation events seem to have a stable trend along the period of study with annual precipitation around 60 mm (1988–2015) with considerable inter-annual variations. The precipitation effect can cause either an increase in the discharge or a decrease this can be noticed in Fig. 3. The fluctuations in the discharge in Sulphur Springs tend to have a decreasing fashion. The decline indicate there is an impact of human activity that contributed into the declination of the spring. The discharge values fluctuation and its tendency to low flow manner makes it difficult to analyze its seasonal signature (Fig. 4).

Therefore, the use of SPI and SDI is useful tool to use to have a close look into the behavior of the precipitation, discharge and there evaluate their seasonality [26]. In Fig. 5, 6, 7 and 8, the SPI and SDI method has be utilized to analyze the drought effect on the Sulphur Springs for the period of 1988 to 2014 using three time intervals 3,6, and 12 for SPI and 3,9 and 12 for SDI. The time periods of seasonal accumulation of precipitation are totally depended on the use of the SPI. The 3 monthly analyses can be utilized in tracking certain cultivation and major impacts on agriculture zones and

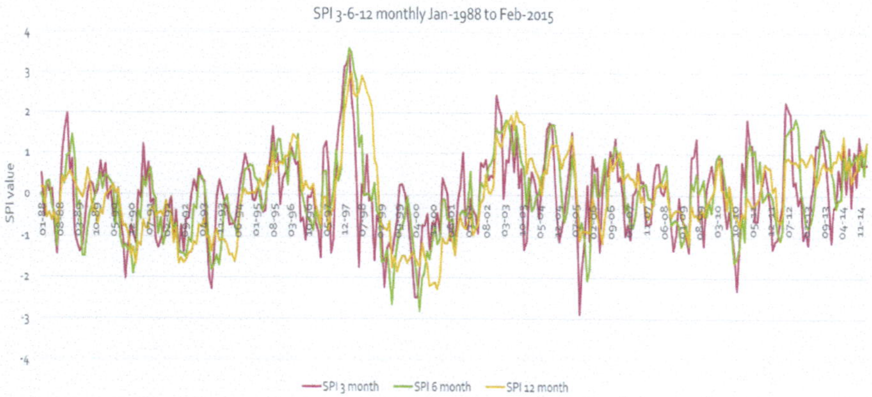

Fig. 4 SPI of the 3, 6, and 12 months for the period of 1988–2015 from the Tampa rainfall gage

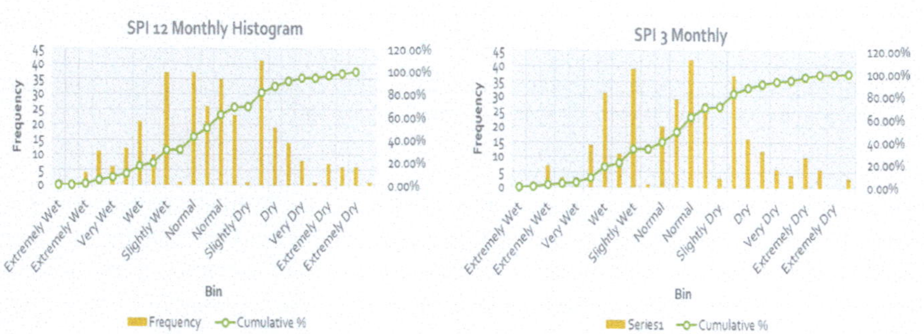

Fig. 5 Frequency analysis for the 3 and 12 monthly SPI

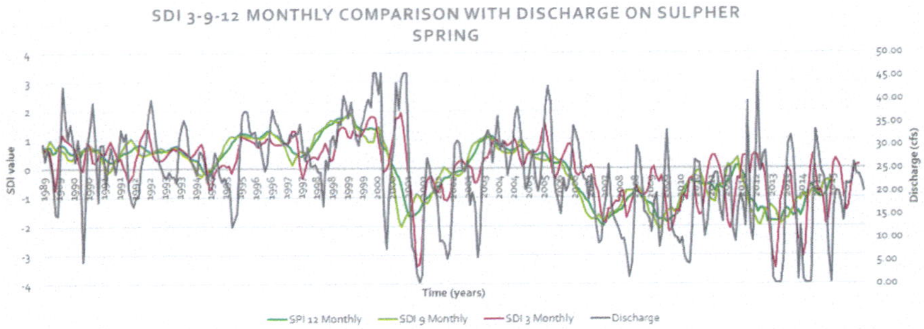

Fig. 6 SDI of the 3, 9, and 12 monthly for the period of 1988–2015 from the Tampa rainfall gage

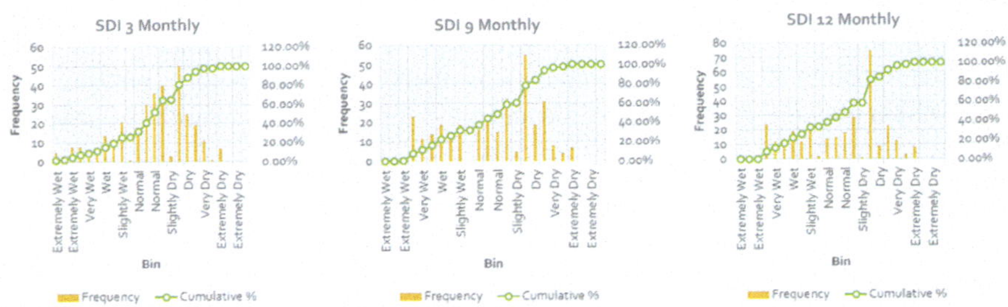

Fig. 7 Frequency analysis for the 3, 9 and 12 monthly SPI

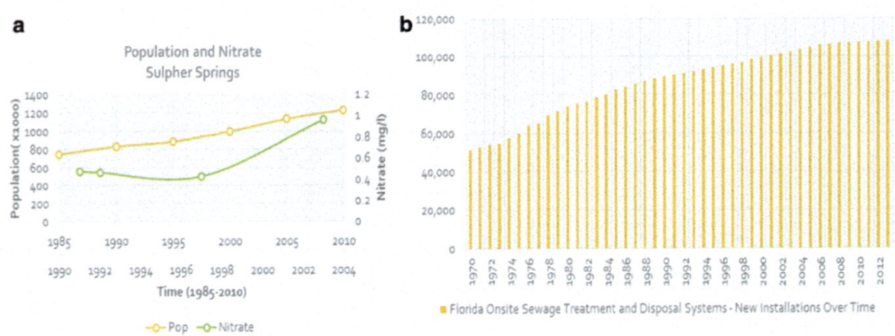

Fig. 8 **a** The population growth along with the increase of Nitrate Concentration in mg/l. **b** Increase of Hillsborough onsite sewer treatment systems—new Installations

plan for mitigation strategies. In agricultural droughts, the availability of soil to the plants and crops takes precedence over precipitation deficits. The capacity of water retention in soil root zone is the most important element in agricultural drought. In the 12-month period SPI, the trends can be used for avoiding wildfire threat, and monitoring water supply. In the Sulphur Spring area, the effect of drought is minor due to the abundancy of rainfall amounts. The accumulations of precipitation during

3, 6, 9 and 12-month period is shown in Fig. 4, 5, 6 and 7. The frequent events occurred tends to be in the slightly dry to slightly wet zones, which reflect the availably of rainfall except for the beginning of the year 98 to 2003.

6.3 Human Impact

With the growth of population, the reflectance of that increase can have an impact on the environment. Human increasing in population will get affected by utilizing a grand vast of space for urbanizing and recreational activates which cause an increase in the impervious surfaces. This effect has an impact on the groundwater replenish springs and aquifers. Since 1960's when the river flows dip low the city need to pump water direct from Sulphur Springs, which cause raise in bacterial counts, such as closed the swimming zone in 1986, and it had been replaced by unconnected pool. Sink holes and underground channels has been clogged and stuffed with undissolved material has caused into decreasing into the amount of recharge in the Sulphur Springs. The baggiest Spring has been clogged in 1950s by a lot of cars dealer, which was turned to a dumpster for the waste into the sinkholes to be completely blocked off. Other sinkholes were clogged because of the increased number of tossed garbage, waste from building materials, and natural debris. In addition, some of the sinkholes was converted to ponds of stormwater detention by the authority of the city for pollution control purposes [27]. Moreover, there is a noticeable increase in the onsite installment of sewage treatment systems in the Hillsborough County. This can be one of the contributing factors that lead to the increase in nitrate on Sulphur Springs. Some type of this installment has gone over a lot of heavy duty activates with time which may has caused some damage into them which led to leakage or spills out. All of this containment gets washed off into the drainage system or directly recharged into the underground aquifers [28]. Moreover, with the growth of population the water demand increases with it especially in the cases of long drought seasons. In Figs. 8 and 9, the data shows that in the highly drought areas such as in California there an increase of pumping from the groundwater compared to areas with water availability. However, between the period of 98 to 2003 drought event in Sulphur Springs it had been noticed that there an increase the groundwater withdrawals as well [29]. This indicates that the County of Hillsborough has a huge demand to compensate for the deficit from by groundwater pumping.

7 Conclusion and Recommendation

Studying the SPI and SDI of the Sulphur Springs area it has recognized that there is a drought event which has occurred between the year of 1998 and 2003. This event has caused a direct increase in the groundwater pumping to meet the demand of water supply, one of the contributing factors in the discharge depletion. However,

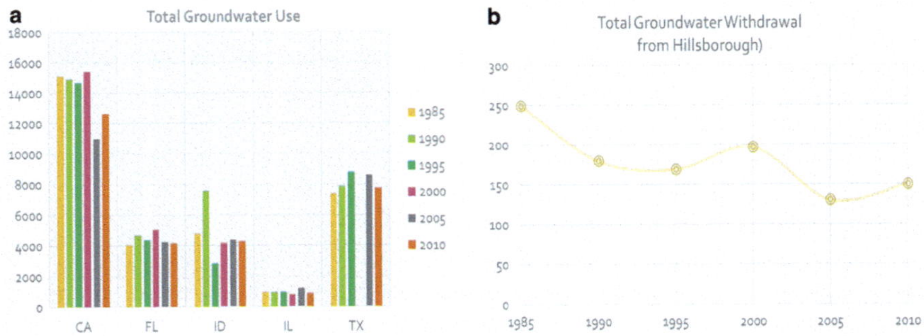

Fig. 9 **a** The total groundwater use in California, Florida, Idaho, Illinois, Texas from 1985 to 2010 **b** The total ground water withdrawal from the Hillsborough county from 198 to 2010 by the USGS water use data base

the anthropologic effect also has caused the spring water quality decreasing if we take the nitrate concentrations as an impact factor on the spring. Also, the land cover change has contributed into transferring those impurities into groundwater aquifers which caused an increase in the contamination as shown in Figs. 2 and 6. Based on the (Environmental, 2000), the points below summarize the historical and current impacts from the human activates as follows:

The pollution levels have increased despite 44% reduction in the flow, resulting in part to increase water pumping from the nearby aquifer. Hydrologists from the SWFWMD found that the main problem of the springs is the clogging of the connected underground channels from the surrounding sinkholes where water is funneled from the surface. Because the sinkholes remained clogged, the flow of water in the spring was reduced by 10 million gallons per day, causing "the salinity of Sulphur Springs to increase proportionally with decreasing of freshwater recharge, clogged sinkholes.

Standardized Precipitation Index SPI is a recommended method to estimate the dry and wet conditions because it uses only the precipitation data, and it is suitable for all type of weather. However, the SPI does not include other parameters such as evapotranspiration, which is important to study the drought conditions.

References

1. Tafazzoli M, Nochian A, Karji A (2019). Investigating barriers to sustainable urbanization. In: International Conference on sustainable infrastructure 2019: leading resilient communities through the 21st Century (pp 607–617). Reston, VA: American Society of Civil Engineers. https://doi.org/10.1061/9780784482650.065
2. Tsai WP, Cheng CL, Uen TS, Zhou Y, Chang FJ (2019) Drought mitigation under urbanization through an intelligent water allocation system. Agric Water Manag 213:87–96. https://doi.org/10.1016/j.agwat.2018.10.007
3. Zhao Y, Weng Z, Chen H, Yang J (2020) Analysis of the evolution of drought, flood, and drought-flood abrupt alternation events under climate change using the daily SWAP index.

Water 12(7):1969. https://doi.org/10.3390/w12071969
4. Aiguo Dai KE (2006) A global dataset of palmer drought severity index for 1870–2002: relationship with soil moisture and effects of surface warming. National Center for Atmospheric Research, Boulder, Colorado, 1117–1130. https://doi.org/10.1002/0471743984.vse8899
5. Dash SK, Sharma N, Pattnayak KC, Gao XJ, Shi Y (2012) Temperature and precipitation changes in the north-east India and their future projections. Global Planet. Change 98:31–44. https://doi.org/10.1016/j.gloplacha.2012.07.006
6. Hayes M, Svoboda M, Wall N, Widhalm M (2011) The Lincoln declaration on drought indices: universal meteorological drought index recommended. Bull Am Meteor Soc 92(4):485–488. https://doi.org/10.1175/2010bams3103.1
7. Mondol MAH, Ara I, Das SC (2016) Drought index mapping in Bangladesh by using Standardized Precipitation Index (SPI) over 1991–2010. https://doi.org/10.20944/preprints201608.0060.v1
8. Abou Zakhem B, Kattaa B (2016) Investigation of hydrological drought using cumulative Standardized Precipitation Index (SPI 30) in the eastern Mediterranean region (Damascus, Syria). J Earth Syst Sci 125(5):969–984. https://doi.org/10.1007/s12665-015-5013-3
9. Katz BG (2004) Sources of nitrate contamination and age of water in large karstic springs of Florida. Environ Geol 46(6–7):689–706. https://doi.org/10.1007/s00254-004-1061-9
10. Lerner DN, Harris B (2009) The relationship between land use and groundwater resources and quality. Land Use Policy 26:S265–S273. https://doi.org/10.1016/j.landusepol.2009.09.005
11. Grigg Neil S (2010) Water, wastewater, and stormwater infrastructure management. CRC Press. https://doi.org/10.1201/9781420032338
12. Weng Q (2015) "Impact of climate zone on impervious surface estimation and mapping. In: Remote Sensing of Impervious Surfaces in Tropical and Subtropical Areas, pp 82–105. CRC Press. https://doi.org/10.1201/b18836-10
13. McIntyre JK, Davis JW, Hinman C, Macneale KH, Anulacion BF, Scholz NL, Stark JD (2015) Soil bioretention protects juvenile salmon and their prey from the toxic impacts of urban stormwater runoff. Chemosphere 132:213–219. https://doi.org/10.1016/j.chemosphere.2014.12.052
14. Walling S, Osborne A, Lee B, Durham R (2014) Residential rain gardens: design, construction, and maintenance
15. FL Department of Environmental Protection. (n.d.). Springs. Florida department of environmental protection. https://floridadep.gov/springs
16. Struhs DB, Mainella FP (2000) Outdoor recreation in Florida: Florida's statewide comprehensive outdoor recreation plan. Florida Department of Environmental Protection, Tallahassee, FL
17. EPA (n.d.) Recreational Water Quality Criteria. OFFICE OF WATER 820-F-12–058
18. Hayashi M, van der Kamp G, Rosenberry DO (2016) Hydrology of prairie wetlands: understanding the integrated surface-water and groundwater processes. Wetlands 36(2):237–254. https://doi.org/10.1007/s13157-016-0797-9
19. US Department of Commerce, N. O. A. A. (n.d.). National weather Service. National Weather Service. https://www.weather.gov/
20. National land cover database—usgs.gov. (n.d.). https://www.usgs.gov/centers/eros/science/national-land-cover-database
21. Southwest Florida water Management District 2018 Estimated (n.d.). https://www.swfwmd.state.fl.us/sites/default/files/medias/documents/2018%20Estimated%20Water%20Use%20Report.pdf
22. Zarch MA, Asadi BS, Sharma A (2015) Droughts in a warming climate: a global assessment of Standardized precipitation index (SPI) and Reconnaissance drought index (RDI). J Hydrol 526:183–195. https://doi.org/10.1016/j.jhydrol.2014.09.071
23. McKee TB, Doesken NJ (1997) The relationship of drought frequency and duration to time scales. In: Proceedings of the 8th Conference on Applied Climatology, pp. 179–184, Anaheim, California

24. Horrocks WK (2010). MODIS Level 1, atmosphere and land data products (LAADS). Retrieved May 1, 2012, from LAADS: http://ladsweb.nascom.nasa.gov/data/
25. Drought Indices MJ (2010). National Drought Mitigation Center. Retrieved May 3, 2012, from with modifications by Dev Niyogi and Umarporn Charusambot Indiana State Climate Office, Purdue University (http://iclimate.org), http://www.drought.unl.edu/whatis/indices.htm
26. Giddings L, SOTO M (2005) Standardized precipitation index zones for México. Atmósfera, 32–55
27. Environmental Protection Department of Florida (2000) Florida's springs: strategies for protection & restoration
28. Lincoln UO (2012) Drought basics. http://drought.unl.edu/DroughtBasics.aspx, http://drought.unl.edu/
29. Lincoln UO (2012) Monitoring tools. Retrieved from NDMC Monitoring Tools Web site: http://drought.unl.edu/MonitoringTools.aspx

Hydrodynamic Modelling for the Chilia—Bystroe Danube Sector: Model Calibration and Validation

Georgeta Tudor, György Deák, Miruna Arsene, Tiberius Marcel Danalache, Bianca Petculescu, Danut Marian Tuca, Edward Bratfanof, and Mohd Remy Rozainy Mohd Arif Zainol

Abstract Morpho-hydrodynamic changes are one of the major problems affecting the abiotic sphere and the ecological status of water bodies. In order to assess the impact of the works carried out in the Bystroe Channel on the hydrodynamic and hydromorphological conditions of the Chilia and Old Stambul Danube branches, solid scientific based tools are required. Using MIKE 3 Flow Model and input data provided by in situ campaigns carried out by the National Institute for Research and Development in Environmental Protection Bucharest, an 8.5 km river length hydrodynamic model was developed, calibrated and validated in two independent river flow conditions. The model shown to provide high—confidence results (>90%) by comparative analysis of modelled versus measured values for water discharge and similarity in water velocity distribution for transversal control sections.

Keywords Hydrodynamic modelling · Boundary conditions · Model calibration · Model validation · Danube river · Bystroe Channel

---

G. Tudor · G. Deák (✉) · M. Arsene · B. Petculescu · D. M. Tuca · E. Bratfanof
National Institute for Research and Development in Environmental Protection Bucharest (INCDPM), 294, Splaiul Independentei, 6th District, Bucharest 060031, Romania

T. M. Danalache
University of Agronomic Sciences and Veterinary Medicine of Bucharest, Bucharest, Romania

E. Bratfanof
Dunarea de Jos University of Galati, Galati, Romania

M. R. R. M. A. Zainol
School of Civil Engineering, Universiti Sains Malaysia, Engineering Campus, 14300 Nibong Tebal, Pulau Pinang, Malaysia
e-mail: ceremy@usm.my

N. Mohamed Noor et al. (eds.), *Proceedings of the 3rd International Conference on Green Environmental Engineering and Technology*, Lecture Notes in Civil Engineering 214,
https://doi.org/10.1007/978-981-16-7920-9_17

1 Introduction

Over time, the Danube River has undergone alteration processes that have affected the dominant natural systems and created structures for economic purposes (navigation, hydropower, agriculture, ports, etc.) that have led to the reduction of floodplains and changes in morphological structures. The Chilia Branch status is particularly important due to the fact that it acts as natural state border between Romania and Ukraine and from it branches the Bystroe Channel, for which intended increased navigability, works have been carried out since 2004 that lead to international intense debate over the adverse transboundary impact on the local hydrodynamic and hydromorphological conditions [1] and habitat loss for fish and birdlife [2, 3].

It is necessary to use techniques and tools [4] to help determine the correct spatio-temporal evolution of hydromorphological properties with impact on water quality [5]. Currently, high-performance software programs are available that ensure the development of numerical models with a high level of confidence in simulating the hydrodynamic and hydromorphological conditions of rivers [6]. With the evolution of technology, a series of programs have been developed related to the hydraulic modeling of water flow [7]. Given that the numerical models take into account all the elements that can influence the water dynamics (flow, bathymetry, riverbed roughness, etc.), by 3D modeling can be calculated hydrodynamic variables and sediment transport, water quality parameters can be determined and trends in riverbed morphology can be monitored [8].

2 Materials and Method

In order to develop a hydrodynamic model for the Chilia-Bystroe area, the first steps carried out consisted of modelling domain delineation by implementing the measured bathymetry values into the dedicated computational grid and setting up the boundary conditions and simulation parameters. Considering the extent of the 3D riverbed bathymetry data set provided by the high resolution multibeam measurements carried out by the National Institute for Research and Development in Environmental Protection Bucharest (INCDPM), the modelling domain covered an 8.5 km length river sector, as seen in Fig. 1.

The in situ campaigns carried out by INCDPM [9] provided data sets (Fig. 2) that were used for setting up the geometric model, boundary conditions (discharge and water level values) and water velocity distributions in control sections for the model calibration and validation.

The developed geometric model for the study area consisted of 17680 triangular elements and 9446 nodes in the horizontal plane and 10 sigma layers with variable thickness (2, 3, 4, 6, 8, 10, 12, 15, 20, 20%) in the vertical plane. The data used for the boundary conditions resulted from Acoustic Doppler Current Profiler (ADCP)

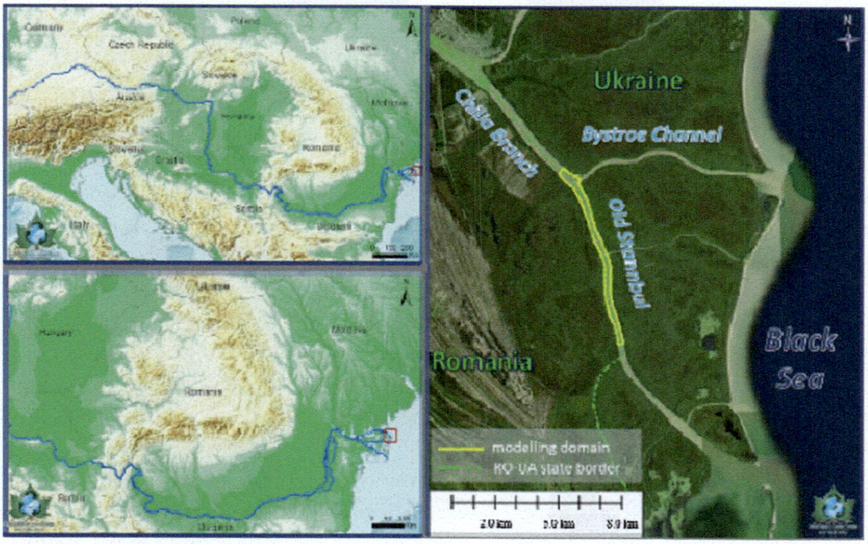

Fig. 1 Position of the hydrodynamic model domain in the Danube river basin

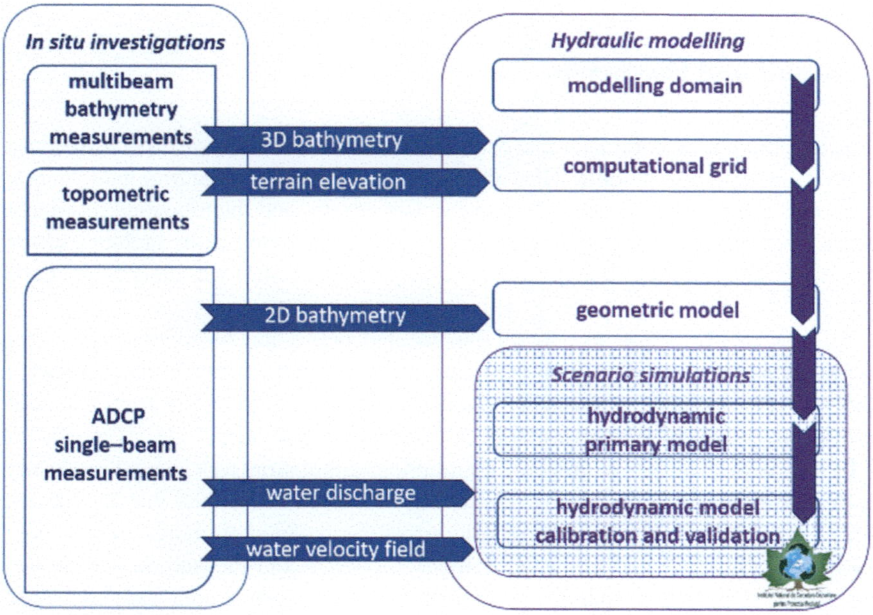

Fig. 2 Hydraulic modelling and measured input data sets (INCDPM 2020)

Table 1 Parameter values measured in situ campaigns used for boundary conditions

Control section	Water discharge values [m^3/s]		Water level values [m]	
	1st set	2nd set	1st set	2nd set
Upstream modelling domain (C1)	1434	2199	0.272	0.677
Bystroe channel access (C2)	658	1036	0.265	0.669
Downstream modelling domain (C3)	776	1163	0.202	0.592

single-beam measurements (water discharge) and from topometric measurements (water level) that were carried out in different hydrological conditions (Table 1).

The initial model was calibrated by gradually varying the parameters values until the modelling results were similar with the in situ measured values for the set flow conditions.

The model validation process consisted in setting up the boundary conditions for another data set recorded in different flow conditions and running the model by using the parameter values determined in the calibration process and compare the resulted modelled values with the measured ones.

3 Result and Discussion

The main simulation parameter that estimates the influence of the riverbed resistance in the Hydrodynamic module (HD) of MIKE 3 Flow Model is the *roughness height (RH)*. For different RH values, 15 test models were run and the modelled discharge values were compared with the measured ones (Fig. 3). With resulted relative errors of −1.38% (C3) and −1.74% (C2), the optimum value for RH was determined to be 0.09 m.

As seen in Fig. 4 for a cross-section located below the bifurcation (Fig. 4a), another step in the model calibration process consisted in the comparative analysis of modelled (Fig. 4b) and measured (Fig. 4c) water velocity distributions for transversal sections. Figure 4 shows the similarity of the two distributions.

In the model validation process, a numerical simulation was carried out using the second data set from Table 1 for the boundary conditions and the determined 0.09 m value for RH. The resulted relative errors for modeled discharge values being below 10% (Table 2) and the water velocity modelled and measured distributions being similar, the hydrodynamic model for the Chilia—Bystroe area was considered validated.

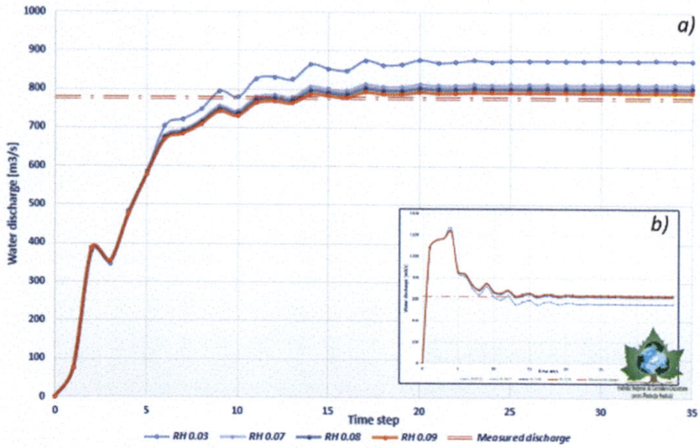

Fig. 3 Modelled discharge versus measured values—model calibration stage for C2 (**a**) and C3 (**b**) control sections

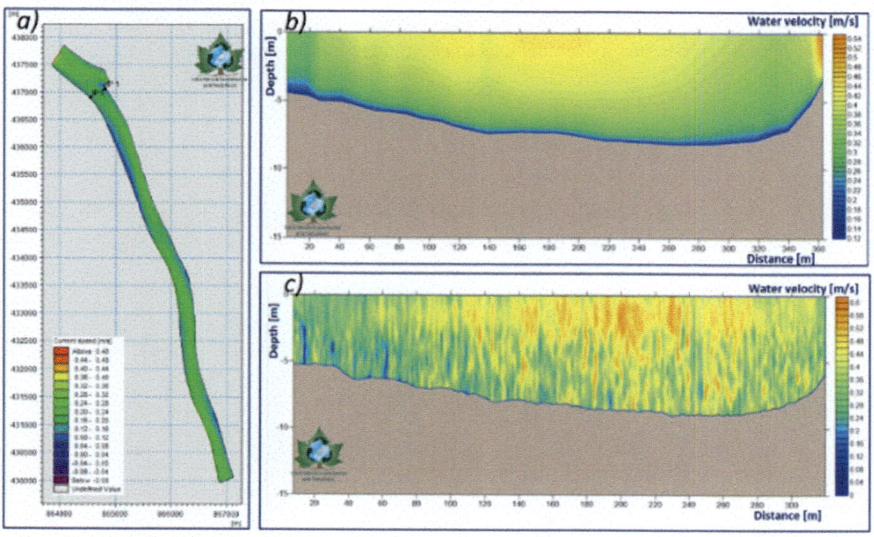

Fig. 4 Modelled water velocity versus measured distribution—model calibration stage

4 Conclusion

Using MIKE 3 Flow Model, a software with international repute on solving challenges in water environments and results of in situ measurement campaigns carried out by INCDPM for the 3D bathymetry on which the geometric model was developed, for the boundary conditions and reference values for the model calibration and validation processes, the hydrodynamic model for the Chilia—Bystroe area had

Table 2 Modelled versus measured values of water discharge in C3 and C2 control sections [m^3/s]

Test model	Control section	Measured [m^3/s]	Modelled [m^3/s]	Relative error [%]	Control section	Measured [m^3/s]	Modelled [m^3/s]	Relative error [%]
1	C3	776	951.839	−22.66	C2	658	479.33	27.15
2			874.716	−12.72			555.98	15.5
3			548.764	29.28			885.221	−34.53
4			694.489	10.5			738.72	−12.27
5			694.583	10.49			742.849	−12.89
6			674.369	13.1			758.902	−15.33
7			812.76	−4.74			617.977	6.08
8			837.771	−7.96			592.975	9.88
9			812.76	−4.74			628.48	4.49
10			802.701	−3.44			628.012	4.56
11			824.295	−6.22			606.449	7.83
12			798.894	−2.95			636.894	3.21
13			791.717	−2.03			653.431	0.69
14			793.795	−2.29			643.477	2.21
15			786.699	−1.38			669.428	−1.74
Model validation		1163	1048.72	9.83		1036	1133.91	−9.45

shown that it provides results with a high degree of confidence (>90%). The model will be updated with new data as it becomes available and it will be used for making short- and medium-term forecasts based on scenarios in order to assess the morpho-hydrodynamic changes for this area with strategic importance and their impact on migration routes and sturgeon habitats.

This paper was developed through Nucleu Programme—Contract number 39 N/2019 carried out with the Romanian Ministry of Research and Innovation support, within the project "*Research on the morphological and hydrodynamic evolution tendencies in the Chilia—Bystroe transboundary area*".

References

1. Tudor G, Deak G, Cïrstinoiu C, Raischi M, Danalache T, Cornateanu G, Bratfanof E (2020) Transboundary river hydrodynamic conditions assessment in sustainable development context. A case study of Danube River: Chilia Branch -Bystroe channel area, IOP Conf. Series: Earth and Environmental Science 616, 012072. https://doi.org/10.1088/1755-1315/616/1/012072
2. INCDPM (2014) Study regarding the updating of the physical-chemical and biological components evaluation from Danube Delta and the adjacent coastline considering the transboundary impact of the Bastroe channel and establishing a transboundary monitoring system coordinated together with the Ukrainian part, in order to analyze the environmental impact on the Danube Delta Biosphere Reserve of the project regarding the construction of the deep sea navigation channel on the Bastroe branch, INCDPM Bucharest, INCDDD Tulcea, INCDM GA Constanţa, INHGA
3. Danalache T, Badilita AM, Deák G, Holban E, Popescu I, Daescu A, Raischi MC, Ghita G, Nicolae CG, Diaconescu S (2017) Assessment of Bâstroe channel possible impact on lower danube sturgeon migration, aquaculture, aquarium, conservation and Legislation. Int. J. Bioflux Soc 10(5):1011–1018
4. INCDPM (2017) PN 16 04 01 13.1: Current trends and approaches to the use of GIS techniques in the analysis of environmental factors, INCDPM
5. Gy D, Tudor G, Nicolae AF, Uriţescu B, C-tin C, Raischi MC, Zamfir S, Badea G, Danalache TM (2019) Hydrodynamic parameters assessment using specialized techniques in the context of low flows. Case study Epurasu Branch—Danube River, AIP Conference Proceedings 2129. https://doi.org/10.1063/1.5118097
6. Acuña GJ, Ávila H, Canales FA (2019) River model calibration based on design of experiments theory. A case study: meta river, Colombia, Water 2019(11):1382. https://doi.org/10.3390/w11071382
7. Sârbu D (2015) Strategies to minimize anthropogenic impact on riverbeds. Dissertation, Technical University of Civil Engineering—Faculty of Hydrotechnics, Bucharest, Romania
8. INCDPM (2015) PN 09 06 01 25.1 Use of numerical simulations to analyse the variation of hydraulic parameters in the context of anthropogenic interventions. Bala branch case study, INCDPM
9. INCDPM (2020) PN 19 43 03 01.3 Establishment of the morpho-hydrodynamic reference status for the Chilia—Bystroe area, INCDPM

Identification and Validation of a Method for Determining the Age of Sturgeons

Tiberius Danalache, Elena Holban, György Deák, Razvan Matache, Raluca Prangate, Monica Matei, Madalina Boboc, and Nor Wahidatul Azura Zainon Najib

Abstract Fisheries research has focused mainly on the dynamics of populations under exploitation in order to be able to inform what measures are needed for the sustainable harvesting of biological resources. A necessary tool for the successful achievement of this objective is to determine the age of the fish. Thus, this article will detail a method for identifying the age-based on scanning electron microscope analysis of dorsal shield and pectoral fins especially the calcified structures by making several sections, which are necessary in age determination and validation.

Keywords Age · Analysis · Sturgeon · SEM

1 Introduction

Determining the age of fish specimens is the basis for quantitative assessment and the forecasted trend of their stocks. Research incorporating this data is not only significant for assessing the state of stocks relative to their historical value, for a reference value of biological significance or for management objectives, but also for assessing the consequences of alternative management scenarios for stocks fish, respectively for fishermen. Therefore, it can be stated that although the analysis of age has been done for over 250 years, currently much faster and more accurate techniques have been developed to determine these data. In our study, the main purpose of efficient testing of a method for estimating the age of sturgeon species is to identify characteristics in correlation with the sexual maturity of specimens, the number of reproductive cycles [1]. Thus, this study consisted in studying previous research in order to identify a method for estimating the age used so far worldwide, respectively

T. Danalache (✉) · E. Holban · G. Deák · R. Matache · R. Prangate · M. Matei · M. Boboc
National Institute for Research and Development in Environmental Protection Bucharest (INCDPM), 294 Splaiul Independentei Street, 6th District, 060031 Bucharest, Romania

N. W. A. Z. Najib
Centre of Excellence, Water Research and Environmental Sustainability Growth (WAREG), Universiti Malaysia Perlis, 02600 Arau, Perlis, Malaysia
e-mail: norwahidatul@unimap.edu.my

N. Mohamed Noor et al. (eds.), *Proceedings of the 3rd International Conference on Green Environmental Engineering and Technology*, Lecture Notes in Civil Engineering 214,
https://doi.org/10.1007/978-981-16-7920-9_18

in selecting a method and its actual testing on samples held by the National Institute for Research and Development for Environmental Protection, Bucharest following the activities of capturing, tagging, releasing, respectively monitoring the sturgeon species found at the level of the Lower Danube [2].

2 Materials and Method

2.1 Materials

A modern technique used, based on scanning electron microscopy, consisted in the use of specialized equipment (including HITACHI SU-70 FE-SEM scanning electron microscope, oven, milling cutter, carbon strips for fixing, QUOORUM SC7620 deposition unit, metal films) from the endowment of the Laboratories Department of INCDPM Bucharest, in order to analyze the biological samples such as the pectoral fins, respectively the dorsal shields, collected from individuals raised in aquaculture environment, from the species *Acipenser stellatus*.

2.2 Method

This research has addressed several studies on age determination in fish species, particularly sturgeon species. Thus, a method was identified and considered to be effective, following efficient testing: the use of scanning electron microscopy.

Recourse to the use of scanning electron microscopy in order to identify the age of sturgeon specimens of the species *A. stellatus*, from aquaculture systems, to not close the vital integrity of wild specimens are currently in conservational decline [3, 4]. A set of biological samples consisting of pectoral fins and dorsal shields was used (Fig. 1). The samples taken from the pectoral fin were prepared in advance by making two cross sections to ensure the necessary thickness for easy reading at the microscopic level. Also, sections of the dorsal shields were previously taken from the

(a) (b) (c) (d)

Fig. 1 Samples preparing procedure; **a**—dorsal shields, calcified pectoral fin; **b**—samples sectioning with the help of a cutter; **c**—samples obtained after grinding; **d**—section of the pectoral fin

(a) (b) (c)

Fig. 2 Samples holding; **a**—pectoral fin samples fixed on the stab with carbon band; **b**—samples fixed with carbon band and paste to ensure conductivity; **c**—QUORUM SC7620 holding unit

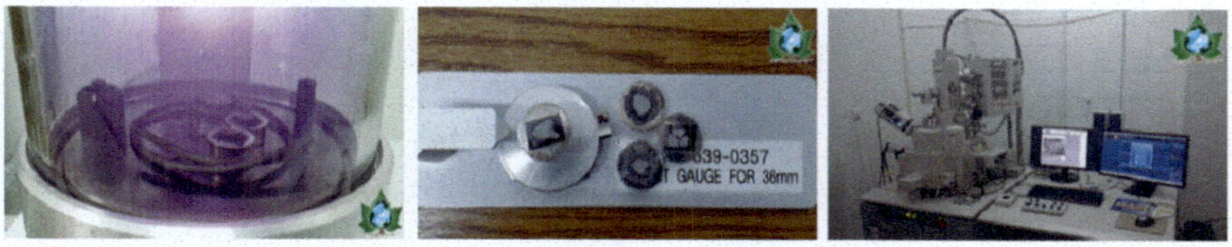

(a) (b) (c)

Fig. 3 Samples preparing for SEM reading; **a**—the deposition process of the Au–Pd metal film; **b**—samples used for SEM reading; **c** HITACHI SU-70 FE-SEM scanning electronic microscope

posterior portion of the head, cleaned and dried for 48 h, in order to fix as samples, following Brennan [5] and Bruch [6] methodology.

Analysis of samples, was focused on calcified structures represented by the pectoral fins, dorsal shields respectively. Sections of pectoral fins were made, which were divided by means of a manual milling cutter, obtaining new samples, which were then sanded on a fine surface, preparing them for a final sanding by means of a diamond paste. After the grinding process, the samples were left in the oven to dry, after which they were fixed to stab by means of a carbon strip, in order to ensure the conductivity necessary for reading the samples using electron microscopy. For a better fixation, the open spaces were completed with a carbon paste. The dry samples were introduced in the QUORUM SC7620 unit, where they were covered with an Au–Pd metal film, in order to ensure a better electrical conductivity (Fig. 2).

However, the last part of the method has been the use of scanning electron microscope (SEM) Hitachi FE-SEM SU-70 (Fig. 3), X-ray spectrometer coupled with energy dispersive (EDS).

3 Result and Discussion

A total of 16 samples from the pectoral fins were analyzed, of which 10 samples without Au–Pd film, respectively 6 samples with film and 5 samples from the dorsal shields (3 samples without Au–Pd film, respectively 2 samples with film). Considering the methodology used (Brennan) [5], the samples that were covered with Au–Pd film did not illustrate growth patterns. Thus, an attempt was made to cover the samples with carbon tape and paste, but due to the low electrical conductivity, these samples did not resist. Of the 10 samples without Au–Pd film, only one representative sample was obtained that could indicate the growth index, thus representing a section of the fibrils that enter the structure of the fin. Considering the fact that in the autumn the individuals of this species tend to accumulate less calcium [7, 8], compared to the spring season, then it can be specified from the analyzes that this specimen is 3.5 years old, more precisely 3 springs and an autumn.

After analyzing the structure of the fibers at the level of the analyzed samples, respectively the chemical composition, they highlighted the fact that the fibers that make up the fin have the shape of circular tubes whose chemical component is based on P elements, respectively Ca and shows various agglomerations in the EDS spectrum on the elementary map. For the dorsal shield was subjected to the same, so as to result predominantly in the chemical composition of the elements are present: Ca, Mg, P (Fig. 4). The dorsal shield is a calcified structure that is found on the surface of the body of a sturgeon coming into direct contact with water and having a faster bioaccumulation capacity unlike the sections of fins, which do not come into

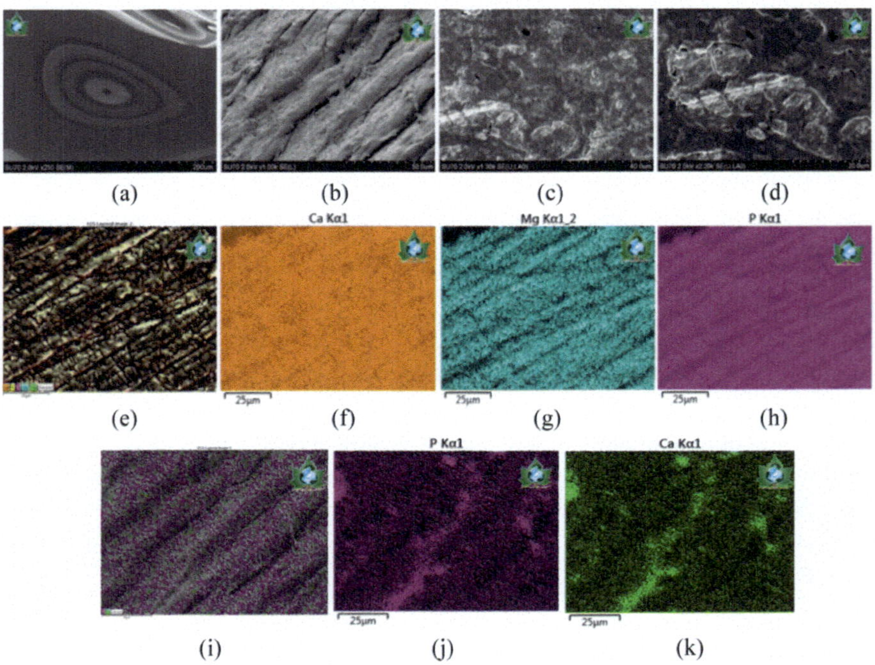

Fig. 4 SEM representations; **a**—growth rings present in the fibers which form the fin; **b**—fiber structure of the pectoral fin; **c**, **d**—fiber structures that make up the dorsal shields of the analyzed specimens; **e**—dorsal shields elemental map; **f**—Ca map within dorsal shields fiber structures; **g**—Mg map within dorsal shields fiber structures; **h**—P map within dorsal shields fiber structures; **i**—pectoral fin elemental map; **j**—P map within pectoral fin fiber structures; **k**—Ca map within pectoral fin fiber structures

direct contact with metals in the environment, bioaccumulation gradually and with a different frequency.

4 Conclusion

Following the analysis of the literature, a method for determining the age and growth of fish was identified: study of seasonal ring formation inside hard parts such as scales and bones. The identification method based on the seasonal growth rings using scanning electron microscopy was performed for the first time, generally used conventional methods such as optical microscopy. Scanning electron microscopy showed that in the case of a specimen of seasonal growth rings possibly present in the fibers that are part of the analyzed bone structure, they indicated that it would be estimated to be 3 and a half years old. In conclusion, this study provides evidence designed to quantify sturgeon age, so by analyzing the biometric data collected during the study period and combined with information from the data INCDPM Bucharest and through practical experiments conducted in INCDPM laboratories. The biological samples used came from specimens raised in intensive aquaculture systems for marketing purposes. In the study, no biological samples from wild specimens were used.

This paper was developed through Nucleu Programme—Contract number 39 N/2019 carried out with the Romanian Ministry of Research and Innovation support, within the project "*Evolution of sturgeon populations and Danube ichthyofauna in the context of the Danube course alterations from recent decades*".

References

1. S. Khan, M. A. Khan, and K. Miyan (2011) Comparison of age estimates from otoliths, vertebrae, and pectoral spines in African sharptooth catfish, Clarias gariepinus (Burchell), Estonian Journal of Ecology, 2011, 60, 3, 183–193 https://doi.org/10.3176/ECO.2011.3.02
2. INCDPM Bucharest, Rapoarte Intermediare 1–17, (2011–2017). Proiect „Monitorizarea impactului asupra mediului a lucrărilor de îmbunătățire a condițiilor de navigație pe Dunăre între Călărași și Brăila, km 375 - km 175", beneficiar Administrația Fluvială a Dunării de Jos R.A. Galați.
3. Deak Gy. et al. (2017) "Action for preserving sturgeon species from Lower Danube River," presented at the International Warsaw Invention Show, Varsovia, Polonia, Oct. 09, 2017.
4. Danalache T. et al. (2018) "Sturgeon migration monitoring on Danube Delta Branches using ultrasonic telemetry and DKMR-01T monitoring system," presented at the 12th ELSEDIMA Conference, Cluj-Napoca, Romania, May 17, 2018.
5. Brennan JS, Cailliet GM (1989) Comparative Age-Determination Techniques for White Sturgeon in California. Trans Am Fish Soc 118(3):296–310. https://doi.org/10.1577/1548-8659(1989)118%3c0296:CATFWS%3e2.3.CO;2
6. Bruch RM, Campana SE, Davis-Foust SL, Hansen MJ, Janssen J (2009) Lake Sturgeon Age Validation using Bomb Radiocarbon and Known-Age Fish. Trans Am Fish Soc 138(2):361–372. https://doi.org/10.1577/T08-098.1

7. F. E. Lux, "AGE DETERMINATION OF FISHES," *Fishery Letters*, vol. 637, pp. 1–10.
8. Chilton DE, Beamish RJ (1982) Age Determination Methods for Fishes Studied by the Groundfish Program at the Pacific Biological Station. Department of Fisheries and Ocean, Ottawa

Properties of Irradiated Bioplastic-A Review

Nurin Najwa Rohidi and Siti Amira Othman

Abstract Nowadays, bioplastic is one the popular research in this world. The benefits of the bioplastic including ease of to degrade and the most important is the contribution in environmental aspect. It helps in reducing the effect of green house that relates with thinning of the ozone layer. Meanwhile in industry, bioplastic is produce extensively to fill the market demand. Currently, exist awareness among people about the use of bioplastic in daily life. In order to enhance the properties of bioplastic, radiation is use to see the difference. This paper review about the properties of irradiated bioplastic.

Keywords Bioplastic · Plastic · Irradiated · Properties

1 Introduction

Among all types of packaging material, plastic was one of the mainly used in the list. Not only that, flexible, lightweight, odorless, great in mechanical strength, hygienic and most important thing is low cost. Bioplastic which is referred to as one of the biodegradable products where being a hero for reducing dependency on fossil fuel. The ability of attract water molecules is needed to pay attention to achieve waterproof or at least near to the hydrophobicity level but even so biodegradability needs to be maintained. Therefore, wettability depends on the water solubility of the bioplastics. In detail, moisture content percentage will be studied in the change of bioplastic weight before and after dried up in the oven. Other than that, water contact angle can be done using a receding angle to ensure the wettability of bioplastics. When it comes to industrial use, gamma radiation and electron beams are generally involved. In this review, the types of plastic and bioplastic will be explained. Hydrophobicity

N. N. Rohidi · S. A. Othman (✉)
Faculty of Applied Sciences and Technology, Universiti Tun Hussein Onn Malaysia, 84600 Pagoh, Johor, Malaysia
e-mail: sitiamira@uthm.edu.my

N. N. Rohidi
e-mail: aw170042@siswa.uthm.edu.my

N. Mohamed Noor et al. (eds.), *Proceedings of the 3rd International Conference on Green Environmental Engineering and Technology*, Lecture Notes in Civil Engineering 214,
https://doi.org/10.1007/978-981-16-7920-9_19

will be the focus on to understanding wettability works. Other than that, radiation and its effect on polymer will be discussed.

2 General Properties of Plastic Material

Plastics have been used for over half a century and it was produced using petroleum-based. In a large production, plastics were available in supermarkets, industries and households as it was very cheap and gave ease for sellers to provide their customers' satisfaction when handling products, especially in the packaging sector. Unlike cupboards, metal and glass, plastics are lighter and have good barrier properties also high stretchability. Almost 40% of plastics were under the packaging sector due to matching characteristics of packaging material. For example, flexible, lightweight, odourless, great in mechanical strength, hygienic and most important thing is low cost. There are five key functions of packaging, the first one protection which to protect from contamination that occur on the product due to the presence of microorganisms, air, moisture, toxins and others that can triggered lifetime of the product. Next, plastic is used for containment to prevent product from spilling. Plastic is also used to inform details of ingredient and manufacturer address. Other than that, utility of use for smoothing transportation, storage process and not to forget advertising about products update. Therefore, resulting in different types of plastics used for satisfying producers' and customers' needs. Plastics have a few major categories, polyethylene (PE), polypropylene (PP), polyester and substituted olefins [1].

Polyethylene can be divided into three more groups, low-density polyethylene (LDPE), high-density polyethylene (HDPE) and linear low-density polyethylene (LLDPE). First of all, PE barrier of water vapour is in great condition but the oxygen barrier is quite low. Compared to the flexibility and stretchability, LDPE is greater than HDPE because HDPE has a higher crystalline density. LLDPE is also not bad when it comes to clarity, toughness and heat sealing. PP also has a weak point where it is used to combine with other molecules such as ethylene–vinyl alcohol (EVOH). PP was rigid and temperature for heat deflection temperature (HDT) was high until being used from cold-chain to heat-treated product.

Polyester is quite different and categorized by crystallinity degree (amorphous and crystallized), both have high HDT and provide good barrier properties in aspects of water and gas. Examples for substituted olefins are polystyrene (PS) and polyvinyl chloride (PVC). PS is good in absorbing shock or shock resistance meanwhile PVC makes sellers satisfied as oxygen requirement of food can be maintained resulting in a good look and freshness of the food [2]. All categories were match in a non-biodegradable team due to unbreakable bond characteristics. Surprisingly it can be oxo-biodegradable if the plastics have stabilization with antioxidants. In order to decrease the degradation period, prodegradants were used.

Pollution will always be linked to the plastics. Emission carbon for the production itself is already a risk to our health system. Every year, the use of plastic bags can reach over millions of digits. The annual production was 78 million tonnes but the

Table 1 Global flows of plastic packaging material in 2013 [15]

14% Collection of plastics for recycle	8%	Cascaded recycle	86% Remaining	14%	Incineration
	4%	Process Loss		40%	Landfill
	2%	Closed loop recycle		32%	Ocean

output after consumption end up to varies of places. From the 78 million tonnes plastics, only 2% of it recycled into a similar quality application (closed-loop), 8% of it undergo cascaded recycle where recycled to lower quality applications. 4% were a loss from the total collection for recycling. Another 86% were total from 14% for incineration, 40% gone to landfill and 32% leaked to the ocean. Degradation does not work on plastics in a short time but long lasts until hundreds of years. Plastic end up on the top of the pollutant lists when it comes to solving the pollution either on land or water (Table 1).

2.1 *Bioplastics*

Bioplastics can be defined as plastic made from biomass sources and it is environmentally-friendly. Bio-based polymers generated from biomass are also classified as bioplastics, despite the fact that they are not biodegradable. The reason certain bio-based do not degrade like other bioplastics due to the ingredient of is petroleum-based. Examples of biomass sources are carbohydrate-based polymers and fat and oil-based polymer. However, starch from carbohydrate-based always got huge attention due to the low cost and easy to obtain stock. Such as corn, potato, wheat and rice. All of them contained two types of molecules, amylose and amylopectin. Amylopectin is the major component present in starch as a crystalline region with range ratio in 75–85%, while the amylose is the opposite present in starch as amorphous region with range ratio in 20–25% [3]. Table 2 shows the amount of amylose exceeds 30%.

Other than cost, and ready stock, starch has been chosen for its zero-toxicity with good extensibility which was similar to plastics. Plus, with pure starch, odourless and colourless were value added characteristics that have been focused in development of bioplastics. However, the properties of starch which is hydrophilic makes it less favourable among consumers as this property threatens the life span of the bioplastics.

Table 2 Amylose and amylopectin ratios in different starches [3]

Starch	Amylose (%)	Amylopectin (%)
Corn	28	72
Potato	20	80
Rice	18.5	81.5

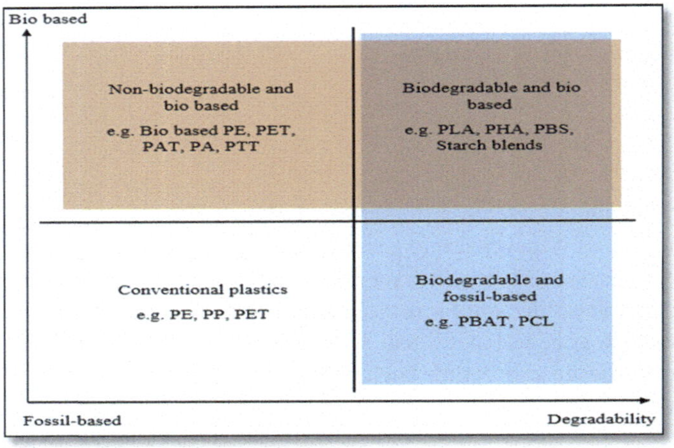

Fig. 1 Categories of plastics [5]

Several studies have shown clear results using different starches and plasticizers to strengthen the mechanical and chemical properties of the bioplastics [16].

Polylactic acid is also in the bioplastic category. Polylactic acid (PLA) was the most glamour among plastic with its biodegradability as it formed from starch fermentation that has been enriched for polymerisation. Degradation of PLA started at 60 °C, but when it has the presence of additives, temperature degradation will increase until it is able to reach 200 °C. Like other bioplastics, PLA is eco-friendly and non-toxic makes it suitable to be applied in the medical field. PLA outperforms other biopolymers in terms of processing ability, such as polyhydroxyalkanoates (PHA). Plus, PLA production does not require energy production like petroleum-based plastic does, PLA only needs about 50% and maximum energy requirement can reach 75% of the total energy that petroleum-based plastic used to consume [4] (Fig. 1).

A study by Van et al. [6] mentioned that the crystallinity level defines the flexibility of bioplastics. Rich in amylopectin brings the struggle in tensile strength resulting in crack and brittle occur. The purpose of their study is to study the effect of the ratio of the amylose and amylopectin on extruded starch and the structure of starch. Materials that have been used were waxy corn-starch which has a high content of amylopectin and amylomaize which has 70% amylose content with 12–13% of moisture content for each sample. The samples were extruded and mixed with glycerol with ratio starch/glycerol: 100/30 (w/w). Parts of the samples were immediately quenched in liquid nitrogen and stored at −22 °C. Another two parts stored in different relative humidities (RH) for 2 weeks. To determine moisture content, an infrared dryer or Gallenkamp vacuum oven is used to measure gravimetrically. Pulling materials was to determine the mechanical properties. The results demonstrate that materials with high amylopectin content are more flexible than materials with high amylose content, but not in as good of a condition in stress tests, resulting in brittleness [17]. The study shows interest only on the starches and put aside the effect of plasticizers and amylose crystallization was undescribed.

An experiment conducted by Amalia et al. [7] that involved corn husk, corn starch, chitosan and acetic acid. The experiment started with different mesh of corn husk. Each of them was added with corn starch with balancing ratio of molar (1:1). Next chitosan inserted but with different of concentrations, followed by acetic acid and sorbitol with amount 2.5 millilitres (ml) and 1.75 ml respectively. Then distilled water was added to the mixture was achieved 70 ml. the mixture heated for 90 min with temperature at 80 °C. From Fourier Transform Infrared (FTIR) Spectroscopy result, hydrogen bonded OH group, C = C, = CH, CH and CO comprised by maize and CO is component that permit starch degradation. Biodegradability of the bioplastics was done by soil burial method for 30 days long to consider the landfill environment. The results only collected with images and lacking of data of weight to prove the biodegradability change. About 70–100% of the whole sample degrade in 21 days after growth of fungal detected at day 14.

A study conducted by Marichelvam et al. [8] to produce a combination of rice and corn starch of bioplastic consists of 5 samples with a total weight 10 g (g) with different weight ratio of rice starch and corn starch added with constant weights of 3 g of glycerol, 1 g of citric acid, 2 g of gelatine and 100 g of water in each samples. Tensile strength tests show that sample 5 with 7 g of rice starch and 3 g of corn starch has the highest tensile strength, 12.5 MPa with Young's Modulus 0.183 GPa and elongation at 6.8%. Plus, the content of moisture for sample 5 the least, 11.7% followed with the water solubility only 11.5%. Curious about the wettability of sample 5, water contact angles were determined and compared with (LDPE) plastic. The results were only 70° for sample 5 and LDPE got around 90° . So far, sample 5 shows hydrophobicity, but when it comes to biodegradation, it degrades with 48.73% in 15 days using soil burial degradation. It can be concluded that an increased amount of rice starch will strengthen the mechanical properties and waterproof characteristics but when compared to other samples in content of moisture and water solubility test, the effect of rice starch amount was not consistent especially for sample 3 and 4 with weight of rice starch 6 g and 6.5 g respectively. Sample 3 and sample 4 readings for water solubility and moisture content should be near with sample 5 reading. It might be because the effect of plasticizers is not equal in every sample.

Another study using potato and yam tubers starches with weighing 2.5 g each, followed by the addition of 25 cm^3 of distilled water, then hydrochloric acid and propan-1,2,3-triol with 2 cm^3 each. The mixtures then boiled for 15 min [9]. After that the mixtures is poured in petri dishes to dry out for 2 days in lamina air flow. Thermal stability of starches was studied by heating the samples in a nitrogen atmosphere in a range temperature 30–500 °C with a rate 20 °C/min. Strength of mechanical of the samples was also carried out with a crosshead speed test. Results show that both have weight loss 50% at 250 °C for yam and 310 °C for potato. Potato has more flexibility compared to yam as yam has better tensile strength with almost three times greater than potato.

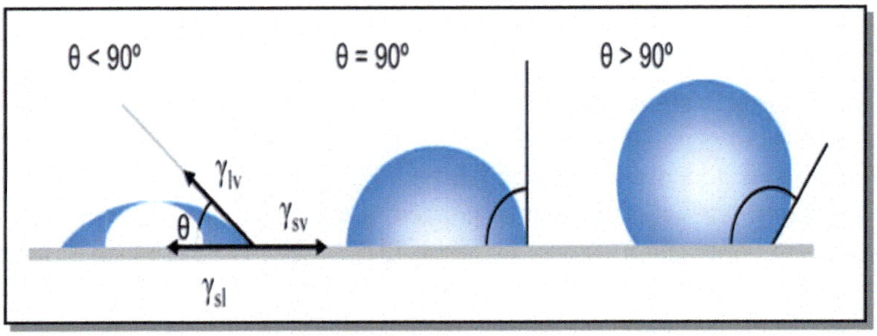

Fig. 2 Contact angle formed by sessile liquid drop [11]

2.2 *Hydrophobicity*

When it comes to plastic, unable to react with water is one of the main characteristic that we will discover first. This is because the polymerization process combining the monomers which are a double bond between two carbon atoms and four pendant groups into a long chain that we called as "polymer" resulting emission of water molecules or hydrochloric acid. However, differ from bioplastic, water permeability is quite high due to the presence of water itself as plasticizer makes some bioplastic easier to attract water molecules. Fortunately, adding other plasticizer will affect relative humidity (RH) where the water content of the material will follow it [10] (Fig. 2).

The lower the RH, the lower the water content will be when the amount of plasticizer increases and vice versa. Therefore, contact angle measurement will be needed to measure hydrophobicity. Sessile droplets are mostly used to examine wettability. Angle of 0° labelled as the most hydrophilic and 180° for superhydrophobic. Other than that, the content of the moisture test can tell the level of hydrophilicity of a material. By weighing mass before and after dry the sample, if there has a huge different then the percentage of moisture content will be high and vice versa.

2.3 *Ionizing Radiation*

Radiation is defined as the emission of energy from a substance and radiation divided by two categories. First is non-ionizing radiation which does not give harm towards us. For example, radio waves, microwaves basically refer to any electromagnetic radiation as they carry not enough energy to emit electrons from an atom. Second is ionizing radiation which can ionize other atoms as it carries enough energy to interact with Coulomb's force.

Ionizing radiation that undergoes direct interaction like alpha radiation and beta radiation. For alpha radiation, high energy alpha particles are involved. Alpha particle production is called alpha decay. Alpha particle is large due to the presence of double positive charge as it is made up of two neutrons and two protons and binds to helium as both of them are identical particles. Since alpha particles are heavy, the distance that they can travel only a few centimetres, even a piece of paper can stop them from traveling more.

Next would be beta radiation, where a free electron or positron is involved. This kind of radiation radiates smaller particles as a single negative charge is carried. Due to its lightness, it can travel or penetrate more than alpha particles and the difference in distance travel is quite big compared to alpha particles until several meters but the limitation of beta particle is aluminium. Gamma-ray, X-ray and neutron radiation are classified under indirect ionization radiation. Both gamma ray and x-ray are known as photon radiation due to the involvement of photons that contain high energy. Distance travel of those two rays can achieve up to 10 m or longer.

3 Effect of Radiation on Polymer

When it comes to industrial use, gamma radiation and electron beams are generally involved. The reason for gamma radiation is because gamma-rays are extreme in penetration and Cobalt 60 (**Co**^{60}) mainly used and often for sterilization purpose. Unlike gamma radiation, electron beams have limitation for penetration. When radiation is exposed to high polymer, different reactions may occur. There are four reaction, abstraction, recombination, disproportionation and polymerization. The effect of radiation leads to crosslinking or degradation towards chemical bonds. When crosslinking occurs, molecular weight affected and increase resulting elastic modulus and tensile strength increase as well and bringing elongation decline. Different from crosslinking, degradation leads to reduction of molecular weight and lessening the viscosity. For instance, polyethylene molecular characteristics affected by radiation were influenced by the thickness of the material, intensity of irradiation and dose. When it comes to starch, radiation can make the starch more soluble as it can undergo oxidative degradation, in fact, the structure of amylopectin might undergo fragmentation [12]. Reaction of irradiation towards oxygen studied by Haji Saied et al. [13] reported that formation of free radicals can occur. By the presence of free radical, crosslinking of the polymer may occur due to the potential possessed by the attacked chain.

According to Gonzalez et al. [14], sterilization processes with dry heat and moist heat method were used along with gamma radiation and ethylene oxide. For irradiation, **Co**^{60} were used with 25 kGy for total dose at rate 5.56 Gys^{-1} and moist heat minutes at 121 °C. Dry heat process refers to drying out microorganisms to death, so the process took 3 h at 150 °C. Meanwhile, for ethylene oxide gas were used for the adsorption 40% of relative humidity with concentration 600 mgl^{-1} at 40 kPa subatmospheric pressure and in range temperature 50–55 °C with 20 h long for contact time. The result shows only dry heat can extend the tensile strength, but in the elongation part it drops while others rise up to achieve maximum. When it comes to capacity of absorption of water, gamma radiation shows the least percentage, the same goes to biodegradation degree. The study successfully figured out the sterilization product does not affect the material properties too much.

Based on the study of ultraviolet (UV) treatment effect of improvement on the film of rice starch properties. Rice was reported to be the second largest production of grain [18]. When rice is chemically modified, mechanical properties can be improved such as swell due to water content and one of the ways is crosslinking by blending either dry or semi-dry way. The starch film prepared with rice starch undergo gelatinization for 10 min at 85 °C, added with glycerol and stirred for 2 min followed by sodium benzoate as photo-initiator. Then the sample was put at 55 °C for 10 to 12 h. The films were held in a desiccator for about 72 h at 55% RH. Before that, a mercury lamp was used to irradiate the samples with wavelength exceeding 290 nm at different duration 10, 20 and 30 min. The ideal flexibility due to presence of photosensitizer was at a concentration range of 3–6% with decreasing elongation from 116 to 67%. Even water vapour permeability at 6% show lowest reading, 4.12 g.mm/m^2.day.kPa. However, samples experienced a decrease in transparency and crystallinity at 14.97%, yet swelling degree were the lowest with high production of gel fraction. Due to the presence of photosensitizer, the samples were having improvement in barrier properties.

4 Conclusions

Recently, bioplastic has gained much attention based on the benefit and irradiated method used. Yet, a lot of research needs to be implemented to make sure the properties of bioplastic improve. With this positive perspective, it is helpful in enhancing the tremendous research relate to bioplastic.

Acknowledgements The authors like to thank University Tun Hussein Onn Malaysia for facilities provided.

References

1. Cantor KM, Watts P (2011) Plastics materials. In: Applied plastics engineering handbook (pp. 3–5). William Andrew Publishing
2. Kim YT, Min B, Kim KW (2014) General characteristics of packaging materials for food system. In Innovations in food packaging (pp 13–35). Academic Press
3. Jambrak AR, Herceg Z, Šubarić D, Babić J, Brnčić M, Brnčić SR, Gelo J (2010) Ultrasound effect on physical properties of corn starch. Carbohydr Polym 79(1):91–100. https://doi.org/10.1016/j.carbpol.2009.07.051
4. Rasal RM, Janorkar AV, Hirt DE (2010) Poly (lactic acid) modifications. Prog Polym Sci 35(3):338–356. https://doi.org/10.1016/j.progpolymsci.2009.12.003
5. Lackner M (2015) Bioplastics—Biobased plastics as renewable and/or biodegradable alternatives to petroplastics. In: Book: Kirk-Othmer Encyclopedia of Chemical Technology (6th edn). Wiley

6. Van Soest JJG, Essers P (1997) Influence of amylose-amylopectin ratio on properties of extruded starch plastic sheets. J Macromol Sci Part A Pure Appl Chem 34(9):1665–1689. https://doi.org/10.1080/10601329708010034
7. Amalia D, Saleh D, Djonaedi E (2020). Synthesis of biodegradable plastics using corn starch and corn husk as the fillers as well as chitosan and sorbitol. In: Journal of physics: conference series (vol 1442, No. 1, p 012007). https://doi.org/10.1088/1742-6596/1442/1/012007
8. Marichelvam MK, Jawaid M, Asim M (2019) Corn and rice starch-based bio-plastics as alternative packaging materials. Fibers 7(4):32. https://doi.org/10.3390/fib7040032
9. Ismail NA, Mohd Tahir S, Yahya N, Wahid A, Firdaus M, Khairuddin NE, Abdullah MA (2016) Synthesis and characterization of biodegradable starch-based bioplastics. In: Materials science forum (vol 846, pp 673–678). https://doi.org/10.4028/www.scientific.net/MSF.846.673
10. Chaléat C, Halley PJ, Truss RW (2014) Mechanical properties of starch-based plastics. Starch Polym 187–209. https://doi.org/10.1016/B978-0-444-53730-0.00023-3
11. Yuan Y, Lee TR (2013) Contact angle and wetting properties. In Surface science techniques (pp 3–34). Springer, Berlin, Heidelberg
12. Khalil SA (2010) Effect of ionizing radiation on the properties of prepared plastic/starch blends and their applications as biodegradable materials. In: Thesis. Ain Shams University
13. Haji-Saeid M, Sampa MHO, Chmielewski AG (2007) Radiation treatment for sterilization of packaging materials. Radiat Phys Chem 76(8–9):1535–1541. https://doi.org/10.1016/j.radphyschem.2007.02.068
14. González ME, Lunati C, Floccari M, Salmoral EM (2009) Effects of Sterilizing Agents on the Biodegradation of a Bioplastic Material. Int J Polym Mater 58(3):129–140. https://doi.org/10.1080/00914030802583726
15. World Economic Forum (2016) The new plastics economy, Rethinking the Future of Plastics
16. Abe MM, Martins JR, Sanvezzo PB, Macedo JV, Branciforti MC, Halley P, Botaro VR, Brienzo M (2021) Advantages and disadvantages of bioplastics production from starch and lignocellulosic components. Polymers 13:2484. https://doi.org/10.3390/polym13152484
17. Lourdin D, Valle GD, Colona P (1995) Influence of amylose content on starch films and foams. Carbohyd Polym 27(4):261–270. https://doi.org/10.1016/0144-8617(95)00071-2
18. Sampad Nandy (2021) India set to harvest record rice, wheat crops in 2020–21. S&P Global

Quality Control & Field Guidelines for Using Non-identical Mixed GPS in Surveying Project: EGYPT Case Study Antennas

Mohamed M. Hosny, Samy Ayaad, Mohamed H. Elwany, and Hassan G. El-Ghazouly

Abstract Nowadays, GNSS accuracy has reached a level where the remaining error sources could be identified as the effect caused by phase center variations (PCV) of the antenna. The correction of PCV effect is generally important for GNSS observations and should be greatly considered when using mixed antenna design. Mixed antenna usage within the same project may degrades the observations accuracy severely. This research shows that using mixed antennas within the same projects may degrades the 3D coordinates accuracy severely. A certain combination and field guidelines for the study case are outlined within this research.

Keywords GNSS · Quality control · Quality assurance · Mixed antennas

1 Introduction

Nowadays, GNSS accuracy has reached a level where the remaining error sources could be identified as the effect caused by phase center variations (PCV) of the antenna. As indicated previously by Seeber et al. [1], the correction of PCV effect is generally important for GNSS observations and should be greatly considered when using mixed antenna design. Mixed antenna usage within the same project may degrades the observations accuracy severely. In addition, it is known that the phase center of a GNSS antenna is not a fixed stable point. For any given GNSS antenna, the direction of the incident satellite signal will affect the antenna phase centers. The problem is considerably important when different antennas are used in

M. M. Hosny (✉)
Civil and Environmental Engineering Department, King Abdulaziz University, Jeddah, Saudi Arabia
e-mail: mohamed.hosny@alexu.edu.eg; mmyuosef@kau.edu.sa

M. M. Hosny · S. Ayaad · M. H. Elwany · H. G. El-Ghazouly
Department of Transportation, Alexandria University, Alexandria, Egypt
e-mail: Samyayaad@alexu.edu.eg

M. H. Elwany
e-mail: elwany@dataxprsM.H

N. Mohamed Noor et al. (eds.), *Proceedings of the 3rd International Conference on Green Environmental Engineering and Technology*, Lecture Notes in Civil Engineering 214,
https://doi.org/10.1007/978-981-16-7920-9_20

measurements. Dawidowicz [2] stated that, ignoring these phase center variations in measurements can lead to serious vertical errors.

The phase center also varies and depends on the received signals. The computed baseline vector is a relative measurement between two antennas. Over short baselines, phase center variations will be cancelled if identical antennas to be used and orientated to the same direction. On the other hand, the effect will not be cancelled if different antennas are used at either end of a baseline [2].

Consequently, using mixed antennas in the same project can become significant for applications requiring the highest attainable precision from GNSS. Therefore, the effect of phase center variations was studied with identically antennas and non-identically antennas with different baselines lengths [3]. The main goal of this study is to identify the effect of non-identical receiver antennas in the same project on the overall accuracy and then to recommend specific guidelines for quality control in using mixed non-identical receiver antennas within the same project in Egypt [4].

2 Outline of Practical Study

As shown in the research work towards the development of Egyptian GNSS guidelines and specifications, the effect of phase center variations with identical and non-identical antennas with different baselines was investigated [4]. Based on the results of the questionnaire conducted during that research, among all participant categories the results indicated that GNSS users, professionals in EGYPT have a great concern on the importance of studying the effect on non-identical mixed GNSS antennas within the same project because it is a common situation in their GNSS projects. Therefore, a practical study was done to indicate the effect of non-identical antennas on the accuracy with different baseline lengths [3].

The analysis was conducted using the observations of four control stations observed as a deferential with ALX1 station, which is located on top of the administration building, Faculty of Engineering, Alexandria University. The four control stations are namely, ST01: Faculty of Engineering Alexandria University, ST02: on top of a high building in Mostafa Kamel Alexandria, ST03: on top of a high building in Kafr EL-Dawar city and ST04: at the Agricultural Road (Abo Hommos city). Each station was occupied by four different types of antennas sequentially with 30 min observation time for each receiver antenna with different baselines. Hence, the effect of non-identically mixed antennas on the accuracy with different baseline lengths could be identified.

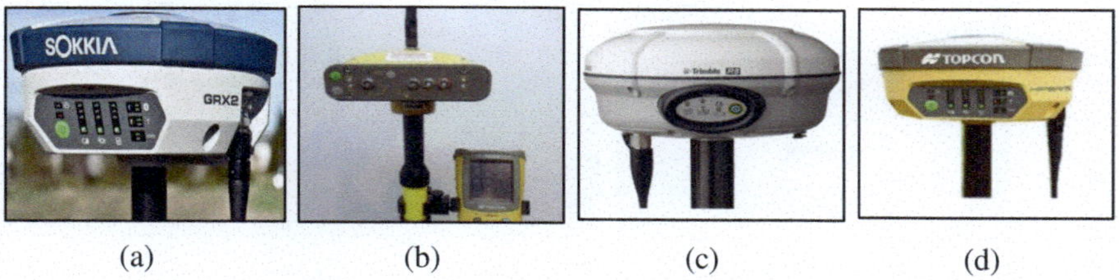

Fig. 1 Receiver Antennas; **a** Sokkia GRX2, **b** Topcon Hiper lite Ga, **c** Topcon Hiper V, **d** Trimble R8

3 Field Observations

In this research, four popular types of GPS receivers, within the Egyptian GNSS industry, were used for data collection, namely: Sokkia GRX2 receiver, Topcon Hiper lite Ga, Topcon Hiper V and Trimble R8. The selected Receiver antennas (as shown in Fig. 1) are coming from different industrial background and had the same specifications of duel frequency and signal tracking capabilities from multi-constellation satellite systems, including GPS and GLONASS [4].

4 Control Points Selection

It is noteworthy that the main objective of this study is the monitor the effect of using mixed antennas on the accuracy with different baseline lengths. Therefore, four control points ST01, ST02, ST03 and ST04 were selected whose locations are shown in Figs. 2, 3, 4, 5 and 6. The data of these points were collected by

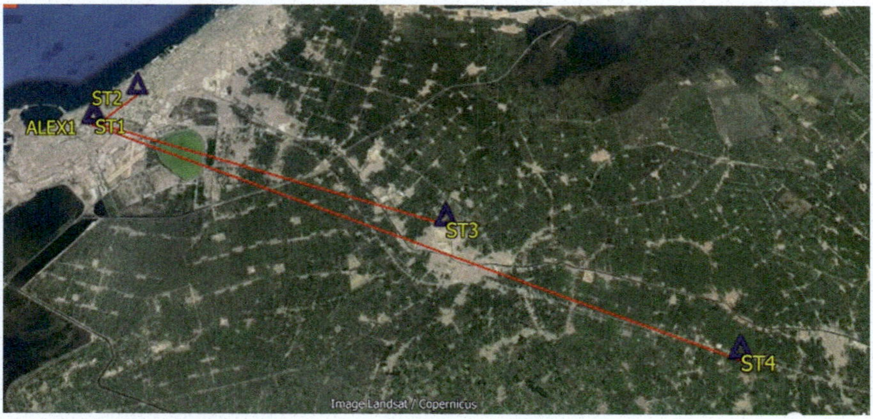

Fig. 2 Control points

Fig. 3 ALX1 and ST01 station

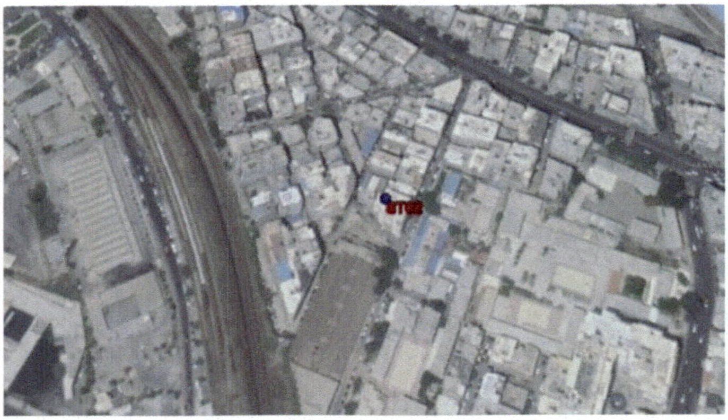

Fig. 4 ST02 station

the four previously selected antennas which have been placed sequentially on these points with 15 s time intervals for observation time, 25 min for each antenna as a differential with ALX1 station (known previously measured and corrected control point-Alexandria University) [4].

5 Procedure and Results

As mentioned previously, the four control points ST01, ST02, ST03 and ST04 were fixed at distances from the master station ALX1, located on top of the administration building, in the Faculty of Engineering Alexandria University. Table 1 illustrates the localization of the four control points and their distances from the Master Station ALX1 (baselines).

Fig. 5 Localization of ST03 station

Fig. 6 Localization of ST04 station

Table 1 Localization and baselines of the control stations

Station	Localization	Baseline length(km)
ST01	Faculty of Engineering - Alexandria	0.16
ST02	Mostafa Kamel - Alexandria	3
ST03	Kafr EL-Dawar city	20
ST04	Agricultural road in Abo Hommos city	35

In order to ensure isolating the effect of mixed antenna usage only, the analysis was done based on the observations conducted at the same conditions on the four control stations. The coordinates of each point were calculated for each receiver antenna as a differential with master point (ALX1). Hence, four coordinates have been calculated for each point because of the sequential change of receivers on each point. To study the effect of non–identical receiver antennas, the calculated coordinates from identical receivers have been considered as the benchmark case for this study, namely: the most accurate (master Sokkia and rover Sokkia). Consequently, the differences in coordinates (dx, dy, dz) between each case and the benchmark case has been calculated. Tables 2, 3, 4 and 5 show the resulted coordinates & differences in meters:

The relationship between the differences in coordinates at each point and the baseline length to study the effect of non–identically on the accuracy. Figures 7, 8 and 9 showing the relationship between the length of baselines and the coordinate's differences in X, Y, and Z.

Figure 7 shows the relationship between the differences in coordinates in **X direction** and the baselines of control points from the master point (ALX1), it can be noticed that the following:

- For the curve of Topcon Hiper V, it was noticed that the difference varies in range from 1 to 4 mm with respect to the baseline that varies in range from 0.16 to 34 km

Table 2 Coordinates and differences (m) of ST01 at baseline (0.16 km)

Master	Rover	X	Y	Z	Differences		
					dx	dy	dx
Sokkia	Sokkia	4,732,205.646	2,723,931.079	3,285,533.389	0	0	0
	TOPCON Hiper V	4,732,205.644	2,723,931.077	3,285,533.386	0.002	0.002	0.003
	TOPCON Hiper Lite	4,732,205.642	2,723,931.076	3,285,533.382	0.004	0.003	0.007
	TRIMBLE	4,732,204.706	2,723,931.908	3,285,533.162	0.94	0.829	0.227

Table 3 Coordinates and differences (m) of ST02 at baseline (3.00 km)

Master	Rover	X	Y	Z	Differences		
					dx	dy	dx
Sokkia	Sokkia	4,721,502.155	2,755,264.082	3,274,830.597	0	0	0
	TOPCON Hiper V	4,721,502.159	2,755,264.08	3,274,830.592	0.004	0.003	0.005
	TOPCON Hiper Lite	4,721,502.069	2,755,264.14	3,274,830.625	0.086	0.058	0.028
	TRIMBLE	4,721,501.969	2,755,264.279	3,274,830.629	0.186	0.197	0.032

Table 4 Coordinates and differences (m) of ST03 at baseline (20.00 km)

Master	Rover	X	Y	Z	Differences		
					dx	dy	dx
Sokkia	Sokkia	4,725,474.682	2,741,567.793	3,280,607.516	0	0	0
	TOPCON Hiper V	4,725,474.679	2,741,567.79	3,280,607.508	0.003	0.003	0.008
	TOPCON Hiper Lite	4,725,474.6	2,741,567.742	3,280,607.466	0.082	0.051	0.051
	TRIMBLE	4,725,474.475	2,741,567.643	3,280,607.488	0.207	0.15	0.028

Table 5 Coordinates and differences (m) of ST04 at baseline (35.00 km)

Master	Rover	X	Y	Z	Differences		
					dx	dy	dx
Sokkia	Sokkia	4721502.155	2755264.082	3274830.597	0	0	0
	TOPCON Hiper V	4721502.159	2755264.08	3274830.592	0.004	0.003	0.005
	TOPCON Hiper Lite	4721502.069	2755264.14	3274830.625	0.086	0.058	0.028
	TRIMBLE	4721501.969	2755264.279	3274830.629	0.186	0.197	0.032

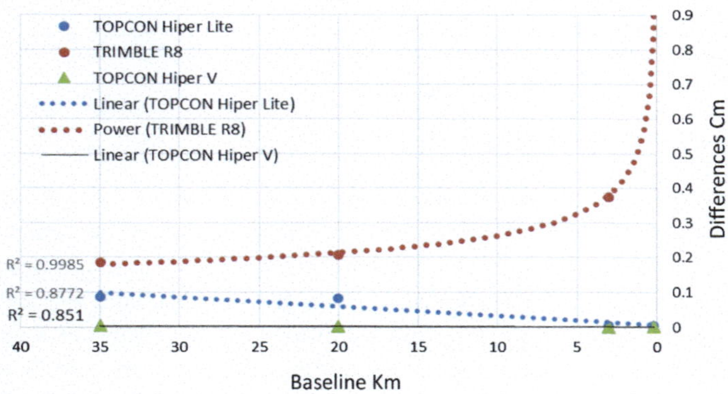

Fig. 7 Differences in X direction

at curve with root mean squared value on chart (R^2) equal 0.851 and noticed that the longer the baseline, the large value of difference coordinates in X direction.

- For the curve of Topcon Hiper lite, it was noticed that the difference varies in range from 4 to 86 mm with respect to the baseline that varies in range from 0.16 to 34 km at curve with root mean squared value on chart (R^2) equal 0.877 and noticed that the longer the baseline, the large value of difference coordinates in X direction.

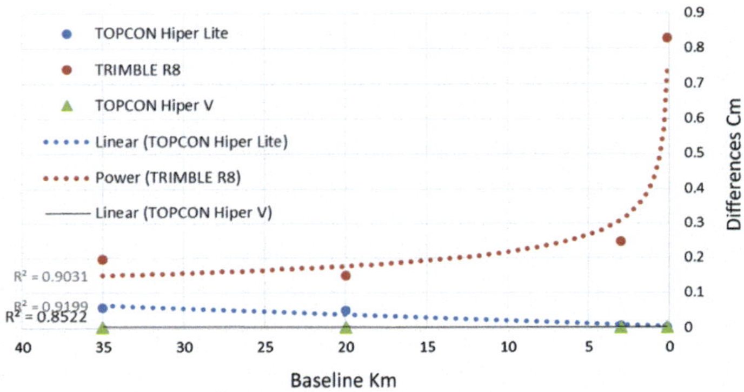

Fig. 8 Differences in Y direction

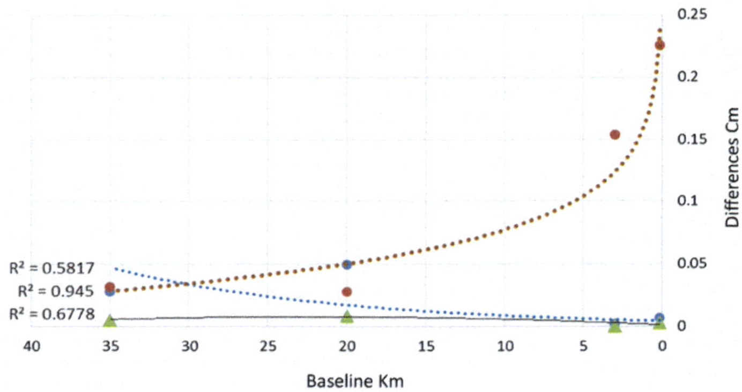

Fig. 9 Differences in Y direction

For the curve of Trimble R8, it was noticed that the difference varies in range from 186 to 940 mm with respect to the baseline that varies in range from 0.16 to 34 km at curve with root mean squared value on chart (R^2) equal 0.998 and noticed that the longer the baseline, the less value of difference coordinates in X direction.

Figure 8 shows the relationship between the differences in coordinates in **Y direction** and the baselines of control points from the master point (ALX1), it can be noticed that the following:

- For the curve of Topcon Hiper V, it was noticed that the difference varies in range from 2 to 3 mm with respect to the baseline that varies in range from 0.16 to 34 km at curve with root mean squared value on chart (R2) equal 0.852 and noticed that the longer the baseline, the large value of difference coordinates in Y direction.
- For the curve of Topcon Hiper lite, it was noticed that the difference varies in range from 3 to 58 mm with respect to the baseline that varies in range from 0.16 to 34 km at curve with root mean squared value on chart (R2) equal 0.919 and

noticed that the longer the baseline, the large value of difference coordinates in Y direction.

- For the curve of Trimble R8, it was noticed that the difference varies in range from 150 to 829 mm with respect to the baseline that varies in range from 0.16 to 34 km at curve with root mean squared value on chart (R2) equal 0.903 and noticed that the longer the baseline, the less value of difference coordinates in Y direction.

Figure 9 shows the relationship between the differences in coordinates in **Z direction** and the baselines of control points from the master point (ALX1), it can be noticed that the following:

- For the curve of Topcon Hiper V, it was noticed that the difference varies in range from 0 to 8 mm with respect to the baseline that varies in range from 0.16 to 34 km at curve with root mean squared value on chart (R^2) equal 0.678 and noticed that the longer the baseline, the large value of difference coordinates in Z direction.
- For the curve of Topcon Hiper lite, noticed that the differences values varies in range from 2 to 51 mm with respect to the baselines that varies in range from 0.16 to 34 km at curve with root mean squared value on chart (R^2) equal 0.582 and noticed that the longer the baseline, the large value of difference coordinates in Z direction.
- For the curve of Trimble R8, it was noticed that the difference varies in range from 28 to 227 mm with respect to the baseline that varies in range from 0.16 to 34 km at curve with root mean squared value on chart (R^2) equal 0.945 and noticed that the longer the baseline, the less value of difference coordinates in Z direction.

From the previous results, it is clear that the curve of Topcon Hiper V for all directions X, Y and Z is the closest to the Sokkia receivers with minimum error differences which means that the software of Topcon Hiper V and Sokkia are the same. The trend of Topcon Hiper V and Hiper lite curves are rising with the increase of baseline in all directions. Conversely, the Trimble curve is a downward direction with the baseline increasing.

6 Analysis

As shown in Figs. 7, 8 and 9, for the curve of Topcon Hiper V, the differences of values vary in range from 1 to 4 mm for X direction, 2 to 3 mm for Y direction and 0 to 8 mm for Z direction with respect to the baselines varying in range from 0.16 to 34 km. In addition, for all directions, when using Topcon Hiper V as a rover, it is clear that the accuracy will be degraded as the baseline gets longer.

For the curve of Topcon Hiper Lite, the differences of values vary in range from 4 to 86 mm for X direction, 3 to 58 mm for Y direction and 2 to 51 mm for Z direction with respect to the baselines varying in range from 0.16 to 34 km. The same was observed as in Topcon Hiper V, when using Topcon Hiper Lite as a rover, it is clear that the accuracy will be degraded as the baseline gets longer.

Finally, for the curve of Trimble R8, the differences of values vary in range from 186 to 940 mm for X direction, 150 to 892 mm for Y direction and 28 to 227 mm for Z direction with respect to the baselines varying in range from 0.16 to 34 km. On the contrary to the other two receivers, when using Trimble R8 as a rover, it is clear that the accuracy will be upgraded as the baseline gets longer.

7 Conclusion

In conclusion, it has been noted that when used for the same project, mixed non-identical receiver antennas lead to a lack of the accuracy. Using Sokkia GRX2 as a master station, it was noticed that the curve of Topcon Hiper V for all directions X, Y and Z is the closest to the Sokkia receivers with minimum differences error which means that the software of Topcon Hiper V and Sokkia are the same. In addition, as the base line increases, the anticipated accuracy of using Topcon Hiper V and Hiper lite will be decreased. Conversely, the Trimble R8 accuracy will be increased as the base line with the master Sokkia GRX2 increases in all directions. Generally, using Sokkia GRX2 as a master station, the non-identical receivers' error that occurs when using Topcon receivers as rover is the smallest while using Trimble receivers as rovers will lead to the worst accuracy. Finally, the produced curves in this research may be used as a guideline for the baseline length selection, the receiver types and the anticipated level of accuracy when applying the same procedure within Egypt.

References

1. Seeber G, Menge F, Völksen C, Wübbena G, Schmitz M (1998) Precise GPS positioning improvements by reducing antenna and site dependent effects. Int Assoc Geodesy Symp 237–244. https://doi.org/10.1007/978-3-662-03714-0_38
2. Dawidowicz K (2014) Phase center variations problem in GPS/GLONASS observations processing. In: Environmental Engineering. Proceedings of the International Conference on Environmental Engineering. ICEE 9, 1. Vilnius Gediminas Technical University, Department of Construction Economics & Property
3. Ayaad S, Elwany MH, El-Ghazouly HG, Hosny MM (2018) Assessment of quality assurance and quality control for GNSS surveying firms—A case study in Egypt. Int J Adv Sci Res Manage 3(3):35–46
4. Ayad S (2018) Assessment of quality assurance and control for global navigation satellite system GNSS surveying: a study towards the Egyptian GNSS guidelines and specifications

SCS Curve Number to Model Flooding in the Upper St. Johns River Using Retrieved Remotely Sensed Precipitation from NEXRAD, and TRMM

Marwan Kheimi

Abstract The economy, environment, and society are heavily impacted by precipitation during floods, and droughts events. Recent technological advancements in satellite-retrieved, and radar precipitation products showed better resolution and coverage that resulted to sustainable data collection. Nevertheless, uncertainty and lack of consistency in information records can portrays a weakness while incorporation retrieved estimate and operational decision making. In this paper, the goal is to evaluate. Hydrologic Modeling System (HEC-HMS) to simulate one flood event to investigate the effect of watershed subdivision. The Model utilized satellite products rainfall estimator (TRMM 3B42) and NEXRAD Stage III in the Upper St. Johns River, located in the Mid-Eastern part of Florida. While taking into consideration extreme event in August 2008. The data were collected from 5 local precipitation gages in the St. Johns River District. Using validation metrics, this data was aggregated and differentiated between in-situ based rainfall, radar, and satellite derived rainfall. The detected rainfall amounts and volume of correctly identified precipitation, NEXRAD Stage provides a better possibility for accurate estimation and variability of precipitation at this high spatial resolution. In fact, the simulation results show less accuracy in predicting flow in the study area, but overestimation in the regions. Although comparable to radar-only simulations in terms of magnitude and time-to-peak, there is no discernible improvement in predicting these quantities.

Keywords Precipitation · Hydrologic modeling · Satellite modeling · Remote sensing

M. Kheimi (✉)
Department of Civil Engineering, Faculty of Engineering-Rabigh Branch, King Abdulaziz University, Jeddah 21589, Saudi Arabia
e-mail: mmkheimi@kau.edu.sa

N. Mohamed Noor et al. (eds.), *Proceedings of the 3rd International Conference on Green Environmental Engineering and Technology*, Lecture Notes in Civil Engineering 214,
https://doi.org/10.1007/978-981-16-7920-9_21

1 Introduction

Reliable and timely precipitation measurements, their integration into datasets and predictive models, and the accuracy of merged information are important to water resources engineering and management. Precipitation is the primary driver of the hydrologic cycle and the main input in hydrometeorological models and climate studies. The accuracy of hydrometeorological predictions depends significantly on the quality of observed precipitation intensity, patterns, duration, and aerial extent. Models, estimates and predictions of watershed, atmospheric, soil moisture, and water availability all begin with a source of measurement data. Hydrometeorological applications have long relied on rain gauge data as the main source of precipitation measurement. However, rain gauges suffer from poor spatial coverage and lack of aerial representation over land, and are unavailable over the oceans. Further, the quality, density, and coverage of gauge networks vary significantly across the globe [1].

Despite the fact that the United States has better coverage, there is a limit to the areas covered by rainfall gauges. Weather radar systems, which supplement rain gauge stations, provide high-resolution precipitation data that is represented in space and time. However, radar installation spatial coverage is insufficient to detect precipitation patterns across the entire United States. In fact, the available radar systems only allow for regional-scale studies in the United States, Western Europe, and a few other places. In sparse gauge areas where is not enough data satellite data plays an important role to estimate hydrologic response in retrospect [2–4]. Even in the United States, many mountainous regions (for example, the Western United States) have poor radar coverage due to significant obstruction [5]. Neary, Habib [6] has used this method in two subbasin of the Cumberland River basin in Middle Tennessee. Neary, Habib [6] evaluated the hourly radar data, radar estimates were compared to corresponding gage observations and calculating the stream flow volume bias, root mean square difference, mean normalized peak error and mean peak timing error at each study sub-basin. His findings after simulation were that the radar precipitation estimate suffers from systematic underestimation. Moreover, Zhang, Wang [7] has simulated two major floods using (HEC-HMS) and evaluated the output result over the study area using peak error, mass balance error, and Nash–Sutcliffe efficiency coefficient. After defining the model basin and sub-basin of the Clear Creek Watershed located in the upper Mississippi River Basin. Results showed a variation of depending on the watershed subdivision.

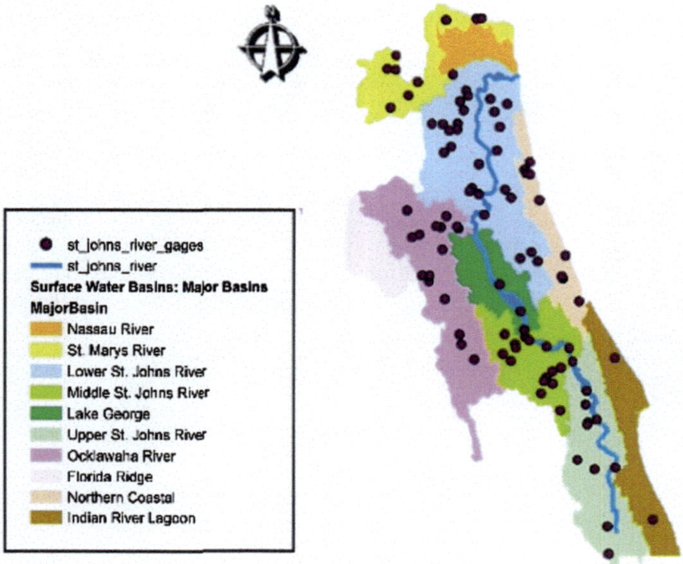

Fig. 1 Map shows the location, river boundaries and stream gages of the St. Johns River from the St. Johns Water Management District (SJWMD), Florida, USA

2 Study Area, Data, and Methodology

2.1 Study Area

The St. Johns River (HUC-8:0,308,101) is located in Mid-Eastern part of the Florida State. St. Johns River originate from the head water of Ft Drum Creek northward to its junction with the Econlockhatchee River, a distance of over 110 miles of length in Fig. 1. The river slope drops 1 foot per 5 miles river length. This slight slope and large drainage allow the Upper St. Johns River and its surroundings to function as storage area, serving as a natural of high and low supply stages. For the purpose of this project the study area is focused on the Upper St. Johns River basin, which is 1733.07 square miles [8].

2.2 Data Acquisition

As an Input for creating the model a collection of hydro-metrological data from the U.S. Geological Service (USGS) and locally from St. Johns Water Management District (SJWMD). Retrieval of physical data from satellite images such as, land cover/land use from the National Land Cover Database (NLCD), digital elevation model (DEM) and soil cover form (SURRGO) shown in Fig. 2. The soil cover and

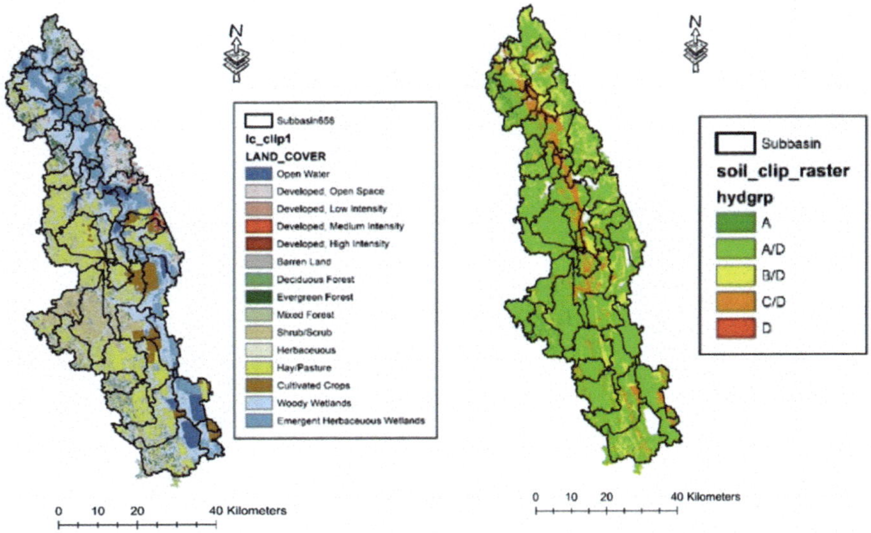

Fig. 2 (Left) Land cover map shows different categories corresponding to each sub-basin from the (NLCD). (Right) Soil grid map from SRROGO for each sub-basin. The void white areas in the maps are waterbodies

the land use/land cover data will be used to find the curve number to be utilized in the model to calculate run-off using SCS curve number method [9]. While the utilization of radar-based, satellite rainfall estimation from TRMM and raingages are simulated to compare the hydrologic predictions of streamflow at outlets of the sub-basin, which are obtained using HEC-HMS.

Precipitation Data

Three sets of daily precipitation data are available for the current study: gage observations, TRMM 3B42 [1], and radar-gage estimates. Gauge data were collected from five recording gages located in the study area of St. Johns Watershed and were provided by the National Climatic Data Center (NOAA) for the study period of August 2008 and selecting one extreme event in the 21/Aug/2008. The quality of gages are susceptible to frequent failure caused by mechanical and electronic glitch. The gage data will be checked for quality by cross-comparison with nearby gages to eliminate outliner and flagged observations (Table 1).

Radar estimates used in this study are stage III precipitation products of the NEXRAD system and obtained from the National Climatic Data Center (NOAA). Stage III products are hourly accumulations of precipitation over a grid 4×4 km^2 that is referred to as a HARP grid [Hydrologic Rainfall Analysis Product [10]]. The Tropical Rainfall Measuring Mission (TRMM) used are precipitation estimate over 0.25°x0.25° pixel size. A daily data was retrieved over the study area which is an accumulation of 3-hourly precipitation.

Table 1 List of station and its location in the Upper St. Johns River basin

Station Id	Station_Name	Lat	Long	Major_Sw_Basin	County
670,294	Seminole Ranch at Mims	28.36	−80.58	Puzzle Lake Unit	Brevard
5,140,560	Bull Creek APT	28.07	−81.00	Jane Green Creek	Osceola
3,990,652	Lake Louise at Bithlo	28.35	−81.06	Econlockhatchee River	Orange
3,200,347	Lake Ashby at Deltona	28.56	−81.06	Deep Creek Unit	Volusia
1,140,474	Osceola Landfill at Geneva	28.47	−81.05	Deep Creek Unit	Seminole

Table 2 Percentage for each allocated Land Cover/Land Use grid

lc_clip1

Rowid	VALUE	COUNT	RED	GREEN	BLUE	LAND_COVER	PERCENTAGE
13	90	1187047	0.729412	0.847059	0.917647	Woody Wetlands	35.14%
11	81	732375	0.858824	0.847059	0.235294	Hay/Pasture	21.68%
14	95	553848	0.439216	0.639216	0.729412	Emergent Herbaceuous Wetlands	16.4%
9	52	316134	0.8	0.729412	0.486275	Shrub/Scrub	9.36%
12	82	155691	0.666667	0.439216	0.156863	Cultivated Crops	4.61%
0	11	138768	0.278431	0.419608	0.627451	Open Water	4.11%
7	42	108378	0.109804	0.388235	0.188235	Evergreen Forest	3.21%
1	21	94981	0.866667	0.788235	0.788235	Developed, Open Space	2.81%
2	22	38093	0.847059	0.576471	0.509804	Developed, Low Intensity	1.13%
10	71	26022	0.886275	0.886275	0.756863	Herbaceuous	0.77%
3	23	12042	0.929412	0	0	Developed, Medium Intensity	0.36%
8	43	5892	0.709804	0.788235	0.556863	Mixed Forest	0.17%
5	31	4818	0.698039	0.678431	0.639216	Barren Land	0.14%
4	24	2691	0.666667	0	0	Developed, High Intensity	0.08%
6	41	1204	0.407843	0.666667	0.388235	Deciduous Forest	0.04%

The Highest percentage of land cover is occupied by the Woody Wetlands with 35% of the total area. While the deciduous Forest occupy 0.04% of the total area of the Upper St. Johns River. As for the soil cover most of the Upper St. Johns River Basin soil is occupied by the hydrologic group A and moderately by group B and C while D has the least percent of coverage (Table 2).

2.3 *Methodology*

HEC-HMS is a numerical model that includes a large number of methods for simulating the behavior of watersheds, channels, and water-control structures, thereby predicting flow, stage, and timing. Individual storm events, as well as continuous precipitation input at minute, hourly, or daily time steps, can be simulated by the HEC-HMS. Also shown is a diagram of the structure of the river basin rainfall runoff process [7]. Watershed precipitation, runoff volume, direct runoff (including overland and interflow), baseflow, and channel flow are all simulation methods. The data are driven by HEC-HMS simulation methods, which are flexible enough to enable the application to watersheds of any size for long and short event analysis to resolve

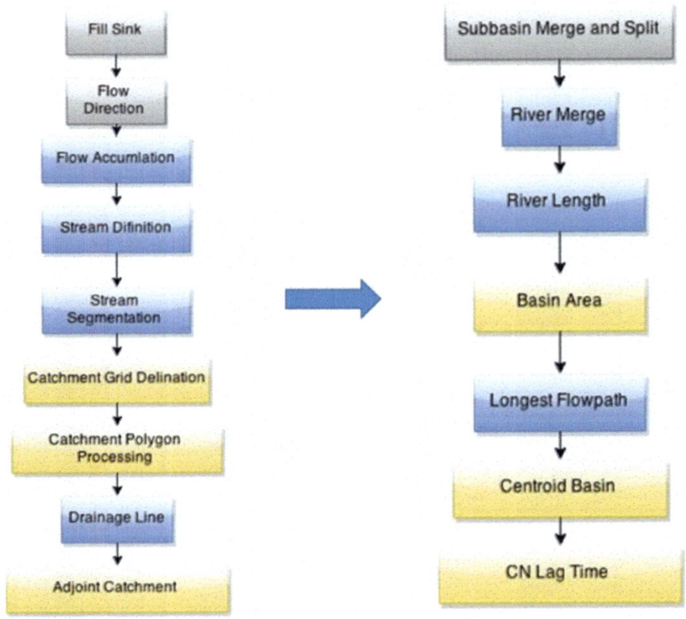

Fig. 3 Flowchart of ArcHydro Methodology and the processing generating the HEC-HMS. The (grey color boxes) represents the pre-processing for mesh data, (blue boxes) represents the stream and streamline process, while the (yellow boxes) represents the catchment output

the model equations with time steps that are suitable for analysis. For various sub-basins sizes, daily precipitation data were used. For such an application, a 24-h time step is used (Fig. 3).

This study focuses on estimating streamflow and comparing peak discharge as a result of different precipitation estimates in HEC-HMS 21 basins for August 2008. These collected data were used in ArcGIS 10.2 and ArcHydro 10.2 preprocessing for computing hydrologic parameters in HEC-GeoHMS. They will be used in HEC-HMS to estimate streamflow runoff. The SCS transform method is used in HEC-HMS to calculate runoff. Basin area, River length, longest flow path, centroid of subbasin, Curve Number (CN), lag time, Time of concentration, and Impervious area are calculated from HEC-GeoHMS data from the US Army Corps of Engineers [11]. The Rainfall-Runoff model input parameters in HEC-HMS. In HEC-HMS, the SCS Transform, SCS Loss, and Muskingum methods were used for the transform, loss, and routing methods, respectively. Furthermore, in this study, the Thiessen weighted precipitation method was used to reflect the spatial characteristics of precipitation during specific hydrological event periods. These include SCS curve number in Fig. 4, SCS unit hydrograph, and base flow estimation methods that are important for calculating water losses, runoff transformation, and base flow rates. The SCS loss model for watershed loss given by:

$$P = \frac{\left(P - (Ia)^2\right)}{P - I_a + S} \quad (1)$$

where P is the total amount of precipitation, I_a is the initial abstraction, which can be set to 0.2 S, and S is the potential maximum retention and function curve number (CN).

$$S = \frac{\frac{1000}{CN}}{-10}(\text{SIUnits})[11] \quad (2)$$

The parameters initial abstraction and CN are required. The SCS unit Hydrograph (UH) rainfall-runoff transformation model is a dimensionless unit hydro-graph U_i expressed as a percentage of peak discharge. U_p for any time fraction $\frac{t}{T_p}$, where T_p is the time to peak. $U_p = \frac{CA}{T_p}$ gives the peak discharge, where C is the conversion constant (2.08 in SI units) and A is the sub-watershed area. $T_p = t$ is the time step in HEC-HMS, and t_p is the time lag defined as the time difference between the center of excess precipitation and the center of UH [11]. The sub-basin lag-time, measured from the centroid of hyetograph to the peak time of the hydrograph, is calculated with following formula [12]:

$$T_{lag} = L^{0.8}\frac{(S + 1)^{0.7}}{1900\sqrt{Y}} \quad (3)$$

where:

T_{lag} lag time in hours.
L hydraulic length of watershed in feet.
S maximum retention in the watershed in inches.

The expositional recession model for baseflow is given by

$$Qt = Q_0 x k_t \quad (4)$$

where Q_0 denotes the initial baseflow and k denotes the exponential decay constant. During the flood event's recessive period. The HEC-HMS employs SCS unit hydrograph and base flow estimation methods to compute water losses, runoff transformation, and base flow rates. According to the Army of Corps HEC-HMS Manual [11], the input data and result are based on an iterative driven runoff solution that is applied using the above equations and user selection. The input data is displayed as a time-series component for each location within the time interval. The approach to model methodology can be divided into four major tasks:

(1) Creating different configuration of sub-watersheds.
(2) Evaluate daily radar data, radar estimates to corresponding gage observations.
(3) Analyzing the parameters and associated relationships.

(4) Analyzing the streamflow at the outlet.

The CN is calculated using the land use and soil data depicted in Fig. 4. The SCS Unit hydrograph has only two parameters: watershed area A and lag time t_L. The time to peak t_p is estimated as a function of rainfall duration D, and the unit hydrograph peak is estimated as follows [13]:

$$T_p = \frac{D}{2} + t_L \text{ in Hours} \tag{5}$$

$$Q_p = 484\frac{A}{t_p} \text{ English units} \tag{6}$$

2.4 Unit Hydrograph Parameters in GIS

The unit hydrograph can compute basin area, channel length, basin centroid, longest flow path, CN basin lag time, and time of concentration. The HEC-Geohash algorithm is used to compute these parameters. DEM preprocessing via ArcHydro is required prior to operating HEC-HMS. The flowcharts in Fig. 3 describes the processing that will be used to generate the input data for HEC-GeoHMS to compute the Hydrologic parameters.

Thiessen polygons are built using the rainfall gage point and basin polygon. Thiessen polygons can be created in ArcHydro using the create Thiessen polygon tool. After constructing the Thiessen polygon, basin polygons with portions of each gage are created by intersecting the Thiessen polygon and basin polygon. HEC-GeoHMS was used to create the sub-basins. The basin is divided into sections. 59 sub-basins as shown in Fig. 4 with an average area of 51km^2.

2.5 Evaluation Procedure

Bias, Mean Error (ME), Correlation Coefficient (CC), and Root Mean Square Error (RMSE) are the statistical methods used to validate these products. They are used to assess the accuracy of each product against the rain gauge ground observations, and compare their percent of agreement against each other and the precipitation ground observing network. The location lacks adequate precipitation observing networks, which are required for water resource management, atmospheric analysis, and natural disaster mitigation. This is especially true in challenging terrain regions like our research location, where metropolitan areas and infrastructure are few. To quantitatively compare the hydrologic simulations of rainfall gage, NEXRAD, and TRMM products some common performance statistics were used. Two variables are defined as the variables' (sample) covariance divided by their (sample) standard deviations

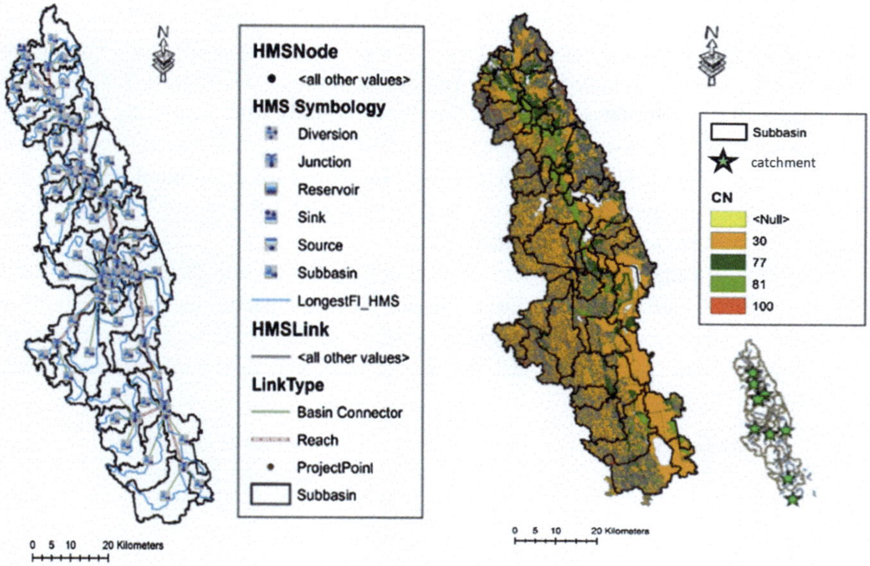

Fig. 4 (Left) Map shows the HEC-HMS legends and each subbasin hydrologic component of the model. (Right) Curve Number based on the corresponding land cover grid values from using the soil data from (NLCD)

Table 3 Comparison matrix for estimated output and observed data

Accuracy matrix	Equation	Range	Perfect score
Bias	$\frac{\sum S}{\sum G}$	0 to $+\infty$	1
Mean Error (ME)	$\frac{1}{N}\sum(S-G)$	$+\infty$ to $-\infty$	0
Correlation Coefficient (CC)	$\frac{\sum_{i=1}^{N}(S_i-\overline{S})(G_i-\overline{G})}{\sqrt{\sum_{i=1}^{N}(S_i-\overline{S})^2\sum_{i=1}^{N}(G_i-\overline{G})^2}}$	$-\infty$ to 1	1
Root Mean Square Error (RMSE)	$\frac{1}{N}\sum(S-G)^2$	0 to $+\infty$	0

where S_i and G_i are the values of S(Simulated output) and G (Observed) for the i_{th} individual and $\overline{S}$ and $\overline{G}$ are average values (Table 3).

3 Result and Discussion

To assess the accuracy of rainfall estimates from reengages, NEXRAD Stage III, and TRMM 3B42 precipitation estimates for the 21st of August 2008 rainfall data was collected and analyzed for daily data for the month of August. Different validation statistics were also evaluated on a daily basis. Bias, Root Mean Square (RMSE), Mean Error (ME), and correlation coefficient are all calculated and compared. Validation matrices vary when compared to daily rainfall accumulation. Figure 5 is a scatter plot of rainfall estimates over the St. Johns River at daily time scale, 0.25° long/lat special resolution (TRMM), and $4 \times 4\ km^2$ (NEXRAD Stage III) from two different remote sensing rainfall products. The comparison shows the agreement between the observed data (raingage) and remotely sensed data (NEXRAD and TRMM), both have similar pattern deducting the rainfall.

Comparing simulated and observed streamflow for all rainfall estimators and raingage in Fig. 6 indicated that neither model preforms particularly well. There is significant disagreement between simulated and observed peaks as shown in Fig. 6. All models failed to predict the last two peaks observed in the stream discharge.

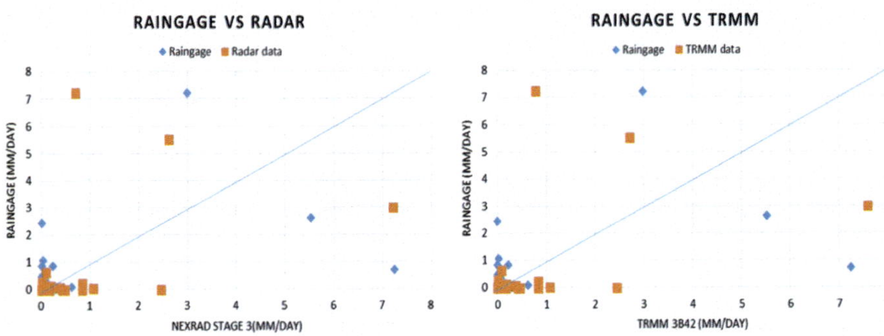

Fig. 5 Comparsion between the Rainfall estimate from NEXRAD Stage 3, Radar and TRMM 3b42

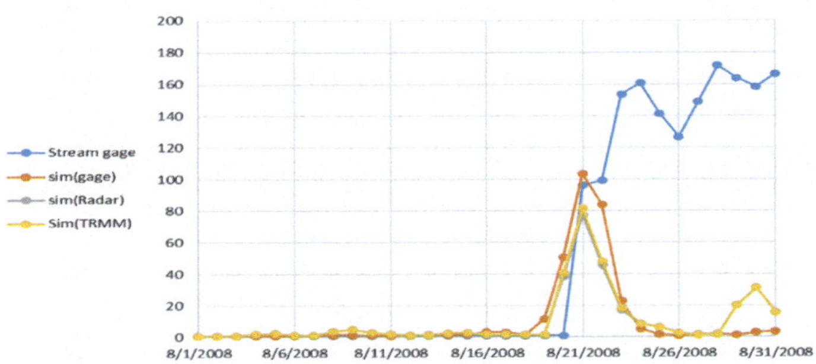

Fig. 6 Simulation output results of sub-basin outlet compared to the nearest streamflow gage

Table 4 Bias and root mean square error between TRMM 3b42 and NEXRAD Stage III

Raingage versus Radar			Raingage versus TRMM		
Bias	RMSE	CC	Bias	RMSE	CC
−17.64	2.31	0.45	7.75	2.34	0.45

Table 5 Validation statistics comparing the performance of daily simulated rainfall estimate

	Raingage versus Stream gage	Radar versus Stream gage	TRMM versus Stream gage	Perfect score
CC	0.16	0.36	0.34	1
Bias	0.20	0.19	0.20	1
RMSE	660.44	357.19	395.48	0
ME	9.25	8.83	9.18	0

This failure may indicate a potential problem with model to the point of direct runoff cannot be generated. Although, the radar model exhibited less bias in the streamflow as it is reflected in Table 4.

The validation analysis of satellite rainfall revealed that all of the products are capable of detecting rainfall on a daily basis. The statistical results for rainfall estimation accuracy show that all of the products either overestimate or underestimate rainfall in the region, but radar has less bias and RMSE than the other rain gauges, which makes it difficult to conclude from this validation study that the product has the best or worst performance as shown in Table 5. All of the products were capable of recording extreme events. As a result, a regional validation study would have provided additional insight into the accuracy of satellite rainfall estimation. These types of findings can be used to suggest that more effort be made to improve the quality of operational Stages and remove systematic biases before using radar data in hydrologic modeling.

4 Conclusion

Two widely available remotely sensed precipitation satellite products were evaluated over the St. Johns River to study the hydrologic response in the watershed located in the mid-center of Florida. The two products are TRMM, NEXRAD Stage III. All the products data is at daily temporal scale and (0.25°) gird resolution courser than NEXRAD Stage III with $4 \times 4\ \text{km}^2$ resolution. Before applying the satellite precipitation data in the model, accuracy of the products was assessed by robust inter-comparison between rain-gauges observations and satellite-retrieved estimates of daily rainfall data. The analysis takes place during the month of August, which has a high rainfall event. The products' accuracy has varied on different days. A higher correlation exists between radar estimates and discharge gage. Overall, the study concludes that some of the current satellite daily rainfall products are potentially usable in hydrologic response analysis over the Upper St. Johns River watershed. However, these findings suggest that there is still room for improvement in these products in order to eliminate errors and improve rainfall estimates. TRMM has a variety of products available, and TRMM 3B42 was used in this study. Because this product is gauge adjusted, it is more likely to perform better than others. The outcomes reflect this as well. However, validation using the TRMM real time (RT) and other products such as GPM, PERSIANN CC, PERSIANN CDR and GsMAP would have revealed whether or not there was any improvement with the bias adjusted product. Future work should include forceful calibration, and more robust simulations may result in improved detection of extreme flood events.

References

1. Huffman GJ, et al.(1997). The global precipitation climatology project (GPCP) combined precipitation dataset. Bulletin of the american meteorological society, 78(1), 5–20. https://doi.org/10.1175/1520-0477(1997)078%3C0005:TGPCPG%3E2.0.CO;2
2. Talchabhadel R et al (2021) Evaluation of precipitation elasticity using precipitation data from ground and satellite-based estimates and watershed modeling in Western Nepal. Journal of Hydrology: Regional Studies 33. https://doi.org/10.1016/j.ejrh.2020.100768
3. Santra Mitra S, Santra A, Kumar A (2021) Catchment specific evaluation of Aphrodite's and TRMM derived gridded precipitation data products for predicting runoff in a semi gauged watershed of Tropical India. Geocarto Int 36(11):1292–1308. https://doi.org/10.1080/10106049.2019.1641563
4. Abdelmoneim H, Soliman MR, Moghazy HM (2020) Evaluation of TRMM 3B42V7 and chirps satellite precipitation products as an input for hydrological model over Eastern nile basin. Earth Systems and Environment 4(4):685–698. https://doi.org/10.1007/s41748-020-00185-3
5. Maddox RA et al (2002) Weather radar coverage over the contiguous United States. Weather Forecast 17(4):927–934. https://doi.org/10.1175/1520-0434(2002)017%3C0927:WRCOTC%3E2.0.CO;2
6. Neary VS, Habib E, Fleming M (2004) Hydrologic modeling with NEXRAD precipitation in middle Tennessee. J Hydrol Eng 9(5):339–349. https://doi.org/10.1061/(ASCE)1084-0699(2004)9:5(339)
7. Zhang H et al (2013) The effect of watershed scale on HEC-HMS calibrated parameters: a case study in the Clear Creek watershed in Iowa. US. Hydrology and Earth System Sciences 17(7):2735–2745. https://doi.org/10.5194/hess-17-2735-2013
8. District SJR.W.M., Upper St. Johns River Basin Project. 2007.
9. Service S, Hydrology, National Engineering Handbook, Supplement A, Section 4, Chapter. 1985.
10. Maidment RSMDR (1999) Coordinate transformations for using NEXRAD data in GIS-based hydrologic modeling. J Hydrol Eng 4(2):174–182. https://doi.org/10.1061/(ASCE)1084-0699(1999)4:2(174)

11. HEC, UA (2000). Hydrologic modeling system HEC-HMS technical reference manual. US Army Corps of Engineers, Davis, CA.
12. USDA S (1972). National engineering handbook, section 4: Hydrology. Washington, DC.
13. Wurbs R and W James (2002). Chapter 12. River basin management. Water Resources Engineering, Upper Saddle River: Prentice Hall.

Study of Oil Palm Frond (OPF) and Oil Palm Trunk (OPT) for Sustainable Development

Noraishah Shafiqah Yacob, Hassan Mohamed, and Abd Halim Shamsuddin

Abstract The depletion of fossil fuel and the harmful environmental emissions such as greenhouse gases (GHG) has led to continuous effort to produce high value-added products from the biomass. With more than 5.8 million hectares of oil palms, Malaysia's oil palm industry has been one of the largest contributors to oil palm biomass. Oil palm production in Malaysia is increasing every year, resulting in an abundance of oil palm waste. Due to these circumstances, Malaysia has the great potential to utilise oil palm biomass in a wide range of applications, including renewable energy. The characteristics of biomass complement the ability to generate reliable energy sources. This study will brief the potential utilisation of the oil palm residues from plantations, which are oil palm frond (OPF) and oil palm trunk (OPT)—the availability, characteristics, and related technologies on biomass conversion.

Keywords Oil palm · Frond · Trunk · Biomass · Renewable energy

1 Introduction

Recent concern for the environment has gained more attention for utilizing the environment-friendly material for energy generation. Researchers have also identified biomass as a potential alternative for energy generation and exploiting the potential of biomass as feedstocks for value-added products. Besides, fossil fuel depletion has created challenges in the energy industry to explore other energy sources [1, 2]. Malaysia has a high opportunity to produce various biofuels from renewable

N. S. Yacob · H. Mohamed (✉) · A. H. Shamsuddin
Institute of Sustainable Energy, Universiti Tenaga Nasional, 43000 Kajang, Selangor, Malaysia
e-mail: MHassan@uniten.edu.my

N. S. Yacob
e-mail: noraishah.shafiqah@uniten.edu.my

A. H. Shamsuddin
e-mail: abdhalim@uniten.edu.my

N. Mohamed Noor et al. (eds.), *Proceedings of the 3rd International Conference on Green Environmental Engineering and Technology*, Lecture Notes in Civil Engineering 214,
https://doi.org/10.1007/978-981-16-7920-9_22

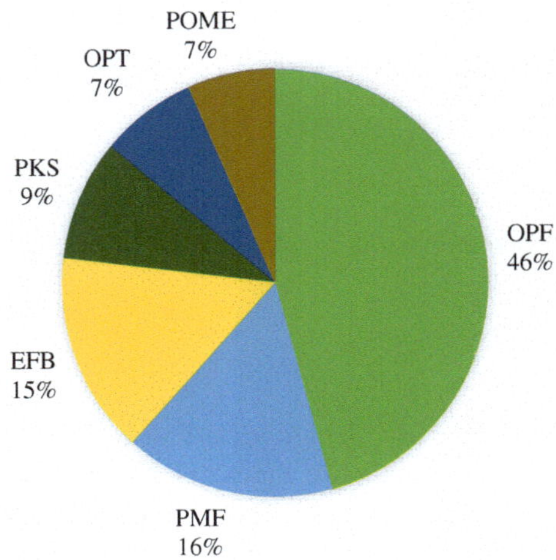

Fig. 1 Availability of oil palm residues in Malaysia in 2017 (dry weight basis) [4]

sources such as palm oil residues due to its abundant resources in the tropical location. Malaysia has also remained one of the world's major palm oil producers for the past decades [2].

The abundance of biomass from the oil palm industry includes oil palm fronds (OPF), oil palm trunks (OPT), empty fruit bunch (EFB), palm mesocarp fibre (PMF) and palm kernel shell (PKS). The availability of oil palm biomass is based on the standard biomass to fresh fruit bunches (FFB) extraction rate [3]. Annual oil palm production is expected to rise to 50 million tonnes by 2030. In 2017, 51.19 metric tonnes (MT) of oil palm biomass wastes were available for replanting, pruning and milling activities in Malaysia, out of 101.02 MT of FFB. From 51.19 Mt of oil palm biomass wastes, 23.39 Mt and 3.74 Mt accounted for OPF and OPT, respectively [4]. Figure 1 shows Malaysia's availability of major oil palm biomass residues (dry weight) in 2017.

As of December 2020 [5], the total oil palm area planted in Malaysia is 5.865 million hectares. Table 1 shows the total area of oil palm plantations in Malaysia's different states by 2020. Approximately 89.2% of the total planted area is mature, which is 5.232 million hectares. Meanwhile, the remaining 10.8% (or 0.633 million hectares) is in the immature stage.

On the other hand, the palm oil industry generates a large quantity of biomass residue that could be used in various sectors. Palm solid residues have several economic benefits, including mitigating environmental pollution, conserving fossil fuels, and reducing reliance on imported fuels. If compared to other renewable energy sources, solid palm waste not only has obvious potential as a fuel source, but it also has a more reasonable cost in terms of energy generation [6]. Therefore, this study provides an overview of the availability of Malaysian oil palm biomass wastes and the usage, particularly for OPF and OPT, will be discussed in detail.

Table 1 Oil palm planted area as of December 2020 (Hectares) [5]

State	Mature	%	Immature	%	Total	%
Johor	688,291	92.9	52,537	7.1	740,828	12.6
Kedah	80,210	89.3	9,572	10.7	89,782	1.5
Kelantan	131,768	78.6	35,831	21.4	167,599	2.9
Melaka	51,672	91.7	4,689	8.3	56,361	1.0
Negeri Sembilan	173,490	91.1	16,972	8.9	190,462	3.2
Pahang	702,163	89.8	80,084	10.2	782,247	13.3
Perak	352,557	90.0	39,211	10.0	391,768	6.7
Perlis	689	99.3	5	0.7	694	0.0
Pulau Pinang	12,540	97.7	289	2.3	12,829	0.2
Selangor	113,911	90.0	12,614	10.0	126,525	2.2
Terengganu	148,244	83.0	30,384	17.0	178,628	3.0
Sabah	1,344,608	87.1	198,446	12.9	1,543,054	26.3
Sarawak	1,431,600	90.3	152,920	9.7	1,584,520	27.0
Total	5,23,743	89.2	633,554	10.8	5,865,297	100.00

2 Oil Palm Biomass

Palm oil is a potentially abundant biomass source in Asia, requiring fewer fertilizers and pesticides than other oil seeds and even being grown in pear soil with less water. In comparison to other vegetable oil producers worldwide, palm oil fruit can reach a height of 25 m and produce fruit all year [1]. The solid biomass wastes from the oil palm industry are generated in the plantation and mill sites [7]. The two residues that are easily accessible at plantation sites are the frond and trunk. Figure 2 shows an illustration of an oil palm tree that consists of frond and trunk [8].

During the harvesting of FFB, about 24% of OPF is collected from each oil palm tree every year. At the same time, OPT is obtained around 70% of all replanting activities. It means that the possibility of OPT availability will increase over time as the plantation grows and replanting occurs throughout the year [9]. Conversely, EFB, PMF, PKS and POME are oil palm residues commonly available at mill sites during palm oil extraction [10]. The quantity of oil palm biomass residues is primarily determined by plantation size and the amount of processed FFB generated from the plantation and mills, respectively [9, 11].

Palm trees typically have a vertical trunk and feathery leaves, approximately 20–40 new leaves known as oil palm fronds growing yearly. During the oil palm fruit harvesting period, the pruning of OPF generates about 44 million tonnes dry weight of OPF yearly basis. However, during the replanting season, up to 15 tonnes of dried OPF and 75 tonnes of dried OPT were chopped and left to decompose in the plantation for nutrient recycling and mulching [7]. Thus, the potential value of these biomass wastes is overlooked for their more profitable usage.

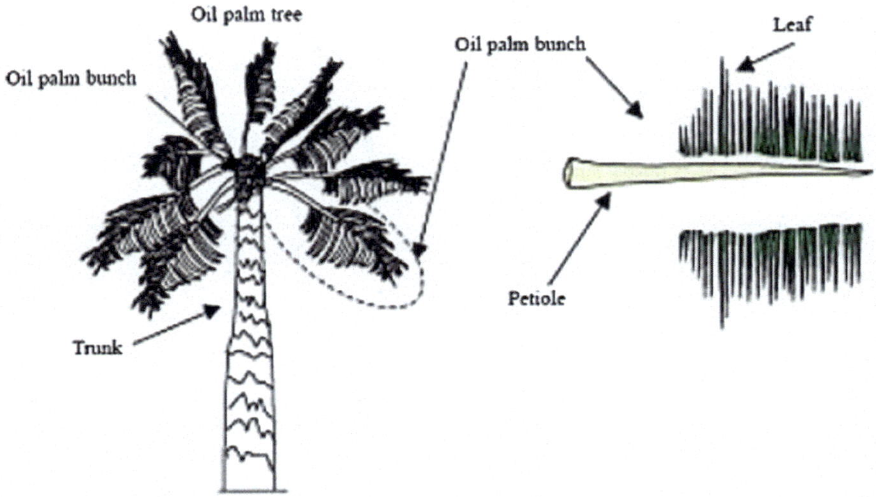

Fig. 2 Oil palm tree [8]

2.1 Composition of OPF and OPT

It is essential to determine biomass composition to identify the availability and suitability of the biomass for thermochemical processes [12]. Cellulose, hemicellulose, and lignin are the main components of oil palm residues. Because of the high content of cellulose and hemicellulose of oil palm residues, it may be converted into simple sugars and used to produce biochemicals or biofuels [9, 10].

The amount of lignin in the cell wall affect the utilisation of oil produced from the palm [4]. The structural components of OPF and OPT that play a significant role in estimating the higher heating value were outlined by previous studies, as illustrated in Table 2. The higher heating value or calorific value of a material is more desirable to use as fuel [13].

Compared to fossil fuels, biomass ability to produce a great fuel is determined by the composition of chemicals and physicals components available. Biomass

Table 2 Structural components of OPF and OPT [9, 10]

Parameter	Oil palm fronds (OPF)	Oil palm trunks (OPT)
Cellulose (% Dry wt.)	30.4	34.5
Hemi-cellulose (% Dry wt.)	40.4	31.8
Lignin (% Dry wt.)	21.7	25.7
Extractives (% Dry wt.)	1.7	3.7
Ash (% Dry wt.)	5.8	4.3

Table 3 Elemental analysis of OPF and OPT [10]

Parameter	Oil palm frond (OPF)	Oil palm trunk (OPT)
Ultimate analysis		
Nitrogen (wt.%)	12.402	0.169
Carbon (wt.%)	48.431	51.408
Hydrogen (wt.%)	10.476	11.816
Oxygen (wt.%)	46.75	51.16
Proximate analysis		
Gross Calorific Value (MJ kg-1)	15.72	17.47
Moisture Content (wt.%)	70.60	75.60
Ash Content (wt.%)	3.37	3.35
Volatile Matter Content (wt.%)	85.10	86.70

also contains less carbon and oxygen, which directly influences the polarity of the molecules and reduces the heat molecule content. The ultimate analysis found on biomass samples shows how it will influence the combustion in the boiler and the environment [12]. The highest nitrogen content is found in OPF (Table 3), which explains why most palm plantations use the OPF for mulching purposes because it can increase soil fertility. Biomass fuel also has more oxygen than coal in terms of chemical properties.

Table 3 also shows that OPF and OPT contain high moisture contents, which require additional pretreatment or energy before being used as fuel. Moisture content reduces the heating value of biomass, which can result in incomplete combustion [13]. The low melting point dissolved ash concentrations can lead to major combustion issues such as slagging and fouling. Nonetheless, the high volatile content in oil palm residues will result in favourable conditions such as low ignition temperature.

Therefore, some pretreatment methods are required to digest and open these resistant structures for further conversion. The chemical compositions of lignocellulosic biomass can be altered by an effective pretreatment method. Previous research suggested that a researcher should determine the lignin content before deciding on a pretreatment method to work on the specific type of oil palm biomass [4].

3 Utilisation of OPF and OPT

It will be a loss of opportunity if the vast amount of biomass residues produced by the oil palm industry are not utilised effectively. Oil palm biomass can be converted into a diverse range of value-added products. For example, OPF is classified as fibrous crop residues which can be processed into pulp [14]. It can also be utilised as a roughage source for ruminants of goats and cattle.

Meanwhile, replanted OPT can be used to make manufactured boards such as plywood. This palm trunk plywood's strength is said to be comparable to the commercial plywood. Nevertheless, in many situations, the trunks are chopped and returned to the ground [15, 16]. There are three different methods for converting biomass to energy: (1) thermochemical process, (2) Chemical and physical process and (3) biological conversions. Combustion, gasification, and pyrolysis are examples of thermochemical conversion methods [12, 17]. Gasification is one of the methods that OPT and OPF usually undergo for energy generation due to their high energy and moisture content [14, 16].

Hydrogen is a synthetic fuel that can replace fossil fuels in almost any application due to the benefits of higher engine efficiencies and zero emissions. Supercritical water technology is the most recent gasification technology for converting oil palm biomass to hydrogen. A previous study discovered that OPT yields more energy and hydrogen gas under identical gasification circumstances than mangrove wood and food wastes. Furthermore, past research has also concluded that palm oil wastes from plantations could be used as an alternate energy source in Malaysia in the future, notably in gasification [7].

Besides, OPT, OPF and EFB could be utilised as bio-fertilizer. For example, shredding and reusing as mulch in plantations, which helps preserve and improve soil fertility, is one of the waste management strategies used during oil replanting for OPF and OPT [4].

4 Conclusion

Efforts to utilise oil palm biomass waste will increase revenue for the operator, industry, and eventually the nation's economy. In addition, the social and economic well-being of rural communities also benefits from the investment of these valuable palm oil wastes for energy generation.

Acknowledgements This study was supported by a grant from the AAIBE Chair of Renewable Energy, Institute of Sustainable Energy, Universiti Tenaga Nasional (UNITEN).

References

1. Kaniapan S, Hassan S, Ya H, Patma Nesan K, Azeem M (2021) The utilisation of palm oil and oil palm residues and the related challenges as a sustainable alternative in biofuel, bioenergy, and transportation sector: a review. Sustainability 13:3110. https://doi.org/10.3390/su13063110
2. Zamri MFMA, Akhiar A, Shamsuddin AH (2019) Talking points of green and sustainable palm oil mill system in Malaysia. Int J Recent Tech Eng 8:6474–6479. https://doi.org/10.35940/ijrte.d5160.118419

3. Loh SK (2017) The potential of the Malaysian oil palm biomass as a renewable energy source. J Energy Con Man 141:285–298. https://doi.org/10.1016/j.enconman.2016.08.081
4. Diyanilla R, Hamidon TS, Suryanegara L, Hussin MH (2020) Overview of pretreatment methods employed on oil palm biomass in producing value-added products: a review. BioResources 15:9935–9997. https://doi.org/10.15376/biores.15.4.diyanilla
5. Information on MPOB (2020). https://bepi.mpob.gov.my/index.php/en/area/area-2020/oil-palm-planted-area-as-at-dec-2020
6. Hosseini SE, Wahid MA (2014) Utilisation of palm solid residue as a source of renewable and sustainable energy in Malaysia. J Renew Sustain Energy Rev 40:621–632. https://doi.org/10.1016/j.rser.2014.07.214
7. Awalludin MF, Sulaiman O, Hashim R, Nadhari WNAW (2015) An overview of the oil palm industry in Malaysia and its waste utilisation through thermochemical conversion, specifically via liquefaction. J Renew Sustain Energy Rev 50:1469–1484. https://doi.org/10.1016/j.rser.2015.05.085
8. Sulaiman SA, Balamohan S, Moni MN, Atnaw SM, Mohamed AO (2015) Feasibility study of gasification of oil palm fronds. J Mech Eng Sci 9:1744. https://doi.org/10.15282/jmes.9.2015.20.0168
9. Dungani R, Aditiawati P, Aprilia S, Yuniarti K, Karliati T, Suwandhi I, Sumardi I (2018) Biomaterial from oil palm waste: properties, characterisation and applications. Palm Oil. https://doi.org/10.5772/intechopen.76412
10. Hamzah N, Tokimatsu K, Yoshikawa K (2019) Solid fuel from oil palm biomass residues and municipal solid waste by hydrothermal treatment for electrical power generation in Malaysia: a review. Sustainability 11:1060. https://doi.org/10.3390/su11041060
11. Aljuboori AHR (2021) Oil palm biomass residue in Malaysia: availability and sustainability. Int J Biomass Renew 2:13–18
12. Sivabalan K, Hassan S, Ya H, Pasupuleti J (2021) A review on the characteristic of biomass and classification of bioenergy through direct combustion and gasification as an alternative power supply. J Physics: Conf Ser 183:012033. https://doi.org/10.1088/1742-6596/1831/1/012033
13. Yacob NS, Mohamed H, Shamsuddin AH (2021) Investigation of palm oil wastes characteristics for co-firing with coal. Adv Research in App Sci and Eng Tech 23:34–42. https://doi.org/10.37934/araset.23.1.3442
14. Shuit SH, Tan KT, Lee KT, Kamaruddin AH (2009) Oil palm biomass as a sustainable energy source: a Malaysian case study. Energy 34:1225–1235. https://doi.org/10.1016/j.energy.2009.05.008
15. Abdullah N, Sulaim F (2013) The oil palm wastes in Malaysia. Biomass Sustain Growth Use. https://doi.org/10.5772/55302
16. Mahlia TMI, Ismail N, Hossain N, Silitonga AS, Shamsuddin AH (2019) Palm oil and its wastes as bioenergy sources: a comprehensive review. Env Sci Pollut Res 26:14849–14866. https://doi.org/10.1007/s11356-019-04563-x
17. Chuah TG, Wan Azlina AGK, Robiah Y, Omar R (2006) Biomass as the renewable energy sources in Malaysia: an overview. Int J Green Energy 3:323–346. https://doi.org/10.1080/01971520600704779

Survey on Constraint in Application and Utilization of Biomass Conversion in Malaysia

Fadhilah Izzati Abdul Rani, Nor Wahidatul Azura Zainon Najib, Nor Ashikin Ahmad, Firuz Zainuddin, and Lucian Laslo

Abstract Biofuels from the conversion process of biomass are gradually becoming essential in the means to protect the environment and as a future substitution to fossil fuel. Recognizing the importance of biomass conversion to biofuel and the fact that Malaysia has an abundance of biomass waste generated, many studies have been conducted extensively to explore its potential as biofuel. Despite having benefit to exploit as renewable energy sources, biomass waste conversion in Malaysia is unexploited largely due to some constrains. Therefore, this study is conducted to identify the challenges faced in developing biomass conversion technology and enhancing the utilization of biofuel as energy source at the level of industrial scale. To achieve this study, the secondary data from previous journal and webpages are collected. This data is used to generate the content of questionnaire survey. Biomass researchers and industrial user are chosen as respondent. The data recorded that technology problems and lack of awareness among industrial user are the main challenges in application of biomass conversion. In term of biomass utilization, both categories of respondents agreed the diversity of biomass is the biggest challenge.

F. I. A. Rani · N. W. A. Z. Najib (✉)
Faculty of Civil Engineering Technology, Universiti Malaysia Perlis (UniMAP), 02600 Arau, Perlis, Malaysia
e-mail: norwahidatul@unimap.edu.my

N. W. A. Z. Najib
Centre of Excellence, Water Research and Environmental Sustainability Growth (WAREG), Universiti Malaysia Perlis, 02600 Arau, Perlis, Malaysia

N. A. Ahmad · F. Zainuddin
Faculty of Chemical Engineering Technology, Universiti Malaysia Perlis (UniMAP), 02600 Arau, Perlis, Malaysia
e-mail: ashikinahmad@unimap.edu.my

F. Zainuddin
e-mail: firuz@unimap.edu.my

L. Laslo
National Institute for Research and Development in Environmental Protection Bucharest (INCDPM), 294, Splaiul Independentei Street, 6th District, 060031 Bucharest, Romania

N. Mohamed Noor et al. (eds.), *Proceedings of the 3rd International Conference on Green Environmental Engineering and Technology*, Lecture Notes in Civil Engineering 214,
https://doi.org/10.1007/978-981-16-7920-9_23

Keywords Biomass conversion · Biofuel · Biomass application · Utilization

1 Introduction

Nowadays, biomass resources have broadly utilized as it has become more and more important due to their economic potential as sources of energy [6]. Other than that, because of Malaysia's rate of energy consumption that also has grown, which owing to the industrialization and population expansion (32.4 million, with an annual growth rate of 1.4% in 2018), might pose a concern in the next 40 years due to the depletion of fossil fuel resources [9]. Despite the fact that a variety of renewable energy options have been identified, biofuel remains one of the most promising options for reducing fossil fuel use by totally or partly eliminating [5] also as the only renewable energy that has an alternative for the production of transportation fuels [7].

Though having the potential as biofuels, the conversion of biomass waste to biofuels in Malaysia was not widely being implemented and used due to problems inherent in biomass conversion. As the time passes, the application and commercialization of biomass waste in Malaysia must grow for benefit of Malaysia's environment and economy. Thus, this study is conducted to identify the challenge in bio-fuel application and utilization by analyzing secondary data and conducting survey to biomass researchers and industrial user in Malaysia.

2 Materials and Method

2.1 Secondary Data

Secondary data from journal articles and web pages were used as the reference to develop questionnaire content and to investigate the actual perspective of biomass researchers and biomass industrial users in empowering the use of biofuels. Secondary data used for the research were data from year 2012 until 2021.

2.2 Questionnaire Survey

The targeted respondent for the questionnaire were biomass researchers and biomass industrial users. The questionnaire was transmitted using the electronic survey Google Form software and distributed via social media platforms such as WhatsApp, Facebook and E-mail. Snowball sampling method was used to minimize human contact. The sample size of respondent was decided using Roscoe's guidelines where 45 respondents was decided to be used for each respondents' categories in the study. The questionnaire data collected were than analyzed using Statistical Package for the Social Sciences (SPSS) tool. The data obtained were based on the result from a descriptive analysis and the outcome of the analysis is presented in mean value.

3 Discussion

3.1 Challenge in Biomass Application

Despite of having benefits as sustainable source of renewable energy, there are still have a lot of challenges to commercialize it. Table 1 shows challenges for the application and commercialization of biomass waste in Malaysia.

According to Derman et al. [1], Hamzah et al. [2] and Raud et al. [8], production cost was the most challenging problem in the application of biomass, as listed in Table 1. However, when associating the data in Table 1 with the surveying data in Table 2, it shows that both respondents categories have different opinion regarding the challenges of biomass application and commercialization. Lack of domestic support was the highest issue as the challenges for biomass researchers with means 4.31, while biomass industrial user chooses lack of awareness on its sustainability with 4.18 of means value. Thus, researchers and government must play an important role to introduce the sustainability of biomass more actively with the publication of research papers or articles and introducing new policies of biomass sustainability to the industries.

However, in spite of the different challenges chosen as the highest challenges, both biomass researcher respondents and biomass industrial user respondent agreed on procurement and feedstock problems as the least challenge for biomass application and commercialization with 3.16 and 3.56 of means value, respectively. Based on the data from Table 2, feedstock issues only being mentioned by How et al. [3], proving

Table 1 Challenges to the commercialization of biomass waste in Malaysia

Biomass type	Challenges
Palm oil, rubber, paddy and MSW [2]	Transportation problems, storage problems, technology problem and drying cost (production cost)
Palm oil (EFB) [1]	Technology problems, cost of feedstock, production cost and environmental awareness
Lignocellulosic [8]	Sustainability problems and high capital and processing costs
Biomass industry [3]	Biomass supply issues, logistics challenges, technological barriers, transportation, financial barriers, social awareness barrier and lack of domestic market support
Sawdust	High cost of production, lack of awareness on its sustainability, lack of ready market and poor packaging

Table 2 Biomass application and commercialization challenges mean

Biomass waste cannot be applied and commercialized in Malaysia widely because	Mean	
	Researchers	Biomass industrial user
High cost of production	4.11	3.86
High cost of transportation	4.04	3.89
Lack of awareness on its sustainability	4.04	4.18
Lack of domestic user support	4.31	3.98
Technology problems	3.84	4.11
Procurement of feedstock problems	3.16	3.56

that procurement and feedstock are the least challenges for biomass application and commercialization.

3.2 *Challenge in Biomass Utilization*

The challenge in the application of biomass waste is typically related to the challenge in the utilization of biomass. Until this moment, many researchers have done research related to the utilization of biomass to produce the final form of biofuels which are solid, liquid and gas. However, despite of all the experimental research and advantages of biomass in Malaysia, the utilization of biofuels from biomass materials at a large scale has still not been fully implemented. This is happening because of some challenges of the utilization. Table 3 lists the challenge of biomass utilization.

From Table 3, it shows that moisture content and density are the most listed challenges of biomass utilization as stated by researchers [2–4, 7]. However, from the questionnaire survey data as presented in Table 4, both respondents categories concluded that the most challenging in the utilization of biomass as biofuels is the complexity and diversity of biomass itself, where the highest means value 3.91 and

Table 3 Challenges to biomass utilization according to different researchers

Research concern	Challenges
Palm, rubber, paddy, wood and MSW [2]	Moisture content and density
Biomass industry [3]	Density and high moisture content
Fuel pellets [7]	Moisture content, particle size, feedstock compositions and density
Biomass utilization [4]	Density, moisture content, complexity, and diversity

Table 4 Biomass utilization as biofuels mean

Challenges of biomass utilization as biofuels	Mean	
	Researcher	Biomass industrial user
Moisture content	3.84	3.80
Density	3.76	3.56
Complexity and diversity	3.91	3.84
Different size and shape	3.40	3.49

3.84 recorded for biomass researcher and 3.84 for biomass industrial user, respectively. This result is corresponded with the statements issued by Irmak [4], in which, complexity and diversity of biomass is the biggest problem in biomass utilization.

On the other hand, respondents of biomass researchers also recorded that different size and shape of biomass waste as the least problem of biomass utilization with a mean value of 3.40, allied to biomass industrial user that recorded density with mean value 3.49. This result shows that the data from survey are matching with the secondary data where it recorded that size as the least problems being stated by Pradhan et al. [7].

4 Conclusion

Biofuels have the potential to substitute the fossil fuels. However, technology problems and lack of awareness on its sustainability has become the challenging factors to its application for industrial user. Both researchers and industrial users agreed that the complexity and diversity of biomass is the biggest challenge in utilization of biomass derived biofuel.

References

1. Derman E, Abdulla R, Marbawi H, Sabullah MK (2018) Oil palm empty fruit bunches as a promising feedstock for bioethanol production in Malaysia. Renew Energy 129:285–298. https://doi.org/10.1016/j.renene.2018.06.003
2. Hamzah N, Tokimatsu K, Yoshikawa K (2017) Prospective for power generation of solid fuel from hydrothermal treatment of biomass and waste in Malaysia. Energy Procedia 142:369–373. https://doi.org/10.1016/j.egypro.2017.12.058
3. How BS, Ngan SL, Hong BH, Lam HL, Ng WPQ, Yusup S, Ghani WAWAK, Kansha Y, Chan YH, Cheah KW, Shahbaz M, Singh HKG, Yusuf NR, Shuhaili AFA, Rambli J (2019) An outlook of Malaysian biomass industry commercialisation: perspectives and challenges. Renew Sustain Energy Rev 113:109277. https://doi.org/10.1016/j.rser.2019.109277
4. Irmak S (2019) Challenges of biomass utilization for biofuels. INTECH 32:137–144
5. Masjuki HH, Kalam MA, Mofijur M, Shahabuddin M (2013) Biofuel: policy, standardization and recommendation for sustainable future energy supply. Energy Procedia 42:577–586. https://doi.org/10.1016/j.egypro.2013.11.059

6. Perea-Moreno MA, Samerón-Manzano E, Perea-Moreno AJ (2019) Biomass as renewable energy: worldwide research trends. Sustainability 11:1–21. https://doi.org/10.3390/su11030863
7. Pradhan P, Mahajani SM, Arora A (2018) Production and utilization of fuel pellets from biomass: a review. Fuel Process Technol 181:215–232. https://doi.org/10.1016/j.fuproc.2018.09.021
8. Raud M, Kikas T, Sippula O, Shurpali NJ (2019) Potentials and challenges in lignocellulosic biofuel production technology. Renew Sustain Energy Rev 111:44–56. https://doi.org/10.1016/j.rser.2019.05.020
9. Rezania S, Oryani B, Cho J, Sabbagh F, Rupani PF, Talaiekhozani A, Rahimi N, Ghahroud ML (2020) Technical aspects of biofuel production from. Processes 1–19

Sustainable Structural Design of Residential Tall Building Based on Embodied Energy and Cost Performance

Muhammad Lutfil Hadi Noraini, Wan-Mohd-Sabki Wan Omar, and Radzi Ismail

Abstract Malaysia is one of the countries having a large number of tall structures. This is one of the engineer's initiatives to maximize land use due to the growing population. This rapid development endangered the environment through energy consumption and carbon emission in Malaysia. This study aims to design high-rise structure buildings using Esteem Software that differ in shape and to offer the most efficient models of high-rise buildings based on embodied energy (EE) and cost performance analysis. These analyses were divided into two categories: EE of a building which was calculated by considering the Hybrid life cycle assessment (LCA) method and the entire cost of the materials utilized in building construction. Furthermore, the study is limited to the cradle to gate phase only. The results demonstrate that as the number of sides increases, so does the value of EE and cost performance. It can be seen that the shape's performance while keeping the same value is significantly related to the number of sides of the shape.

Keywords Tall building · Embodied energy · Cost performance · Life cycle assessment · Building shape

1 Introduction

The United Nations Sustainable Development Goals (SDGs) have urged everyone to take care of the world by protecting the environment to ensure prosperity and peace in 2030. In Malaysia, an emission intensity levels basis is calculated as a ratio of greenhouse gas (GHG) emissions to the country. Malaysia's emission intensity

M. L. H. Noraini · W.-M.-S. Wan Omar (✉)
Water Research and Environmental Sustainability Growth (WAREG), Universiti Malaysia Perlis, 02600 Arau, Perlis, Malaysia
e-mail: wansabki@unimap.edu.my

R. Ismail
School of Housing, Building and Planning, Universiti Sains Malaysia, 11800 Pulau Pinang, Malaysia
e-mail: radzi@usm.my

N. Mohamed Noor et al. (eds.), *Proceedings of the 3rd International Conference on Green Environmental Engineering and Technology*, Lecture Notes in Civil Engineering 214,
https://doi.org/10.1007/978-981-16-7920-9_24

level is 1.3 tonnes CO_2 above the global average and the energy sector became major contributor amounting 0.73 tonnes CO_2. Residential tall building construction is considered as the main contribution environmental impact. Therefore, carefully designing residential tall buildings and taking embodied energy and cost performance into account will have a positive impact, particularly on the environment. Due to the rapid development, many techniques such as the method of construction have been compared either industrialized building system (IBS) method or the traditional method. Buildings were identified as the main contribution to the embodied energy and carbon emission. This is due to the embodied energy (EE) being significantly required during the construction phase and ultimately contributed to environmental impact. Life cycle assessment (LCA) is a systematic methodology to evaluate process loads and environmental impacts over their whole life cycle from cradle to grave. In ISO 14040 for instance, LCA is introduced to researchers to solve any problem regarding to environmental impact. A study carried out by Jiao et al. [1] has taken into account the shape of a structure in order to determine the quantity of EE within the building without considering the cost that is highly essential throughout the project's life cycle and it is also something which will influence a project's success or failure. Therefore, this study aims to design tall building structures considering both EE and cost performance at cradle to gate system boundary.

2 Methodology

2.1 Design Model

An architectural concept is the meaning and reason to the end product (the completed building or structure) and is very crucial step to decide the basic shape they need to refer in designing process. Lotteau et al. [2] has claimed that a building shape affected the EE as well as the cost of performance. In the designing process, the load factor is a value that is larger than one where it is multiplied by the nominal load to provide appropriate structural safety. Engineers can use varying design load factors for structural columns based on the number of floors they support, according to the design code. As the number of floors supported increases, the factor typically decreases. The Eurocodes (EC) (EN 1990–1993) require a higher level of protection for high-rise structures. In these codes, the design action in the structure is calculated using the favourable and unfavourable of the permanent and variable actions. Each of structure's design action for dead load (Gk) and live load (Qk), such as column, beam and slab is standardized as 1.35 Gk + 1.5 Qk. There are three types of shape configuration for high rise building designs as shown in Fig. 1. All structure components of building are designed according to Eurocode 2 BS EN 1992: Design of Concrete Structure. The factor of safety for Gk and Qk are 1.35 and 1.5 respectively. For Esteem software attributes, the design parameter used are based on BS 8110:1997 as given in Table 1.

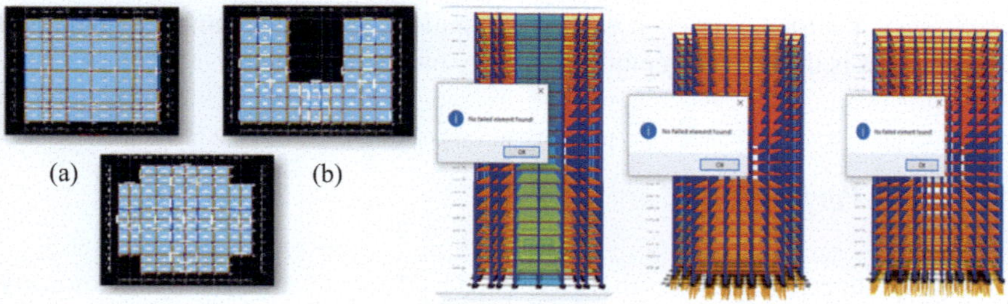

(c) (a) (b) (c)

Fig. 1 Shapes of high-rise building models using Esteem software: **a** Square shape, **b** U-shape and **c** cross shape

Table 1 Design parameter for structural component of building

Design criteria	Value
Characteristic strength of concrete, FCK	30 N/mm^2
Characteristic strength of steel, FYK	500 N/mm^2
Unit weight of concrete	25 kN/m^3
Design life	50 Years
Exposure class	XC2

2.2 *Embodied Energy*

The potential for environmental impact reduction through the use of demolition debris and recycled materials in building construction has been quantified by minimizing the embodied energy and carbon of construction materials [3, 4]. In the architecture criteria, for instance, the building heights as well as recycled materials significantly influenced the total embodied energy and carbon from the choice of structural types and construction materials. Embodied energy per unit of construction material, part or device is calculated as the quantity of non-renewable energy. It is calculated in megajoules (MJ) or gigajoules (GJ) per unit weight (kg or tonne) or area (m^2). Although it is complex and requires various data points in the process of measuring EE, all the calculations in this project that involve the calculation of EE have used the same standard to determine EE intensities on each material. For instance, EE intensities of concrete, plywood for formwork and steel are 1.944 MJ/kg, 33.056 MJ/kg and 72.753 MJ/kg respectively. This method was applied from Wan Omar, et al. [5] where the material quantities were multiplied with EE intensity of materials as shown in Eq. 1.

$$\text{Total EE} = \sum_{n} \text{Q} \times \text{EEI} \tag{1}$$

where EE is the total embodied energy of a building, Q is the number of materials used and EEI is the value of embodied energy intensities.

2.3 Cost Performance

The high quality characteristics in IBS have reduced maintenance and operation costs. Abdul Kadir, et al. [6] claimed that many contractors are reluctant to participate in any IBS project because of the cost, despite the overall cost savings it provides. Therefore, this research studies the design of tall building with different configuration to the cost performance. All the calculations of cost performance have used the same standard to determine the price of each material as calculated using Eq. 2. For instance, the price of concrete (RM 245.00/m^3), plywood (RM 17.48/m^2) for formwork and steel (RM 3000.00/tonne) are multiplied with each quantity of material to identify the cost performance. The price for the transport and installation of various building supplies to work sites in Malaysian Klang Valley. Moreover, all the cost data included different standard specifications of material which are frequently used in Singapore and Malaysia.

$$\text{Total CP} = \sum_{n} \text{Q} \times \text{Cost per material} \tag{2}$$

where CP mean is the total cost performance of a building, Q is the number of materials used and cost per material is the price of each material.

3 Result and Discussion

3.1 Embodied Energy

Table 2 shows the EE and cost performance of conventional and IBS methods. The result of EE is based on the different shape of a residential tall building. The entire value of embodied energy in GJ for both the conventional and IBS techniques of construction was determined based on the findings. According to the results, the lowest value of total embodied energy for a 20-story residential structure is 4.62 GJ/m^2 for the conventional method and 2.88 GJ/m^2 for the IBS method. As the number of sides increases, so does the value of embodied energy. The highest value of EE energy is the cross shape where the value of EE consume by this building is 5.11 GJ/m^2 and 3.38 GJ/m^2 for conventional and IBS respectively. The value is obtained by dividing the total value of EE by the sum of areas on all floors of the building. The results showed that the EE intensity for a square shape building gives the lowest yield value than other shapes.

Table 2 Total EE intensity and total cost performance for residential tall buildings

Shape of the building	No. of side	Building design system	EE Intensity (GJ/m^2)	Cost performance (RM million)
Cross	12	Conventional	5.11	5.11
		IBS	3.38	3.40
U-shape	6	Conventional	4.90	4.89
		IBS	3.17	3.18
Square	4	Conventional	4.62	4.60
		IBS	2.88	2.88
Square-Shape (square footing)	4	Conventional	–	4.42
		IBS	–	2.69

3.2 Cost Performance

The total cost performance is determined by adding all of the value of manufacturing cost, and formwork cost. According to data in Table 2, the lowest total cost for a 20-story residential structure for a conventional method is RM 4.42 million whereas the IBS method cost is RM 2.69 million. The cross shape of tall building design has highest total cost of building material, similar trend with EE intensity. The total cost for conventional building design is RM 5.11 million and IBS building design is RM 3.40 million. As the number of sides increased, the value of cost performance also increased as revealed by previous research [2].

4 Conclusion

In summary, the cross-shaped structure has the worst influence on the environment of all the buildings among the model that has been designed by Esteem software. This type of shapes does not only has highest EE intensity, but it also has the highest performance cost. This proves the higher number of sides of the shape of the building, the higher the EE and the cost performance on the building. While the best model that has the least impact on the environment and gives the most positive impact in managing cost performance among other models is a square-shaped building that uses a rectangular pile as a foundation and is built by IBS.

References

1. Jiao Y, Lloyd CR, Wakes SJ (2012) The relationship between total embodied energy and cost of commercial buildings. Energy Build 52:20–27. https://doi.org/10.1016/j.enbuild.2012.05.028
2. Lotteau M, Loubet P, Sonnemann G (2017) An analysis to understand how the shape of a concrete residential building influences its embodied energy and embodied carbon. Energy Build 154:1–11. https://doi.org/10.1016/j.enbuild.2017.08.048
3. Wan Omar WMS (2018) A hybrid life cycle assessment of embodied energy and carbon emissions from conventional and industrialised building systems in Malaysia. Energy Build 167:253–268. https://doi.org/10.1016/j.enbuild.2018.02.045
4. Crawford RH, Bontinck P-A, Stephan A, Wiedmann T, Yu M (2018) Hybrid life cycle inventory methods—A review. J Clean Prod 172:1273–1288. https://doi.org/10.1016/j.jclepro.2017.10.176
5. Wan Omar WMS, Doh JH, Panuwatwanich K, Miller D (2014) Assessment of the embodied carbon in precast concrete wall panels using a hybrid life cycle assessment approach in Malaysia. Sustain Cities Soc 10:101–111. https://doi.org/10.1016/j.scs.2013.06.002
6. Abdul Kadir MR, Lee WP, Jaafar MS, Sapuan SM, Ali AAA (2006) Construction performance comparison between conventional and industrialised building systems in Malaysia. Struct Surv 24(5):412–424. https://doi.org/10.1108/02630800610712004

Uncertainty in River Hydraulic Modelling: A Review for Fundamental Understanding

Mohd Aliff Mohd Anuar, Mohd Shalahuddin Mohd Adnan, and Foo Hoat Lim

Abstract River hydraulic modelling is a vital component to perform preferable analysis. Mostly river hydraulic modelling was performed flood analysis and water resources analysis. The hydraulic engineer or modeller needs a better technical understanding on the simulated the model. However, river hydraulic modelling frequently encountered model stability or uncertainty problem during computation. The error for model stability was will notice by the modeller. Hence, the modeller needs to solve the error for the purpose to simulate the model without uncertainty interruption. In this paper an attempt has been made to review the research work done on the stability occurred in river hydraulic modelling. This research review shows that hydraulic geometry and hydraulic parameter input are the main role of occurrence model stability.

Keywords River modeling · Hydraulic · Stability

1 Introduction

The River hydraulic modelling is includes hydrological data and hydraulic data to setup the model. River hydraulic model are the most widely used application to simulate the water flow movement. River hydraulic modelling is the model that simulate water movement through the river cross section in the basin by using derived equation such as St Venant's equation (Shallow water equation) [1–3]. River hydraulic model can be classified into one (1) dimensional (1D) and two (2) dimensional (2D) model [4, 5]. Therefore, to simulate the water path motion for basin can be model into 1D and 2D model depend on data availability and the outcome result.

M. A. Mohd Anuar · M. S. Mohd Adnan (✉) · F. H. Lim
Faculty of Civil Engineering and Built Environment, Universiti Tun Hussein Onn Malaysia, 86400, Johor Parit Raja, Batu Pahat, Malaysia
e-mail: shalahuddinadnan@uthm.edu.my

Water Resources and Coastal Department, Angkasa Consulting Services, 47620 Subang Jaya, Malaysia

N. Mohamed Noor et al. (eds.), *Proceedings of the 3rd International Conference on Green Environmental Engineering and Technology*, Lecture Notes in Civil Engineering 214,
https://doi.org/10.1007/978-981-16-7920-9_25

Stability in the modelling is a instability numerical model which causes certain type of numerical error growing and resulting the computational are unable to continue and proceed simulation [5]. There are several features affected model stability for instance flow, manning, cross section, roll wave etc. [5–8]. The river geomorphology also contribute to the stability during applying river section in the model, the river reach will control the flow by geometry of the river due to sedimentation [9–11].

However, model and data accuracy needs to be analysed thoroughly. The model accuracy is to ensure the degree of closeness of the numerical solution to the real condition. The model accuracy is depends on limitation of the model, geometry data and hydrological data [5, 12]. Thus, the model stability can be analysed and during data collection.

Therefore, this research paper was focuses to review the hydrodynamic model stability problem encounter during simulation. Based on the review from other research, the causes of stability can be determined in the earlier stage, which is during data preparation. The stability can be controlled and handled once the modeller identified the problem that will encounter before model the river.

2 Factor Contribute to Model Stability

2.1 *Cross Section Spacing Geometry*

River cross section is the main hydraulic geometry input in modelling. All hydraulic modelling software such as HEC-RAS, InfoWorks ICM, MIKE 11, XPSWMM and etc. need a cross section data as main input to compute the simulation. One of the factors, which affect stability of hydraulic model, is cross section data [5, 13]. The interval of the cross section cannot place far apart to each cross section, it will cause numerical damping of flood wave and make the model unstable [5, 14]. Moreover, if the interval between cross section is closes to each other, it may cause wave steepening and make model instability [5, 14, 15]. However, the first step before model the river is to decide the cross section spacing using Fread and Samuel equation [16, 17].

2.2 *Time Step and Manning Value*

Every hydraulic modelling needs to set the timestep to proceed the simulation. The selection of time step also important because it can contribute as a factor of model instability [18]. The enhanced stability achieved by the new formulation comes at no significant additional computational cost and, in fact, the model performance can benefit from the small time steps value because it can reduce instability [6]. The model instability also can be influence by Manning values. The using of low

Manning value can causes lower depth of water, high in velocity and the occurrence of supercritical flow; meanwhile if the Manning is high it will increase the stage locally [5, 19].

2.3 Structures-Weir, Bridges, Culvert

Most river hydraulic model will encounter uncertainty during hydraulic computation when the hydraulic structure was added in the model. In river hydraulic modelling the structure input cannot be avoided to match the existing condition of the river. However, the instability of the model will be increase once the structures were added or model during simulation. The hydraulic structure such weir, gate, bridges, culvert, spillway levee and etc. can cause oscillation and disturb the stage in the river [20, 21].

2.4 Initial and Low Flow Condition

Any time-marching (unstable) solution needs initial conditions for flows and depths in cross-sectional areas in a river network. Initial flow and depth values must be given to each part of the river network, and these values should be consistent with the inflow boundary conditions at the starting time to avoid instability [3, 22]. In the 1D case, a significant simplification of truth is the presumption that flow is strictly 1D. In the 2D case, at least when modelling channel flow, the assumption that the vertical velocities are negligible may be incorrect [23]. Flow data used as boundary conditions for all models is one of the main uncertainties [23]. Pool and riffle sequences and shallow depths can contribute to severe model instability in low flow conditions. Low flow modelling can be challenging because of the limited volume of water often flowing through the reaches [5, 24]. Due to the limited amount of water often flowing through the reaches and the instability problems the low flows can cause in the HEC-RAS modelling method, low flow modelling can be difficult [24].

2.5 Roll Wave and Steep Stream

Steep stream is the next factors that cause instability in hydraulic modelling. When the slope of river channel is high, the Froude numbers tend to approach 1.0, which means critical depth. Thus, the hydraulic model becomes unstable due to high velocity and rapid changes in depth [5, 25]. Stability can be maintain by ensure the slope of river bed is small compared to friction coefficient, it may be violated in steep channels and, under suitable conditions, roll-waves will be formed [26]. The Vedernikov number is also used to determine if the uniform flow is stable as a criterion [2]. Roll waves are quasi-periodic finite amplitude waves produced in a channel due to uniform steady-state flow instability [27]. Table 1 shows the previous studies conducted on river hydraulic modelling.

Table 1 Previous studied on river hydraulic modelling

Authors	Title	Findings: Factor present in model stability
Brunner	HEC-RAS river analysis system—User's manual version 5.0	Cross section, time step, structures, low flow, steep stream [20]
Pelletier	Reply: uncertainties in the single determination of river discharge: a literature review	Cross section [13]
Haron et al.	Assessing river stability and hydraulic geometry of fluvial river in Malaysia	Cross section [11]
Fread and Lewis	Selection of δ x and δ t computational steps for four-point implicit nonlinear dynamic routing models	Cross section [16]
Samuels	Backwater lengths in rivers	Cross section [17]
De Almeida	Improving the stability of a simple formulation of the shallow water equations for 2-D flood modeling	Time step, manning [6]
Wingate and Read	The maximum allowable time step for the shallow water α model and its relation to time-implicit	Time step [18]
Zidan	A Review Of Friction Formulae In Open Channel Flow	Manning [19]
Brandimarte and Di Baldassarre	Uncertainty in design flood profiles derived by hydraulic modelling	Structure [21]
Betsholtz	Potentials and limitations of 1D, 2D and coupled 1D-2D flood modelling in HEC-RAS	Initial, Low flow [23]
Sharkey	Investigating instabilities with HEC-RAS unsteady flow modeling for regulated rivers at low flow stages	Initial, low flow [24]
Ebrahimian and Nuruddin	Runoff estimation in steep slope watershed with standard and slope-adjusted curve number methods	Roll wave, steep stream [25]
Que and Xu	The numerical study of roll-waves in inclined open channels and solitary wave run-up	Roll wave, steep stream [26]
Boudlal	Stability of regular roll waves	Roll wave, steep stream [27]

3 Identifying Model Stability Problem

The model stability can be detected by identified certain criteria that occurred on the model. The problem will be recognise when the program suddenly stop the simulation while running. Besides that, there are errors were listed regarding hydraulic or hydrology model set up. When the simulation model proceed to the maximum number of iteration for the time step coupled with errors also the method to detect the stability problem. Then, the simulation model will encounter oscillation during computed stage and flow hydrograph. Furthermore, there a lot of warning message will appear during model validation before the simulation started [4, 5, 24].

4 Conclusion

This paper briefly review the instability occurred in river hydraulic modelling. Most river hydraulic modelling using Saint Venant equation. The model stability complication can be divided into two, which are hydraulic geometry and hydraulic parameter input. Moreover, factors of model stability also encounter during computation the simulation also were highlighted. However, to fix and correct the error, which cause model instability need a further discussion as different model software will have different technique.

Acknowledgements The authors express gratitude to Research Management Centre (RMC) UTHM for awarded Post Graduate Research Grant Vot. No. H352. Universiti Tun Hussein Onn Malaysia for the facility and Ministry of Higher Education Malaysia.

References

1. Teng J, Jakeman AJ, Vaze J, Croke BFW, Dutta D, Kim S (2017) Flood inundation modelling: a review of methods, recent advances and uncertainty analysis. Environ Model Softw 90:201–216. https://doi.org/10.1016/j.envsoft.2017.01.006
2. Barker B, Johnson MA, Noble P, Rodrigues LM, Zumbrun K (2018) Stability of viscous st. venant roll-waves: from onset to infinite-froude number limit. J Nonlinear Sci 1–38. https://doi.org/10.1007/s00332-016-9333-6
3. Yu C, Liu F, Hodges BR (2017) Consistent initial conditions for the Saint-Venant equations in river network modeling. Hydrol Earth Sys Sci 4959–4972. https://doi.org/10.5194/hess-21-4959-2017
4. DHI (2003) MIKE 11-a modelling system for rivers and channels—User guide. https://www.mikepoweredbydhi.com/products/mike-11. Accessed 4 May 2021
5. Brunner G (2016) HEC-RAS river analysis system—User's manual version 5.0. United State: US Army Corp of engineer. Institute for Water Resources, Hydrologic Engineering Center. https://www.hec.usace.army.mil/. Accessed 17 Apr 2021
6. De Almeida GAM, Bates P, Freer JE, Souvignet M (2012) Improving the stability of a simple formulation of the shallow water equations for 2-D flood modeling. Water Resour Res 48:1–14. https://doi.org/10.1029/2011WR011570

7. Bessar MA, Matte P, Anctil F (2020) Uncertainty analysis of a 1D river hydraulic model with adaptive calibration. Water 12(2). https://doi.org/10.3390/w12020561
8. Samuel P, Skeels C (1990) Stability limits for preissmann scheme. J Hydraul Eng 116(8):997–1012. https://doi.org/10.1061/(ASCE)0733-9429(1990)116:8(997)
9. Demissie M, Borah DK, Admiraal D (1997) Review of hydrologic and hydraulic studies and unsteady flow routing of the 1993 flood in the upper mississippi river system. https://core.ac.uk
10. Garde RJ (2005) River morphology. New Age International, India, New Delhi
11. Haron NA, Yusuf B, Sulaiman MS, Ab Razak MS, Abu Bakar SN (2019) Assessing river stability and hydraulic geometry of fluvial river in Malaysia. Int J Integr Eng 11(6):214–223. https://doi.org/10.30880/ijie.2019.11.06.023
12. Julien PY, Simons DB (1984) Analysis of hydraulic geometry. Relationships in alluvial channels. https://mountainscholar.org. Accessed 5 June 2021
13. Pelletier PM (1989) Reply: uncertainties in the single determination of river discharge: a literature review. Can J Civ Eng 15(5):834–850. https://doi.org/10.1139/I88-109
14. Knight DW (2000) Derivation of routing parameters from cross-section survey. https://eprints.hrwallingford.com/441/1/SR568
15. Haron NA, Yusuf B, Sulaiman MS, Razak SA, Nurhidayu S, Bakar A (2019) Assessing river stability and hydraulic geometry of fluvial river in Malaysia. Int J Integr Eng 11(6):214–223
16. Fread DL, Lewis JM (1993) Selection of δ x and δ t computational steps for four-point implicit nonlinear dynamic routing models In: Abstract of national conference on hydraulic engineering
17. Samuels PG (1989) Backwater lengths in rivers. Proc Inst Civ Engrs Part 2 87:571–582. https://doi.org/10.1680/iicep.1989.3779
18. Wingate BA (2004) The maximum allowable time step for the shallow water α model and its relation to time-implicit differencing. Mon Weather Rev 132:2719–2731. https://doi.org/10.1175/MWR2816.1
19. Zidan ARA (2015) A review of friction formulae in open channel flow. Int Water Technol J 5(1):43–57
20. Brunner GW, Problems when performing an unsteady flow analysis. https://www.nws.noaa.gov. Accessed 2 May 2021
21. Brandimarte L, Di Baldassarre G (2012) Uncertainty in design flood profiles derived by hydraulic modelling. Hydrol Res 43(6):753–761. https://doi.org/10.2166/nh.2011.086
22. Matić BB, Simić Z (2016) Comparison of pre and post development low flow conditions for Drina river. Procedia Eng. 169:284–292
23. Betsholtz A (2017) Potentials and limitations of 1D, 2D and coupled 1D-2D flood modelling in HEC-RAS. http://lup.lub.lu.se/student-papers/record/8904721. Accessed 21 March 2021
24. Sharkey JK (2014) Investigating instabilities with HEC-RAS unsteady flow modeling for regulated rivers at low flow stages. Desertation, University of Tennessee, Knoxville
25. Ebrahimian E, Nuruddin AAB, Amin M, Soom BM, Sood AM, Neng LJ (2012) Runoff estimation in steep slope watershed with standard and slope-adjusted curve number methods. Pol J Environ Stud 21(15):1191–1202
26. Que Y, Xu K (2006) The numerical study of roll-waves in inclined open channels and solitary wave run-up. Int J Numer Methods Fluids 50:1003–1027. https://doi.org/10.1002/fld.1102
27. Boudlal A, Liapidevskii V (2005) Stability of regular roll wave. Russia, Novosibirsk

Urban Sustainable Development Solution for Irrigation Systems of Green Spaces Located on Buildings Based on a 3D Catalyst Immobilized on Fixed Support

Iasmina-Florina Burlacu, György Deák, Lidia Favier, Irina Urloiu, and Salwa Mohd Zaini Makhtar

Abstract Green spaces located on buildings are an important part of urban environment, improving the quality of life through their aesthetic and functional advantages. However, they consume large amounts of resources from cities' water supply systems. Moreover, rapid urbanization and population growth amplifies the issue on clean water supply, increasing the necessity for efficient solutions enabling wastewater treatment and reuse. This paper focuses on demonstrating of an innovative wastewater treatment method, based on reusable catalyst materials on fixed supports, and their potential application to supply the irrigation scheme of the green spaces located on buildings. The investigations were conducted on a highly persistent organic compound, namely Clofibric Acid, frequently identified in urban wastewater in Europe. The obtained results confirmed that the 3D Hybrid Catalyst has a degradation efficiency of the compound up to 45% in visible light and up to 92% after 7 cycles of reuse, under UV irradiation.

Keywords Urban green spaces · Wastewater reuse · Reusable 3D catalytic materials · AOPs wastewater treatment · Sustainable development

I.-F. Burlacu · G. Deák (✉) · I. Urloiu
National Institute for Research and Development in Environmental Protection Bucharest (INCDPM), 294, Splaiul Independentei, 6th District, 060031 Bucharest, Romania

L. Favier
Ecole Nationale Supérieure de Chimie de Rennes, CNRS, ISCR – UMR6226, F-35000 Rennes, France
e-mail: lidia.favier@ensc-rennes.fr

S. M. Z. Makhtar
Centre of Excellence, Water Research and Environmental Sustainable Growth (WAREG), Universiti Malaysia Perlis, 02600 Perlis, Malaysia
e-mail: salwa@unimap.edu.my

N. Mohamed Noor et al. (eds.), *Proceedings of the 3rd International Conference on Green Environmental Engineering and Technology*, Lecture Notes in Civil Engineering 214,
https://doi.org/10.1007/978-981-16-7920-9_26

1 Introduction

Green spaces located on buildings offer multiple benefits in urban areas, having not only aesthetic value, but also functional—thermal and phonic insulation, increase the energy efficiency and improve the air quality. Currently, the amount of water for their irrigation, which varies between 0.5 and 20 L/m^2 per day, is mainly provided by the water distribution network [1]. Although some research projects (e.g., NaWaTech, SuperGreen) [2, 3] investigated the irrigation of green spaces with wastewater, there are still gaps regarding the impact of this practice on plants, soil and human health. Some studies [4, 5] show that the long-term use of untreated water leads to increased salinity and soil porosity. Additionally, the accumulation in the soil of surfactants present in detergents may affect plants health [6, 7].

The current trend towards the expansion of urban areas and the rapid growth of the population, which generates, proportionally, increasing amounts of wastewater, has led to the development of new methods and technologies in recent years [8]. Advanced Oxidation Processes (AOPs) have emerged as innovative methods with high potential for emergent pollutants degradation, without generating secondary metabolites. Among the different types of AOPs, photocatalytic oxidation has shown a promising capacity for wastewater treatment, with the ability to eliminate persistent organic pollutants. Zinc oxide (ZnO) is highlighted in the literature as one of the ideal candidates, due to its many advantages, such as wide availability, low cost and efficiency in removing pollutants from water. Several studies have reported the degradation efficiency of water contaminants in the presence of ZnO of about 90–100% for dyes (e.g., Basic Red, Methyl blue) and pharmaceutical compounds (e.g., Caffeine) [9–11].

Currently, one of the major technological barriers in engineering applications is the recovery of photocatalytic materials following the treatment process [11, 12]. In numerous experimental studies [9, 13], photocatalysts are used in the form of particles in suspension, which leads to a difficult recovery process and the need for an additional treatment step. Research in the field of AOPs has made it possible to overcome these limitations by immobilizing particles with catalytic activity on support materials, thus obtaining heterogeneous catalysts [14]. The aim of this study is to demonstrate the potential for the application of immobilized ZnO on a fixed support in an integrated system for the removal of emerging pollutants from wastewater and its reuse for irrigation of green spaces on buildings. The pharmaceutical compound Clofibric Acid (CA) was selected to demonstrate the photodegradation efficiency of the proposed catalyst, having the following characteristics: CAS Number 882 09 7, molar mass 214.65 g/mol, pK_a 4.2 at 295 °K, water solubility 583 mg/L at 20 °C and log K_{ow} 0.76 at pH 7.5. CA has been widely found in European waters, having a complex structure and high persistence in the environment, which means that biological degradation is insufficient for its elimination [15]. The material can be applied in the system because its recovery and reuse are practical, having high oxidation efficiency of organic pollutants in both UV and Visible spectrum. Thus, this article aims to contribute to the sustainable development of urban areas through the efficient management of wastewater, in the context of pressures generated by climate change.

2 Experimental

2.1 Materials and Methods

The 3D Hybrid Catalyst was obtained by (I) deposition of ZnO on the porous-support surface, followed by (II) the autoclaving process, for fixing the oxide layer. Zinc oxide confers photocatalytic properties, presenting activity in the UV and visible spectrum. The support material was synthesized by capitalizing, in varying proportions, some waste—glass waste from fluorescent tubes, eggshells and agricultural fertilizer Epsom salt ($MgSO_4$). The applicability of the 3D Hybrid Catalyst material developed in wastewater treatment is supported by the fact that (a) it has a high efficiency in degrading pollutant compounds both in the UV and visible spectrum; (b) recovery of the material after use is practical and does not require an additional treatment step; (c) the material can be reused several times, with a high efficiency of pollutant removal; (d) is a cost-effective option and uses wastes that are found on a large scale as raw materials.

The 3D Hybrid Catalyst was tested by photodegradation of the Clofibric Acid (CA). A solution containing CA (5 mg/l CA) was prepared by dilution in ultrapure water and placed in contact with the catalyst (1,75 g/L ZnO concentration). The initiation of photocatalytic process was performed in the presence UV irradiation, respectively Visible spectrum. Observations on the degradation efficiency of the catalyst were made for 8, respectively 24 h.

3 Results and Discussions

3.1 **3D Hybrid Catalyst** *Degradation Efficiency of the Organic Pollutants*

To evaluate the potential application of the material in wastewater treatment, the photocatalytic performance in the UV and visible spectrum were investigated for the degradation efficiency of CA, a compound frequently identified in municipal water treatment plants. The catalyst reuse is crucial to determine its reliability for long-term use in practical applications. The 3D Hybrid Catalyst was tested to evaluate the degradation efficiency of CA over seven successive cycles of reuse for 8 h. The results showed that the ZnO retained its photocatalytic activity after seven successive experimental stages, as illustrated in Fig. 1. A slight decrease was observed after the

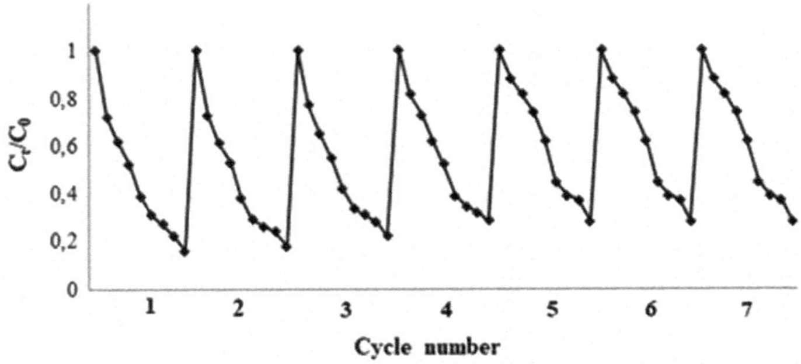

Fig. 1 3D catalyst material reuse capacity in the degradation efficiency of CA for 8 h of UV radiation

Fig. 2 Photocatalytic degradation efficiency of the studied compound in the presence of the 3D Hybrid Catalyst under Visible light for 8 h

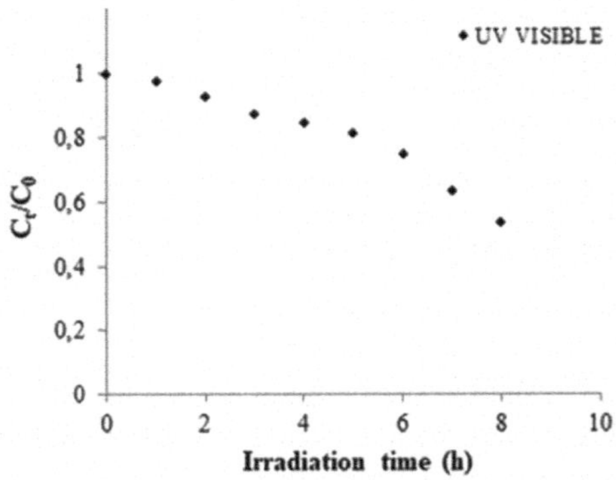

third use. However, after 7 cycles, the decomposition efficiency was maintained at high values, up to 92%.

To evaluate the degradation efficiency of 3D Hybrid Catalyst of CA in the presence of visible light, observations were made for 8 h. As shown in Fig. 2, the efficiency of ZnO removal photocatalytic activity maintained an upward trend throughout experimental stages. After an irradiation time of 8 h, the degradation efficiency was up to 45%. For both experiments (UV and Visible spectrum), the photocatalytic degradation trends are described by the Modify Power equation $C_t/C_0 = ab^t$, where the parameters varies as it follows: the regression coefficient is in the range 0.968 and 0.994, a and b coefficients varies between $a = 1.018$–4.31 and $b = 0.946$–0.808. The standard error is between 0.0057 and 0.14.

4 Conclusions

The 3D Hybrid Catalyst was successfully employed for the degradation of Clofibric Acid, being described by the Modify Power equation. The obtained results indicate a removal efficiency up to 45% of the CA in Visible spectrum and up to 92% under UV irradiation after 7 reuse cycles. Thus, they confirm the applicability potential of the 3D hybrid material in a system of purification and reuse of wastewater for irrigation of green spaces on buildings. These results offer not only a new perspective on the application of catalytic materials for the elimination of emerging pollutants from wastewater, but also a simple and very efficient way that contributes to the sustainable development of urban areas.

References

1. Prodanovic V, Hatt B, McCarthy D, Zhang K, Deletic A (2017) Green walls for greywater reuse: understanding the role of media on pollutant removal. Ecol Eng 102:625–635. https://doi.org/10.1016/j.ecoleng.2017.02.045
2. Masi F, Bresciani R, Rizzo A, Edathoot A, Patwardhan N, Panse D, Langergraber G (2016) Green walls for greywater treatment and recycling in dense urban areas: a case-study in Pune. J Water Sanit Hyg Dev 6(2):342–347. https://doi.org/10.2166/washdev.2016.019
3. Boano F, Caruso A, Costamagna E, Fiore S, Demichelis F, Galvão A, Pisoeiro J, Rizzo A, Masi F (2020) Assessment of the treatment performance of an open-air green wall fed with graywater under winter conditions. ACS ES&T Water 1(3):595–602. https://doi.org/10.1021/acsestwater.0c00117
4. Boyjoo Y, Pareek VK, Ang M (2013) A review of greywater characteristics and treatment processes. Water Sci Technol 67(7):1403–1424. https://doi.org/10.2166/wst.2013.675
5. Ababsa N, Kribaa M, Ouldjaoui A, Tamrabet L, Addad D, Hallaire V (2019) Long-term effects of wastewater reuse on hydro physicals characteristics of grassland grown soil in semi-arid Algeria. J King Saud Univ Sci. https://doi.org/10.1016/j.jksus.2019.09.007
6. James DTK, Surendran S, Ifelebuegu AO, Ganjian E, Kinuthia J (2016) Grey water reclamation for urban non-potable reuse challenges and solutions: a review. In: 7th ICSBE. Kandy, Sri Lanka, pp 16–18
7. More RP, Thakare SB (2017) A study on effect of kitchen wastewater irrigation on quality of soil and crop growth. Int J Sci Res 6(6):1989–1992
8. Dumitru FD, Moncea MA, Baraitaru AG, Panait AM, Olteanu MV, Deak G (2020) Effect of mesoporous silica with TiO2/ZnO nanocomposites in wastewater treatment. Rev Rom Mater 50(1):3–9
9. Yashni G, AlGheethi A, Maya Saphira Radin Mohamed R, Nor Hidayah Arifin S, Abirama Shanmugan V, Hashim Mohd Kassim A (2020) Photocatalytic degradation of basic red 51 dye in artificial bathroom greywater using zinc oxide nanoparticles. Mater Today-Proc 31(1):136–139. https://doi.org/10.1016/j.matpr.2020.01.395
10. Lam SM, Kee MW, Sin JC (2018) Influence of PVP surfactant on the morphology and properties of ZnO micro/nanoflowers for dye mixtures and textile wastewater degradation. Mater Chem Phys 212:35–43. https://doi.org/10.1016/j.matchemphys.2018.03.002
11. Sacco O, Vaiano V, Matarangolo M (2018) ZnO supported on zeolite pellets as efficient catalytic system for the removal of caffeine by adsorption and photocatalysis. Sep Purif Technol 193:303–310. https://doi.org/10.1016/j.seppur.2017.10.056

12. Di Mauro A, Farrugia C, Abela S, Refalo P, Grech M Falqui L, Impellizzeri G (2020) Ag/ZnO/PMMA nanocomposites for efficient water reuse. ACS Appl Bio Mater 3(7):4417–4426. https://doi.org/10.1021/acsabm.0c00409
13. Rogé V, Guignard C, Lamblin G, Laporte F, Fechete I, Garin F, Lenoble D (2018) Photocatalytic degradation behavior of multiple xenobiotics using MOCVD synthesized ZnO nanowires. Catal Today 306:215–222. https://doi.org/10.1016/j.cattod.2017.05.088
14. Burlacu IF, Deák G, Marcu E, Serre I P, Balloy D, Halin DSC (2020) Circular economy-waste reuse into a spongy oxide material with photocatalytic activity for a sustainable development. IOP Conf Ser: Earth Environ Sci 616(1):012068 (IOP Publishing). https://doi.org/10.1088/1755-1315/616/1/012068
15. Caracciolo AB, Topp E, Grenni P (2015) Pharmaceuticals in the environment: biodegradation and effects on natural microbial communities. Rev J Pharm Biomed Anal 106:25–36. https://doi.org/10.1016/j.jpba.2014.11.040

Green Technologies

A Comprehensive Evaluation of Pozzolanic Activity of Ancient Brick Powders Wastes—BPW in Cement Based Materials

Mihaela-Andreea Moncea, György Deák, Florina-Diana Dumitru, Roshazita Che Amat, and Norlia Mohamad Ibrahim

Abstract The recovery potential of bricks wastes in obtaining environmental friendly materials is sustained by their pozzolanic properties, which can be exploited within a Portland cement—brick powder waste (BPW) system, thus reducing the amount of waste as well as the consumption of raw materials and CO_2 emissions from cement factories. These properties were intensively studied and reported in the literature, for their assessment often being used direct analytical methods to highlight the presence of $Ca(OH)_2$ and its subsequent reduction in abundance with time as the pozzolanic reaction proceeds. Since it is more difficult to quantify the pozzolanic activity of calcined clay in a Portland cement /brick powder mixture, for the present work classical determinations, such as thermo-gravimetric analysis (TGA), scanning electron microscopy (SEM) were combined with the individual assessment of BPW in terms of pozzolanicity in order to study the variation of the Ca^{2+} respectively OH^- ions concentration within a BWP—lime saturated solution. SEM analyses showed CH and CSH formation and the DTA curve highlighted a more intense peak around 500 °C after 28 days of hydration, where the $Ca(OH)_2$ decomposition take place. The pozzolanicity test results showed a pronounced decrease of Ca^{2+} concentration after 28 days.

Keywords Brick powder waste · Portland cement · Land field recovery · Compressive strength

M.-A. Moncea (✉) · G. Deák · F.-D. Dumitru
National Institute for Research and Development in Environmental Protection Bucharest (INCDPM), 294, Splaiul Independentei Street, 6th District, 060031 Bucharest, Romania

R. C. Amat · N. M. Ibrahim
Sustainable Environment Research Group (SERG), Centre of Excellence Geopolymer and Green Technology (CEGeoGTech), Universiti Malaysia Perlis, 01000 Kangar, Perlis, Malaysia
e-mail: roshazita@unimap.edu.my

N. M. Ibrahim
e-mail: norlia@unimap.edu.my

N. Mohamed Noor et al. (eds.), *Proceedings of the 3rd International Conference on Green Environmental Engineering and Technology*, Lecture Notes in Civil Engineering 214,
https://doi.org/10.1007/978-981-16-7920-9_27

1 Introduction

Nowadays the civil industry is in continuous development, many ancient buildings are demolished and new ones are built. Together with the comfort that a new house or a new office provides, there are many environmental problems created by solid wastes resulted through construction and demolition works (C&DW) such as environmental pollution, landscape deterioration and consumption of landfill capacity. At the European Union level, the wastes from C&DW represent one third of the total generated wastes, the average rate of collection being below 50% of what should be gathered [1]. A recent study reported that in the past years approximately 14% from the total wastes is represented by C&DW from urban areas with a strong demand for construction activities, the target imposed by legislation for their recycle and reuse at the level of 2020 year being 70% [2].

Among C&DWs the highest recycling potential is for ceramic wastes (bricks, tiles etc.), which through appropriate processing can be used as aggregates in concrete, mineral addition for clinker production or supplementary materials for construction industry [3, 4]. Thus, the use of bricks wastes in construction materials not only has a positive impact on the environment but also could improve the properties of mortar/hardened concrete due to their pozzolanic properties [5–7]. The pozzolanic properties are possessed by many natural or artificial siliceous materials, with no binding properties but capable to react with $Ca(OH)_2$ produced by cement hydration to form calcium silicate hydrate (CSH), reducing thus the porosity and improving the materials durability [8]. The pozzolanic character of different secondary materials, including ceramic wastes, was intensively studied and reported in the literature [4, 9]. The performed researches often use direct analytical methods, (XRD, TGA, SEM) to highlight and monitor the presence of $Ca(OH)_2$ and its subsequent reduction in abundance with time as the pozzolanic reaction proceeds. Since it is more difficult to quantify the pozzolanic activity of calcined clay in a Portland cement/brick powder mixture, for the present work classical determinations were combined with the individual assessment of waste powder which aimed to study the variation of the Ca^{2+} respectively OH^- ions concentration in a lime saturated solution.

2 Materials and Methods

2.1 Materials

The microstructural characteristics of different cementitious compositions influenced by pozzolanic addition were studied and the reactivity of the ancient brick powder waste (BPW) towards $Ca(OH)_2$ was determined. Considering previous results [5], the present paper highlights the hardening microstructure of cementitious materials obtained by 80% ordinary Portland cement (OPC) and 20% BPW, with w/b ratio of 0.45, compared to the etalon. The used cement, CEM I 42.5, presented a specific surface area, Ssp of 3793 cm^2/g. The BPW was prepared by crushing and milling ancient bricks from construction wastes up to advanced fines (Ssp = 3476 cm^2/g), with grain sizes ranging between 0.01 and 100 μm.

2.2 *Microstructure Particularities*

The microstructure investigations were performed on binder paste, cured up to 28 days in close vials at room temperature. For the resulted specimens, SEM investigations were performed, using a Hitachi SU-70 scanning electron microscope. Additionally, the influence of the cement substitution with BPW on the hydration and hardening processes was evaluated by Differential Thermal Analysis (DTA) analysis, using a NETZSCH STA-TG analyzer, on cement pastes hardened for 28 days by heating the samples up to 1000 °C, at a rate of 10 °C/min in nitrogen atmosphere.

2.3 *Pozzolanicity Test*

In order to assess the pozzolanicity of ancient bricks powder wastes a simple approach was practiced by using lime saturated water. The lime saturated solution was prepared by dissolving $Ca(OH)_2$ in 2 L of distilled water, until it saturated, by continuously stirring for three hours at room temperature. When the saturation concentration was achieved the solution was filtered and mixed with 20 g of BPW and further maintained for up to 28 days. The initial concentration of Ca^{2+} and OH^- from lime saturated water was determined.

3 Result and Discussion

The microstructural modifications which intervene through the BPW addition in the cementitious specimens were investigated by SEM analyses and the results are presented in Fig. 1. Portlandite (plate-like shape formations) and CSH (flakes-like structure) are the major hydration products of the OPC for both curing terms (Fig. 1a and b). Additionally, needles-like structure are visible, which indicates the presence of ettringite. The presence of BPW within cement composition favors the internal structure densification, reducing thus the porosity. The BPW powder determines supplementary CSH formation (Fig. 1c), which are more visible after 7 days of hydration as interconnected crystal network from which hexagonal CH crystals are arise in the shape of bundles (Fig. 1d).

In Fig. 2 three significant endothermic peaks can be observed, available for the two strengthen terms. The first peak, in the temperature interval of 55–150 °C, corresponds to the both physically bound water from pores and the water from calcium

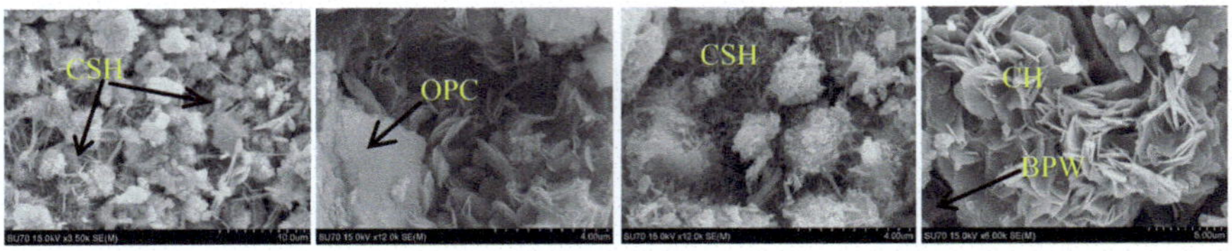

Fig. 1 SEM images of the OPC specimen hydrated for 1 (**a**) and 7 days (**b**) and OPC + BPW respectively, hydrated for 1 (**c**) and 7 days (**d**)

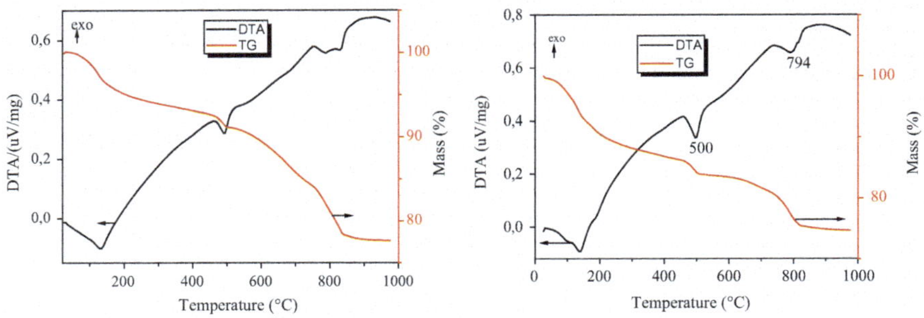

Fig. 2 DTA-TG curves of the OPC + BPW hydrated for 2 days (**a**) and 28 days (**b**)

silicate hydrates (CSH gel dehydration). The peak temperature of the CSH gel decomposition increases with hydration time from 130 °C for 2 days (Fig. 2a) to 140 °C for 28 days (Fig. 2b).

The second endothermic peak corresponds to the $Ca(OH)_2$ decomposition, its temperature increase with the hydration time, reaching 500 °C for 28 days samples. The last endothermic peak ranges between 783 °C for 2 days and 794 °C for 28 days corresponding to calcite decomposition. In order to validate the pozzolanic properties of BPW, its reactivity was assessed in lime saturated water and the test results are presented in Fig. 3.

There are clearly distinguished three variation segments of Ca^{2+} and OH^- ions concentration. In the first 2 h (the first variation segment), the Ca^{2+} concentration decrease with 15% as to the initial value of 424 mg/l. Forward, up to 24 h, the slope of the concentration curve is slightly upward, the difference between the extreme values of this segment being only 3%. This behavior could be correlated with alkali-calcium-ion exchange process in the aqueous solution, shifting to higher alkalinity and lower Ca ion concentrations, a reaction pathway which leads to decrease the Ca ions solubility [10]. However after 1 day, there are significant differences between the curves allure in terms of lime fixation, so that at the end of the test (after 28 days) the BPW fixed around 93% of the total available lime (the final value of Ca^{2+} ions concentrations was 32 mg/l). The OH^- ions concentration curve follows the same variation with that of calcium meaning that the BPW pozzolanicity is verified in the OPC + BPW system which indicates the formation of Ca–Si/Al-hydrates.

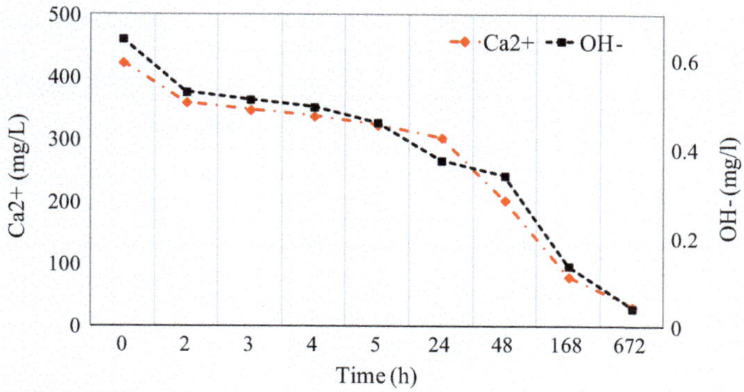

Fig. 3 Ca^{2+} and OH^{-} ions variation from lime saturated water with BPW addition

4 Conclusions

The pozzolanic properties of ancient BPW were studied. SEM analyses revealed the presence of portlandite and CSH as major hydration products as well as ettringite. The presence of BPW within cement composition determines the structure densification. DTA–TG analyses highlights three significant endothermic peaks: between 55 °C and 150 °C, corresponding to both physically bound water from pores and CSH gel dehydration, around 500 °C the $Ca(OH)_2$ decomposition take place and at 783 °C for 2 days and 794 °C for 28 days respectively the calcite decomposition is produced. The pozzolanicity test showed that after 28 days the BPW fixed around 93% of the total available lime indicating the Ca–Si/Al-hydrates formation.

1. References

1. Viitor Plus (2021) Recycling map. https://hartareciclarii.ro/homepage/stiati-ca/stiati-ca-deseurile-din-constructii/. Accessed 24 June 2021
2. Mihai FC (2019) Construction and demolition waste in Romania: the route from illegal dumping to building materials. Sustainability 11:3179
3. Sabir B, Wild S, Bai J (2001) Metakaolin and calcined clays as pozzolans for concrete: a review. Cem Concr Compos 23(6):441–454
4. Awoyera PO, Akinmusuru J, Moncea A (2017) Hydration mechanism and strength properties of recycled aggregate concrete made using ceramic blended cement. Cogent Eng 4(1). https://doi.org/10.1080/23311916.2017.1282667
5. Deák G, Panait AM, Moncea MA, Dumitru FD, Boboc M, Matei M, Olteanu MV (2019) Influence of water to binder ratio on mechanical properties of blended cements with brick powder waste. AIP Conf Proc 2129:020091
6. Cheng H (2016) Reuse research progress on waste clay brick. Procedia Environ Sci 31:218–226
7. Letelier V, Tarela E, Moriconi G (2017) Mechanical properties of concretes with recycled aggregates and waste brick powder as cement replacement. Procedia Eng 171:627–632
8. Hea C, Osbzckb B, Makovicky E (1995) Pozzolanic reactions of six principal clay minerals: activation, reactivity assessments and technological effects. Cem Concr Res 25(8):1691–1702

9. Heidari A, Hasanpour B (2013) Effects of waste bricks powder of Gachsaran company as a pozzolanic material in concrete. Asian J Civ Eng 14(5):755–763
10. Pourkhorshidi AR, Najimi M, Parhizkar T, Hillemeier B, Herr R (2010) A comparative study of the evaluation methods for pozzolans. Adv Cem Res 22(3):157–164

A Review of Natural Fiber Concrete for Radiation Shielding

Yusrina Mohd Yusof and Siti Amira Othman

Abstract As we knew, shielding is important as a protection for radiation. There is many materials used in fabrications of shielding including lead, water and natural fibre. Furthermore, various reinforced concrete has been developed around the world to improve the qualities and performance of concrete in construction and industry. Concrete is known to be strong in compression but weak in tension, which is why reinforcing is utilised to help with this. One technique to improve concrete's performance, particularly in terms of energy absorption capacity, is to incorporate fibres as reinforced concrete. Natural fibre is used to increase the concrete's strength and longevity. As a result, the performance of concrete as a radiation shielding material is automatically improved, and natural fibres are a good reinforcement material in concrete. Natural fibres as reinforcement are the safest and most environmentally friendly option because they cause no harm to the environment, are inexpensive, and may be consumed at any time. As a result, natural fibre is employed as a reinforcement in concrete to improve the shielding material's effectiveness. This paper contains reviews of the past studies undertaken by researchers over the past years or decades which has the similarity either preparation of specimens, machines or techniques used.

Keywords Cogon grass · Concrete · Radiation · Shielding

1 Introduction

All living species and humans on the planet are constantly exposed to ionizing radiation from natural radionuclides through their respiratory and digestive systems, as well as indoor regions and man-made radiological sources [1]. Everyone is aware of the dangers of radiation exposure. Despite all of radiation's drawbacks, there are some significant advantages, particularly in the study of certain biological problems and

Y. M. Yusof · S. A. Othman (✉)
Faculty of Applied Sciences and Technology, Universiti Tun Hussein Onn Malaysia, 84600 Pagoh, Johor, Malaysia
e-mail: sitiamira@uthm.edu.my

N. Mohamed Noor et al. (eds.), *Proceedings of the 3rd International Conference on Green Environmental Engineering and Technology*, Lecture Notes in Civil Engineering 214,
https://doi.org/10.1007/978-981-16-7920-9_28

the treatment of cancer. The use of rays such as x-rays and gamma rays is increasing in a variety of fields, including medical applications and the food processing industry. We must take precautionary measures to lessen the effects of radiation, as we cannot completely eliminate the risk of radiation in our daily lives. According to Azeez et al. [2], three general safety rules exist are the duration of exposure, distance and shielding. Shielding is the most important precaution to take, even if materials like lead and iron are effective ray-proofing devices, and their technical and economical usage is confined to specific locations.

The main goal for radiation protection is to reduce the intensity of the radiation to an appropriate amount (J). Costs, power and size are secondary considerations. If enough thickness is put, almost any substance can minimize radiation. The use of substances requiring extreme thickness, like basic concrete with 146 lb/ft^3 could be unaffordable due to area limits [3]. Moreover, according to Sayyed et al. [4], it has been well recognized, lead (Pb) is one of the standard shielding materials used to protect both the environment and devices from the adverse effects of ionizing radiation. Nonetheless, the use of lead can be discouraged due to other risks such as toxicity, transport and storage frequency, high cost of processing and adverse effects on the human body.

2 Radiation Shielding Concrete

Construction materials, building products and structural components are widely used as alternate gamma-ray shielding materials to lead in a wide variety of uses of nuclear science and technology. Construction materials have been of concern as sufficient shielding materials. Concrete is one of the most frequent and reasonably priced radiation shielding materials utilized in radioactive sources and radiation generating facilities. In comparison to other building materials, concrete has a strong shielding capacity against radioactive radiation and a long lifespan [5]. Concrete is known to be an outstanding and flexible shielding material as it is commonly used in the shielding of nuclear power stations, particle accelerators, research reactors, hot cells and health facilities [6].

Many researchers discovered and have been reported the advancement of radiation shielding concrete. A significant number of studies have been performed to examine the linear attenuation coefficient of radiation and the durability of the concrete made with various building materials and the shielding characteristics. According to McAlister [7], the attenuation coefficient is an amount that indicates how quickly electromagnetic radiation penetrates the material. The attenuation coefficient is most often described in terms of the unit area per mass (cm^2/g). The attenuation coefficient and material density may be used to measure the emission of gamma radiation through the chosen thickness of the shielding material or the thickness of the shielding material to reach the desired attenuation amount. Gamma attenuation coefficients are inversely proportional on gamma radiation and are directly proportional to the atomic number of the element from which the shielding substance is made. To improve radiation shielding effectiveness, altering the ratio of formulation in goods with additions of various specific densities can result in concrete with smaller diameters and bigger attenuation coefficients. Several research on the efficacy of radiation shielding for concrete dopants with particular additions have been published.

2.1 *Linear Attenuation Coefficient*

Because gamma-rays are uncharged and have no mass, they can quickly penetrate matter, according to Medhat [8], making shielding against them extremely difficult. The incoming photon energy is necessary for gamma ray interaction. The linear attenuation coefficients are frequently used to evaluate the performance of any sort of radiation shielding material [9]. The linear attenuation coefficient (μ) is the most essential quantity that characterises the penetration and diffusion of gamma radiation in the medium. It is defined as the likelihood of radiation interacting with a material per unit length of the path. The fraction of attenuated incident photons per unit thickness of a substance is also known as the linear coefficient of attenuation (μ). The fraction of photons per unit thickness of the material separated from a monoenergetic beam is reflected here. The part of the beam that is transmitted is known as the complement.

The photon energy incident, the number of atoms or an atomic number, and the density of the shielding materials all influence the size of linear attenuation coefficients. Because these linear attenuation coefficients are density dependent, they are given as mass attenuation coefficients (/), which is the linear attenuation coefficient per unit mass of the material, according to Medhat [8]. According to AKA [10], the linear attenuation coefficient of a material can be computed and measured experimentally using Beer-law. Lambert's.

Beer-rule Lambert's states that $I = I_o e^{-\mu x}$, where x is the sample thickness, I_o is the number of counts that represent the intensity of gamma-ray photons at a specific energy without attenuation, I is the gamma-ray counts that penetrated the absorber with attenuation in the sample, and cm^{-1} is the linear attenuation coefficient.

2.2 *General Properties of Concrete*

According to Machaka et al. [11], there is no question that the most commonly used material in the building industry is concrete. In 2009, it was reported that every year about 10 billion tons of concrete was poured into various forms of construction around the world. Azeez et al. [2] said that concrete composed of Portland cement, sand, aggregates composed of either blocks, gravel or others, and water, is among the most important materials used in the construction of commercial buildings.

Today, ordinary concrete (density approximate 50 kg/m^3) usually used in surface and ortho-voltage radiotherapy rooms. According to Varun Vikas Rai et al. [12], while modern concrete used today has adequate strength, the key issue is to reduce production costs, material use and environmental hazards. As a result, various materials are partly integrated or substituted in concrete to boost its efficiency, quality and energy savings.

However, because concrete is strong in compression but weak in tension, it is necessary to strengthen it in order to maximise tensile strength. Steel bars are commonly employed as alternatives to boost tensile strength, although natural fibres clearly assist in achieving enough strength. Natural fibers are those obtained from plants, animals or some other type of life that is long and slender. Various types of natural fibers are used in concrete technology as an alternative or partial replacement material for cement, aggregates and reinforcement materials. Such fibers are used to improve post cracking resistance, high energy absorbing properties, fatigue strength and other mechanical properties.

3 Natural Fibre

Natural organic fibres have been found in archaeological artifacts, showing that they were used for simple purposes. Poles, chains, records, and rudimentary suspension bridges were all made from natural fibres, and many of these relics are still in use today. Various cultures have developed new fibre sources, techniques, and applications, frequently utilising their qualities.

Natural fibres were brought back to life in the 1990s as reinforcement applications, despite a temporary stoppage of use due to the growing interest in polymeric products during World Wars I and II. Originally, recycled plastic fibres were used to meet the expanding need for flexible end-of-life items without regard for the environment. The impact of this out-of-scale growth on the ecosystem is now clear, raising awareness of the importance of using sustainable resources such as natural fibres [13].

Natural fibers are not plastic or manmade in simple language. It can come from plants, animals or minerals. In recent decades, much attention has been given to use natural fiber, both renewable and non-renewable such as oil palm, sisal, flax and jute, to build up durable materials [14]. Easier access, light in weight, the durability of raw materials, cheaper, high basic strength and steepness are several advantages of natural fibers. For many countries natural fibers are inexpensive and can be found in the local area. Fibers are like strings that can be used for different purposes. The use of natural fibers decreases the weight of the part by 10%, although the prices of the commodity are 5% below the equivalent steel reinforced material and at the same time the energy used to manufacture is 80% much smaller [15].

According to Jamir et al. [16], cellulose shall be the mixture of plant fibers, while protein-compound animal fibres. Bast, leaf, hard fiber, seed, fruit, wood, cereal straw and other fibers of grass are included in plant fibers.

Cellulose is indeed a very crystalline structure that comprises 80% of the crystalline regions. Hemicellulose is composed of strongly ramified cellulose-associated polysaccharides after the removal of pectin. Lignin becomes amorphous, tightens the cell walls and serves as a cellulose protective shield. A combination of rigid cellulose embedded in a soft lignin and hemicellulose matrix is the structure of cellulose fibre. Natural fibres are sometimes known as cellulosic or lignocellulosic fibres. Natural fibres are composites that naturally create a primary bioengineering material that differs from basic materials including cellulose, lignin, hemicellulose, wax, minerals, and other materials [17]. The fibers of vegetables normally have similar morphology. They consist of a large number of fiber cells formed by main, secondary and tertiary cell walls and lumens. A medium lamellae (ML), composed of lignin and hemicellulose, unites each fibre cell. The fiber cell numbers, the cell wall thickness, and the fiber cross-section are the distinctions among the different fiber types. Each fibers therefore has different mechanical characteristics and behaviour [18].

3.1 Natural Fibre Reinforced Concrete

It is widely acknowledged that one approach for improving the performance of concrete is to incorporate fibres into the mix. Natural fibres are frequently employed instead of waste solutions as a re-inforcing replacement [19]. Fiber insertion into the concrete matrix has long been recognised as a method of improving strength absorption [20]. However, for environmental reasons, it is strongly advised that natural fibres be used instead of synthetic fibres. Several academic research have looked into natural fibres that can be used to replace concrete, steel, polypropylene, and fibreglass. Steel, glass, and synthetic fibres come from a non-renewable source and are very expensive materials with considerable energy consumption in terms of the environment. Natural fibres, on the other hand, are a renewable resource that may be used in a variety of applications in almost every country. Natural fibres are a good alternative to hazardous plastics for the environment and can assist to reduce pollution.

These do have low cost, great mechanical properties and low energy consumption in processing. It is also possible to enhance sustainability by reducing building waste by using these materials in construction work [21]. The use of natural fibers in concrete construction is suggested as different kinds of such fibers are locally sourced and abundantly available. It is no new practice to use fibres such as straw or horsehair to increase the strength and longevity of fallen material, such as bricks and plaster. Natural fiber suitable for concrete construction and readily available in developed countries [22].

According to Zakaria et al. [23], rising demand from the construction industry is sustainable development with higher intensity. The use of natural fibres in concrete reinforcement is more likely to ensure an increase in concrete quality while minimising environmental effect and maximising the use of available natural resources. In conjunction with this, several researchers have used the fiber and the yarn as a concrete strengthening material very successfully. These fibres can help cement-based composites with post-splitting tolerance, high energy absorption, and increased fatigue resistance. The major weaknesses of synthetic fiber use are fairly expensive, health and environmental risks. Organic, low-cost, environmentally beneficial, and widely available natural fibres, on the other hand, are made from natural resources such as the coconut tree, banana tree, cotton, and jute. Numerous research have been conducted to determine the extent to which mechanical and physical behaviour can be improved. In recent years, there have been numerous initiatives to include natural fibres in order to increase energy efficiency, economics, and eco-friendliness flavour.

3.2 *Performing Chemical Treatment on Natural Fibre*

Many researchers studied natural fibers to be used in cement, mortar and concrete as a lightweight building material. Natural fibers, however, have biodegradable characteristics; the cement matrix's alkali condition contributes to fiber longevity issues. In case the fibers were not managed to withstand the impact of the alkaline attack, this gradually leads to a decrease in fibers strength and weight. Furthermore, the poor attachment between natural fibers and the cement matrix decreases the impact of fibers to increase the current composite production. Additionally, multiple treatment methods have been investigated to improve the toughness of vegetable fibers in the concrete mix. One of these is the chemicals treatment that improves the adhesion of the fiber surface and the cement matrix and thus, to reduce the moisture absorption by fiber, increase the surface sturdiness or roughness of the fiber, remove waxes and oils from the surface of the fiber and, above all, increase the durability of the fibers in the concrete composite.

The aim of the surface treatment is generally to improve both the surface morphology and the chemical composition of the fibre. The surface treatment varies accordingly from one fibre to another. Natural fibers are harvested, refined, and treated by chemical agents based on their chemical properties [24]. The following types of chemical treatment include silane, acrylate, acrylic acid, triazine, fatty acid derivatives (oloyl chloride), sodium chloride and fungal, malesated coupling agents, permanganate, acrylonitrile and acetylation grafting, stearic acid, peroxide. Alkali treatment, also known as Mercerization, is a common chemical procedure often used by researchers in fiber treatment.

According to Mohammed et al. [14], also studied the effect of alkaline treatment on surface properties of natural Iranian harvested fibers. The study has shown that alkaline treatment removes a variety of chemical components on the surface of fibres, including uranic acid (hemicellulose) and aromatic molecules released from partial depolymerisation of lignin. The chemical components of non-wood fibers are affected more effectively. The improvement in non-wood fiber crystallinity contributes to just a small rise in softwood fibers. Thus, by alkaline treatment, the basic contact of the fibers can also be enhanced dramatically as well as the fiber wettability. Alkali treatment reduces cellulose-exposure components such as hemicellulose, lignin, pectin, fat, and wax, and enhances surface roughness to strengthen interfacial adhesion, according to Pickering et al. [25]. The structure of cellulose is similarly affected by alkali treatment, with minor treatments leading to an increase in cell crystallinity, which is thought to be owing to the removal of components that could disrupt cellulose crystallinity. Treatment with alkaline resulted in improved fibre quality.

According to Li et al. [26], the major change made to hydrogen-bonding in the network structure, thus rising surface roughness, by alkaline treatment is an important improvement. This extracts a variety of lignin, wax and oils that cover the exterior of the cell wall of fiber, depolymerises cellulose and reveals crystallite longitude. Aqueous sodium hydroxide (NaOH) addition to natural fibers allows hydroxyl group ionization into alcohol.

The alkaline processing therefore affects the degree of polymerization, cellulosic fibril and the extraction of lignin and hemicellulosic compounds. Fibers are soaked for a certain amount of time in NaOH solution in alkaline water. The jute and sisal fibers are treated with 5% aqueous NaOH solution from 2 to 72 h at room temperature. There is also attempted to treat flax fiber in similar therapies. It is stated that 2% alkaline solution was for degummation and defibrillation of individual fibers for 90 s at a pressure of 200 C and 1.5 MPa.

4 Cogon Grass/"*Lalang*" Grass

Grasses usually can grow anywhere. The first thing we can think of is that weeds are growing rapidly and abundantly in the open fields. We do not generally think grasses like canes and bamboo plants are in the family of grass. The grass family comprises 635 genera's and 9000 species. After legumes, orchids and hybrid floral flora, cogon grass is the fourth largest plant family. The grass is a member of the Graminea family, a large monocotyledoneae family, a single flowering monocots class of plants with a single cotyledon of seed leaf [27].

Imperata cylindrica, often known as cogon grass or kunai grass, is a plant in the Imperata family. A rhizomatous, persistent pan-tropical grass that thrives in the damp and shallow, severely dry seasons that are common in Southeast Asia, and can grow in a wide range of soil conditions [28]. The grass, rough and annual, can be found in tropical and subtropical regions all over the country. In more than 73 nations, cogon grass is now found. Cogon grass is most visible in Malaysia as luxurious stalls of youthful-green grass, typically in the full sun, growing along the sides of roads [29]. Cogon grass is one of the ten most destructive weeds worldwide as it can populate, grow and displace ideal vegetation effectively [30].

Mohd Kassim et al. [31] stated that cogon grass, which is also known as Jap grass, blady grass or spear grass, alang-alang and lalang, belongs to the family of Poaceae (Gramineae). The grass is regarded as a perturbing weed in the world's tropical and subtropical zones. Furthermore, the ability of cogon grass to spread and replace favourable plants overestimated any soil erosion. In comparison, the rough edges of mature cogon grass leaves and silicon bodies around the leaves make it unsuitable for animals to chew the leaves and to graze wildlife. Cogon grass is widely grown in Malaysia, both on the road and in open areas. That is because of the ability of this grass to thrive in dry unproductive areas due to the strong roots that can draw water and minerals from the soil compared to other plants which might have difficulty surviving. Normally cogon grass is approximately 1 m long, with transparent and rough margins of the base.

In Malaysia, cogon grass is a plant with no commercial value that spreads fast via seeds and enormous rhizomes. Cogon grass contributes to greenhouse waste without any need [32]. Cogon grass is typically made up of three primary components consists of cellulose, hemicellulose, and lignin. This is plentiful and rarely used to render building structures such as composites strengthened by natural fibers. The cogon grass is around 40% high in cellulose [24]. Lately cogon grass is used in the Philippines' neighbouring country as a possible export material. The unregulated growth of these grasses, however, will constitute a challenge to the life of the farmers. Cogon grasses are like quiet thieves who gradually snatch the land to such a point that they decrease the crop quality. This violent rhizomatous perennial is typically thought of as pernicious plant because of its potential to disperse, colonize, spread, and then clash with and transfer preferable plants, and disturb habitats over a wide range of environmental conditions [27].

5 Conclusion

Currently, more eco-friendly materials are in high demand around the world, thus most researchers are concentrating their efforts on developing new materials that will improve the efficiency of product environmental quality. Concrete is one of the most common structural building materials nowadays. Concrete has a tremendous density, yet it has an extremely weak structure and a short lifespan. To strengthen the strength of traditional concrete, certain natural fibres were added.

The usage of cogon grass as an alternative fibre will increase resource utilisation and minimise global deforestation demand. The Imperata Cylindrica, or cogon grass, is one of the worst weeds on the planet, thanks to its capacity to quickly conquer, expand, and disturb good vegetation. There has been a lot of research into using different natural fibres as an additive in concrete to increase its strength. Moreover, a number of researchers have reported on the development of radiation shielding concrete. Unfortunately, there are few studies on the use of natural fibre in concrete as a radiation shielding material while simultaneously improving its performance. As a result, several research studies and literature have been gathered and referenced to in order to undertake this research.

Acknowledgements The authors would like to thank the Faculty of Applied Sciences and Technology for facilities provided.

References

1. Çakıroğlu M (2016) Investigation of radiation shielding properties of polypropylene fiber reinforced shotcrete. Acta Phys Polonica A 129(4):705–706. https://doi.org/10.12693/aphyspola.129.705
2. Azeez AB, Mohammed KS, Sandu AV, Al Bakri AMM, Kamarudin H, Sandu IG (2013) Evaluation of radiation shielding properties for concrete with different aggregate granule sizes. Revista de Chimie-Bucharest-Original Edition 64:899–903
3. English RJ (2020) Radiation shielding: high density concrete for shielding radiations below 1 MeV: design, construction, and application 129(2)
4. Sayyed M, Tekin H, Kılıcoglu O, Agar O, Zaid M (2018) Shielding features of concrete types containing sepiolite mineral: comprehensive study on experimental, XCOM and MCNPX results. Results Phys 11:40–45. https://doi.org/10.1016/j.rinp.2018.08.029
5. Horszczaruk E, Sikora P, Zaporowski P (2015) Mechanical properties of shielding concrete with magnetite aggregate subjected to high temperature. Procedia Eng 108:39–46. https://doi.org/10.1016/j.proeng.2015.06.117
6. Gencel O, Bozkurt A, Kam E, Korkut T (2010) Determination and calculation of gamma and neutron shielding characteristics of concretes containing different hematite proportions. Ann Nucl Energy 38(12):2719–2723. https://doi.org/10.1016/j.anucene.2011.08.010
7. McAlister DR (2012) Gamma ray attenuation properties of common shielding materials. University Lane Lisle, USA
8. Medhat M (2009) Gamma-ray attenuation coefficients of some building materials available in Egypt. Ann Nucl Energy 36(6):849–852. https://doi.org/10.1016/j.anucene.2009.02.006
9. Agar O, Tekin H, Sayyed M, Korkmaz M, Culfa O, Ertugay C (2019) Experimental investigation of photon attenuation behaviors for concretes including natural perlite mineral. Results Phys 12:237–243. https://doi.org/10.1016/j.rinp.2018.11.053
10. AKÇA B (2016) Measurement of linear attenuation coefficients of compounds of some essential major elements. J Multidiscip Eng Sci Technol 3:5003
11. Machaka M, Basha H, Abou Chakra H, Elkordi A (2014) Alkali treatment of fan palm natural fibers for use in fiber reinforced concrete Hisham Basha. Eur Sci J 1010:1857–7881
12. Varun Vikas Rai P, Pavanesh C, Niveda J, Shetty R, Poornima D (2018) An experimental study on mission grass concrete and coconut fibre reinforced concrete. Int Res J Eng Technol (IRJET) 05(05):3731–3741
13. Díaz-Ramírez G, Maradei F, Vargas-Linares G (2019) Bagasse sugarcane fibers as reinforcement agents for natural composites: description and polymer composite applications. Revista UIS Ingenierías 18(4):117–130. https://doi.org/10.18273/revuin.v18n4-2019011
14. Mohammed L, Ansari M, Pua G, Jawaid M, Islam M (2015) A review on natural fiber reinforced polymer composite and its applications. Int J Polym Sci 2015:1–15. https://doi.org/10.1155/2015/243947
15. Balasubramanian, Chandrashekaran J, Selvan S (2015) Experimental investigation of natural fiber reinforced concrete in construction industry. Int Res J Eng Technol (IRJET) 02(01):179–182
16. Jamir MRM, Majid MSA, Khasri A (2018) 8—Natural lightweight hybrid composites for aircraft structural applications. In: Woodhead publishing series in composites science and engineering, sustainable composites for aerospace applications. Woodhead Publishing, pp 155–170. https://doi.org/10.1016/B978-0-08-102131-6.00008-6
17. Bhattacharyya D, Subasinghe A, Kim KN (2015) Chapter 4: natural fibers their composites and flammability characterization. In: Multifunctionality of polymer composites. Elsevier
18. Fidelis MEA, Pereira T, Gomes O, de Andrade Silva F, Toledo Filho R (2013) The effect of fiber morphology on the tensile strength of natural fibers. J Market Res 2(2):149–157. https://doi.org/10.1016/j.jmrt.2013.02.003
19. Mahzan S, Fitri M, Zaleha M (2017) UV radiation effect towards mechanical properties of natural fibre reinforced composite material: a review. IOP Conf Ser: Mater Sci Eng 165:012021. https://doi.org/10.1088/1757-899x/165/1/012021

20. Merta I, Tschegg E (2013) Fracture energy of natural fibre reinforced concrete. Constr Build Mater 40:991–997. https://doi.org/10.1016/j.conbuildmat.2012.11.060
21. Sanjay MR, Madhu P, Jawaid M, Senthamaraikannan P, Senthil S, Pradeep S (2017) Characterization and properties of natural fiber polymer composite. J Clean Prod 172(2018):566–581. https://doi.org/10.1016/j.jclepro.2017.10.101
22. Sethunarayanan R, Chockalingam S, Ramanathan R, Natural fiber reinforced concrete. Transp Res Rec 1226:57–60
23. Zakaria M, Ahmed M, Hoque M, Islam S (2016) Scope of using jute fiber for the reinforcement of concrete material. Text Cloth Sustain 2(1). https://doi.org/10.1186/s40689-016-0022-5
24. Setyawan P, Sugiman S (2015) Mechanical properties of cogon grass (Imperata Cylindrica) fiber reinforced unsaturated polyester composites
25. Pickering K, Efendy M, Le T (2016) A review of recent developments in natural fibre composites and their mechanical performance. Compos Appl Sci Manuf 83:98–112. https://doi.org/10.1016/j.compositesa.2015.08.038
26. Li X, Tabil L, Panigrahi S (2007) Chemical treatments of natural fiber for use in natural fiber-reinforced composites: a review. J Polym Environ 15(1):25–33. https://doi.org/10.1007/s10924-006-0042-3
27. Tagal JA, Cataytay JD (n.d.) Production and evaluation of cement-bonded particle board using cogon grass as constituent
28. Haque M, Barman D, Kim M, Yun H, Cho K (2015) Cogon grass (Imperata cylindrica), a potential biomass candidate for bioethanol: cell wall structural changes enhancing hydrolysis in a mild alkali pretreatment regime. J Sci Food Agric 96(5):1790–1797. https://doi.org/10.1002/jsfa.7288
29. Mohd Kassim AS, Mohd Aripin A, Ishak N, Hairom NHH, Fauzi NA, Razali NF, Zainulabidin MH (2016) Potential of cogon grass (Imperata cylindrica) as an alternative fibre in paper-based industry. J Eng Appl Sci 11:2681–2686
30. Jumaidin R, Khiruddin M, Asyul Sutan Saidi Z, Salit M, Ilyas R (2020) Effect of cogon grass fibre on the thermal, mechanical and biodegradation properties of thermoplastic cassava starch biocomposite. Int J Biol Macromol 146:746–755. https://doi.org/10.1016/j.ijbiomac.2019.11.011
31. Mohd Kassim AS, Mohd Aripin A, Ishak N, Zainulabidin M (2015) Cogon grass as an alternative fibre for pulp and paper-based industry: on chemical and surface morphological properties. Appl Mech Mater 773–774:1242–1245. https://doi.org/10.4028/www.scientific.net/amm.773-774.1242
32. Ibrahim S, Baharuddin S, Ariffin B, Hanafiah M, Kantasamy N (2018) Cogon grass for oil sorption: characterization and sorption studies. Key Eng Mater 775:359–364. https://doi.org/10.4028/www.scientific.net/kem.775.359

Comparing the Physical Properties of Coal Bottom Ash (CBA) Waste and Natural Aggregate

Syakirah Afiza Mohammed, Mohamed Reyad Alhadi Ahmad, Norlia Mohamad Ibrahim, Nur Liza Rahim, Roshazita Che Amat, and Salmi Samsudin

Abstract Coal bottom ash (CBA) is a co-combustion product material, which may cause hazards to human health and the environment. Rapid growth in technology causes the increase of CBA waste and this situation led to a waste disposal crisis. Reuse waste material as an alternative material instead of natural materials can led to sustainable and environmentally friendly construction. The main objective of this study is to determine the physical properties of CBA and its suitability to be used as replacement material in civil construction. The physical properties test conducted in this research were aggregate impact value test, aggregate crushed value test, flakiness and elongation test. The results show that the ability of CBA to resist sudden shock and repeated load was lesser than natural aggregate (NA). The differences of aggregate impact value (AIV) and aggregate crushing value (ACV), between NA and CBA were 50.4% and 48.9%, respectively. In addition, CBA has higher amount of flaky

S. A. Mohammed (✉) · M. R. A. Ahmad · N. M. Ibrahim · N. L. Rahim · R. C. Amat
Faculty of Civil Engineering Technology, Universiti Malaysia Perlis, 02600 Arau, Perlis, Malaysia
e-mail: syakirahafiza@unimap.edu.my

N. M. Ibrahim
e-mail: norlia@unimap.edu.my

N. L. Rahim
e-mail: nurliza@unimap.edu.my

R. C. Amat
e-mail: roshazita@unimap.edu.my

N. M. Ibrahim · N. L. Rahim · R. C. Amat
Sustainable Environment Research Group (SERG), Centre of Excellence Geopolymer and Green Technology (CEGeoGTech), Universiti Malaysia Perlis, 01000 Kangar, Perlis, Malaysia

S. A. Mohammed
Water Research Group (WAREG), Universiti Malaysia Perlis (UniMAP), 02600 Arau, Perlis, Malaysia

S. Samsudin
Stesen Janakuasa Sultan Azlan Shah, TNB Janamanjung Sdn Bhd, Jalan Semarak Api, Teluk Rubiah, 32040 Seri Manjung, Perak, Malaysia
e-mail: salmis@tnb.com.my

N. Mohamed Noor et al. (eds.), *Proceedings of the 3rd International Conference on Green Environmental Engineering and Technology*, Lecture Notes in Civil Engineering 214,
https://doi.org/10.1007/978-981-16-7920-9_29

and elongated particles compared to NA. The flakiness index value for NA and CBA were 7.12% and 26.10%, respectively while the difference value of elongation index between NA and CBA was 37%. However, even though the properties of CBA were not as good as NA, the results for ACV and the flakiness index of CBA meet the minimum requirement of Jabatan Kerja Raya (JKR) Standard Specification which indicates that CBA has potential to be used in civil construction.

Keywords Coal bottom ash · Waste · Physical properties · Natural aggregate

1 Introduction

In Malaysia, every state has its own quarry and natural aggregate (NA) production activities. NA production in Selangor is the biggest among all followed by Perak, Sarawak, Johor, Sabah, Negeri Sembilan, Terengganu, Pulau Pinang, Kedah, Pahang, Kelantan, Melaka, and Perlis [1]. The factors contribute to the production of NA are the population growth, the urbanization and the economic growth which led to the high demand in construction activities. The impact of the accelerating of these factors is feared to affect the ecological balance. Therefore, in considering the high demand of the NA to be used in construction industries, researchers were seeking for alternative material as a replacement for better or similar construction performance as produced by NA. Currently, the common waste materials used in civil construction are recycled concrete [2], recycled asphalt [3], steel slag [4], palm oil clinker [5] and glass [6].

On the other hand, due to the developing interest for electricity from coal sources, high amount of coal bottom ash (CBA) waste was produced. This condition has caused additional of landfill which led to the detrimental of land use. Therefore, considering the problem with the disposal ash pond in terms of cost and environmental impact, the need to recycle it has become a necessity than a desire. A limited research work is reported on the utilization of CBA. Previous studies on the utilization of coal ash waste as material in the construction field have focused more on fly ash than bottom ash. However, recent study reported that bottom ash possesses some desirable engineering properties that enable to be used as a construction material [7]. In this study, the physical properties of CBA waste were investigated. Then, physical properties of CBA waste and NA were compared to determine the potential of CBA waste to be used in civil construction industry.

2 Materials and Method

Material used in this research is NA and CBA which was collected from Janamanjung Power Plant in Perak. Figure 1 shows the coarse and fine particles of CBA. By observation, the CBA appearance is similar with NA which has darker color.

(a) (b)

Fig. 1 **a** Coarse CBA . **b** Fine CBA

2.1 Aggregate Impact Value

Sample was sieved and dried in the oven for 4 h, cool down to room temperature, and fill in the cylinder measure, tamped on each layer 25 times, stacked off the surface of the sample to the upper level, and weighed. Then the cylinder was placed at the base of the machine and the hummer fall freely 15 times at the height of 38 cm. The sample was extracted, sieved and weighted.

$$\text{AIV} = \frac{\text{Weight of sample passing sieve 2.36 mm}}{\text{Weight of sample}} \times 100 \quad (1)$$

2.2 Aggregate Crushed Value

Prior testing, the sample was sieved and dried for 4 h (W1). Samples was filled in the cylinder measurement and tamped 25 times on each layer, stack off the surface of the sample to the upper level and weighed. The sample loaded consistently in 10 min. The load was then released, and the crushed material was removed and sieved. Material passing through sieved 2.36 mm was weighted (W2).

$$ACV = \frac{W2}{W1} \times 100 \quad (2)$$

2.3 Flakiness and Elongation

The sample (CBA and NA) was oven dried and sieved. The flaky and elongate material was separated using thickness gauge. The flaky and elongate material passing the gauge to an accuracy of at least 0.1% of the test sample was weighted. The flakiness and elongation index were then calculated.

$$Flakinessindex = \frac{Weight\,of\,flaky\,particles}{Total\,weight\,retained} \times 100 \tag{3}$$

$$Elongation\,index = \frac{Weight\,of\,elongate\,particles}{Total\,weight\,retained} x100 \tag{4}$$

3 Result and Discussion

3.1 Aggregate Impact Value

The aggregate impact value provides a relative measure of the resistance of an aggregate to sudden shock or impact. Resistance of the aggregates to impact is also known as toughness. The aggregates used in civil construction should have sufficient toughness to resist their disintegration as the impact load can break the aggregate into smaller pieces which subsequently may lead to the structure failure.

The result of aggregate impact value of NA and CBA waste was presented in Fig. 2. The result shows that the aggregate impact value of NA is 17.2% which can be classified as strong material while CBA is 34.7% which can be classified as weak materials. The results show that the ability to resist sudden impact or shock load of the NA is better than CBA.

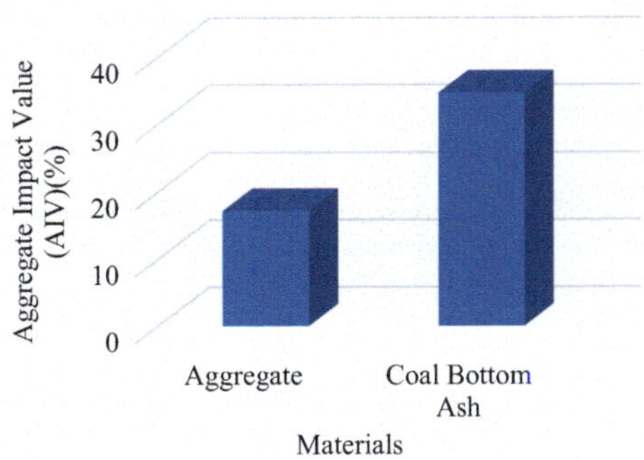

Fig. 2 Aggregate impact value (%) of NA and CBA waste

Table 1 Aggregate crushing value (%) of NA and CBA waste

Properties	Types of materials		JKR Limit	
	NA	CBA	Building work	Road work
ACV (%)	18.5	36.2	≤ 40	≤ 30

3.2 Aggregate Crushing Value

The aggregate crushing value measure the resistance of a material to crush under gradually applied compressive load. In other word, the crushing value of aggregates indicates its strength. Lower crushing value is recommended as it indicates a lower crushed fraction under load which offer a longer service life and an economical performance.

Table 1 shows the aggregate crushing values of NA and CBA waste. The results shows that the aggregates crushing value of CBA is poorer compared to NA. The ACV for NA is 18.5% while ACV for CBA is 36.2%. According to JKR Standard Specification for building work, the ACV value of CBA is in the allowable range as the value is not exceeding 40% [8]. However, the ACV value of CBA is not meet the requirement of JKR Standard Specification for road works as the value is exceeding 30% [9].

3.3 Flakiness and Elongation

Flakiness and elongation index test gives the percentage of flaky and elongate aggregate present in the total aggregate samples. In concrete production, flaky and elongated aggregates caused more water needed to produce workable concrete compared to smooth and rounded compacted aggregate. While in pavement construction, too much of flaky and elongated aggregates may lead to the failure of the pavement due to their random position under repeated loading and vibration.

Figure 3 shows the flakiness and elongation index of NA and CBA waste. The flakiness and elongation index of NA is 7.12% and 23.76%, respectively. For CBA, the flakiness and elongation index are higher, 26.10% and 37.85, respectively. These value meets the requirement of JKR standard except elongation index for CBA which the value exceeding 30%.

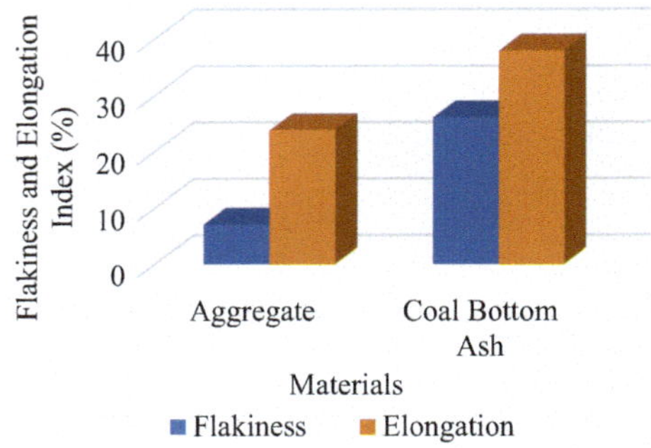

Fig. 3 Flakiness and elongation index (%) of NA and CBA waste

4 Conclusion

In conclusion, the physical properties of CBA were not as good as NA. CBA has lesser ability to resist the sudden shock or impact compared to NA. In addition, the ability of the CBA to resist the repeated load also poorer compared to NA. However, the crushing value of the CBA is meet the minimum requirement of JKR Standard Specification for building works and it indicates that CBA has potential to be used in concrete production. Higher flaky and elongate particles of CBA might be influenced the performance of the concrete, pavement or any civil construction.

Acknowledgements The authors wished to acknowledge the RESMATE 9001-00623 grant provided by Universiti Malaysia Perlis for the financial support.

References

1. Ismail S, Hoe KW, Ramli M (2013) Sustainable aggregates: The potential and challenge for natural resources conservation. Procedia Soc Behav Sci 101:100–109. https://doi.org/10.1016/j.sbspro.2013.07.183
2. Nwakaire CM, Yap SP, Yuen CW, Onn CC, Koting S, Babalghaith AM (2020). Laboratory study on recycled concrete aggregate-based asphalt mixtures for sustainable flexible pavement surfacing. J Clean Prod, 262, 121462. https://doi.org/10.1016/j.jclepro.2020.121462
3. Devulapalli L, Kothandaraman S, Sarang G (2020). Effect of rejuvenating agents on stone matrix asphalt mixtures incorporating RAP. Constr Build Mater, 254, 119298. https://doi.org/10.1016/j.conbuildmat.2020.119298
4. Pang B, Zhou Z, Xu H (2015) Utilization of carbonated and granulated steel slag aggregate in concrete. Constr Build Mater 84:454–467. https://doi.org/10.1016/j.conbuildmat.2015.03.008
5. Babalghaith AM, Koting S, Sulong NHR, Karim MR, Mohammed SA, Ibrahim MR (2020) Effect of palm oil clinker (POC) aggregate on the mechanical properties of stone mastic asphalt (SMA) mixtures. Sustainability 12(7):2716. https://doi.org/10.3390/su12072716
6. Adhikary SK, Rudzionis Z (2020) Influence of expanded glass aggregate size, aerogel and binding materials volume on the properties of lightweight concrete. Materials Today: Proceedings 32:712–718. https://doi.org/10.1016/j.matpr.2020.03.323
7. Yoo BS, Park DW, Vo HV (2016) Evaluation of asphalt mixture containing coal ash. Transp Res Procedia 14:797–803. https://doi.org/10.1016/j.trpro.2016.05.027

8. Jabatan Kerja Raya Malaysia JKR (2014) Standard Specification for Building Works. Concrete works
9. Jabatan Kerja Raya Malaysia JKR (2008) Standard Specification for Road Works. JKR/SPJ/rev2005.

Effect of Bottom Ash and Limestone on the Optimum Binder Content in Hot Mix Asphalt (HMA)

Nur Liza Rahim, Syakirah Afiza Mohammed, Noor Aina Misnon, Nurhidayah Hamzah, Roshazita Che Amat, Norlia Mohamad Ibrahim, Christina Remmy Entalai, and György Deák

Abstract One of the most effective and simplest methods to minimize waste as well as reduce the environmental problems associated with waste disposal is by utilizing waste materials as a cement replacement in hot mix asphalt (HMA) mixtures which can provide the same or better stability as the conventional method. Fillers play an important role in the stability and strength of the pavement by filling voids between the aggregate particles in the performance of the HMA mixture. This research investigated the effect of the utilization of different types of filler (bottom ash and limestone) on the optimum binder content of HMA. Flow, stability, stiffness, air void in

---

N. L. Rahim (✉) · S. A. Mohammed · R. C. Amat · N. M. Ibrahim · C. R. Entalai
Faculty of Civil Engineering Technology, Universiti Malaysia Perlis, Kompleks Pusat Pengajian Jejawi 3, 02600 Arau, Perlis, Malaysia
e-mail: nurliza@unimap.edu.my

S. A. Mohammed
e-mail: syakirahafiza@unimap.edu.my

R. C. Amat
e-mail: roshazita@unimap.edu.my

N. M. Ibrahim
e-mail: norlia@unimap.edu.my

N. L. Rahim · R. C. Amat · N. M. Ibrahim
Sustainable Environment Research Group (SERG), Center of Excellence Geopolymer and Green Technology, Universiti Malaysia Perlis, 01000 Kangar, Perlis, Malaysia

N. A. Misnon
Department of Civil Engineering, Faculty of Engineering, Universiti Pertahanan Nasional Malaysia, Kuala Lumpur, Malaysia
e-mail: nooraina@upnm.edu.my

N. Hamzah
School of Civil Engineering, College of Engineering, Universiti Teknologi MARA, 40450 Shah, Alam, Selangor, Malaysia
e-mail: nurhidayah0527@uitm.edu.my

G. Deák
National Institute for Research and Development in Environmental Protection Bucharest (INCDPM), Splaiul Independentei Street, 6th District, 294 Bucharest, Romania

N. Mohamed Noor et al. (eds.), *Proceedings of the 3rd International Conference on Green Environmental Engineering and Technology*, Lecture Notes in Civil Engineering 214,
https://doi.org/10.1007/978-981-16-7920-9_30

mix (VIM) and void filled with bitumen (VFB) were determined using the Marshal Method test in order to determine the optimum binder content of HMA for all mineral filler. The results of the Marshall test for each filler have been compared with the JKR standard specification. The optimum binder content for bottom ash, limestone and Ordinary Portland Cement (OPC) was 5.42%, 5.65% and 5.54%, respectively. All values of mineral filler used meet the JKR standard specification, where the range is between 4 and 6%. From the result achieved, the bottom ash has the lower optimum binder content value compared to the limestone and OPC. When the lower binder content is used in the bituminous mixture, the cost for pavement construction will be reduced.

Keywords Hot mix asphalt · Mineral filler · Optimum binder content · Limestone · Bottom ash

1 Introduction

Transportation infrastructure is currently one of the essential infrastructures for increasing society's mobility for economic activity. Road is very important in our daily activities because it enables people to move from one location to another location. Maximum road maintenance is required since the current road users are increasing in using the highways and roads. The main step of improving the performance of an existing road starts with ensuring the correct design of the mix and the use of appropriate materials [1].

Modifying asphalt mixture has been studied for a long time. There were various methods have been developed to improve the asphalt mixture in order to improve the efficiency and consistency of hot mix asphalt (HMA) by modifying the HMA constituting ingredients. There are four major asphalt production cost categories: materials, plant production, trucking and lay down (i.e., construction). Materials are the most expensive production cost category, comprising about 70 percent of the cost to produce HMA. Thus, precautions must be taken when selecting the types of ingredient materials and the major constituents in the mixture during the HMA mix design process. This enables the mix designer to achieve the required mixture property.

Mineral fillers consist of inert minerals added to asphalt mixtures to improve the density and strength of the mixture. The example of mineral fillers is rock dust, slag dust, hydrated lime, hydraulic cement, fly ash, loess, or other suitable mineral matter. The effect of the added mineral fillers on HMA is to fill voids between aggregate and reduce the voids in the mixture. Filler has various purposes, which fill voids and reduce optimum asphalt content and increase stability [2]. Besides, fillers can reduce the use of expensive binder material, increase some properties of the blend material and are also environmentally friendly.

Nowadays, the road pavement or asphalt surface is exposed to higher loading due to the increasing numbers of transportation or vehicles on the road. Thus, the failures of pavement such as deformation or rutting, fatigue, cracking and skidding problems have increased significantly over the year due to the increase in road traffic, which is proportional to the insufficient degree of maintenance [3]. The road pavement has low durability and workability. The problems regarding road pavement can lead to any unfortunate incident [4]. For these several decades, lots of testing and research has been done to find the most suitable alternatives material to improve the road pavement. One of the common alternative ways taken into concern is by manipulating filler materials in the asphalt mix design. The modification of asphalt mix design materials used can improve the behavior of mixes which provide a better service life of the road pavement and strengthen the mix.

In order to achieve the better construction of pavement in Malaysia, a mineral filler such as limestone and fine bottom ash can be used as a cement replacement in optimum binder content of hot mix asphalt [5]. The overall objective of this paper is to investigate the effect of the utilization of different types of filler on the optimum binder content of hot mix asphalt. One of the most effective and simplest methods to minimize waste as well as reduce the environmental problems associated with waste disposal is by utilizing waste materials as a cement replacement in HMA mixtures. So, this research aims to reduce the use of Ordinary Portland Cement (OPC) by replacing it with waste material (bottom ash and limestone) that can give the same strength as the conventional method.

2 Materials and Method

2.1 Material

Hot mix asphalt (HMA) basically consists of two main components, which are aggregate and asphalt binder. However, the aggregates are often classified into fine and coarse aggregates and filler percentages due to their particle size [6]. The characteristics of aggregates are very important to the performance of HMA pavements. Aggregate shape properties, such as form, angularity, and surface texture, highly influence the performance of HMA. A significant proportion of mineral aggregate is present in a bituminous mixture, approximately 95% of the weight and 80% of the volume [7].

2.1.1 Aggregates

In this study, the aggregate used is the fine aggregate (sand). Based on the standard specification for road works published by Jabatan Kerja Raya in 2008, fine aggregates would be non-plastic and free foam clay, loam, aggregations of material, vegetative and other organic matter and other deleterious substances. For the mix design, the wearing course (ACW14) is used. The particle size distribution of aggregate of wearing course (ACW14) must follow the standard specification for road work.

2.1.2 Bitumen

In this study, the bitumen grade used is 60/70 since it is widely used in Malaysia. In order to produce the mix design and achieve the optimum binder content, different value of the bitumen content required must follow the standard specification for road works. For this study, the standard bitumen content used for the wearing coarse (AC14) is between 4.0–6.0% [8]. Bitumen used in this research has been obtained from Pens Industries Sdn Bhd.

2.1.3 Limestone

Limestone is defined as a sedimentary rock composed primarily of calcium carbonate (C_aCO_3) in the form of the mineral calcite. The category of limestone used is extracted from natural minerals by mechanical sieving after quarry operation. The limestone used in this research has been obtained from Pens Industries Sdn. Bhd. The quarry is located at Jalan Bukit Ayer, Batu Pahat, Perlis.

2.1.4 Bottom Ash

Bottom ash (BA) used in the research has been obtained from TNB Janamanjung Sdn. Bhd., a thermal power plant at Perak, Malaysia, as shown in Fig. 1. The BA

Fig. 1 Samples of Bottom Ash (BA)

particles are relatively coarse and have angular particles with very porous surface textures. The particle sizes range from as coarse as fine gravel to fine sand (<10 mm to 75 μm) [9]. The BA has been sieved to remove the coarser particle. Subsequently, the BA has been ground to obtain finer particle size; sufficient grinding can improve the pozzolanic activity of the bottom ash [10].

2.2 Method

This research focused on the Marshall properties that were prepared in the laboratory using different types of mineral fillers that passed through 0.075 mm (all P-200) sieve, namely limestone, bottom ash and Ordinary Portland Cement (OPC) as control filler. The optimum binder content and varied mixing characteristics of criteria were discovered using the Marshal Mix design technique. In the Marshall Mix design test, the following criteria are taken into account: stability, flow, stiffness and air void in mix (VIM) and void in aggregate filled with bitumen (VFB), as shown in Fig. 2.

The bitumen penetration test is based on ASTM D5. The purpose of the penetration test used in this study is to determine the uniformity of a bitumen material where was expressed as the depth a standard needle can penetrate a sample material within ten millimeters under a specified time in 5 s, 25 °C of temperature and loading of the needle assembly is 100 g. Three samples of penetration value of bitumen were

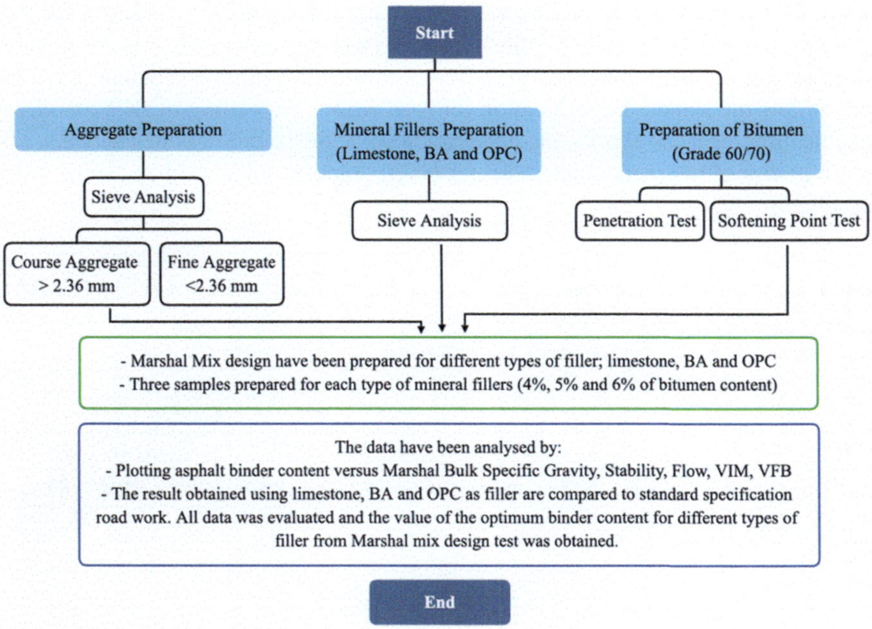

Fig. 2 Summary of Methodology

tested to get the mean value. Determination of three points on each of the sample's surfaces must more than 10 mm from the edge of the penetration tin and not below 10 mm apart from the other points were done.

A total of 1200 g of the sample aggregate were collected and heated in the oven until constant mass. The aggregate was heated to a temperature of 150 °C to 190 °C. Bitumen mix was added in increments of 1.0% starting from 4%, 5% and 6%. Limestone, bottom ash and OPC used as mineral filler were mix with different binder content. Three samples were prepared for each mineral filler with different binder content by compacting the samples to 75 blows on both sides of the sample by using Marshall Compactor [11]. Then, the sample was demoulded after 24 h left at room temperature. The laboratory testing results will be based on the Jabatan Kerja Raya (JKR) standard specification testing of road improvements. The bituminous mixture properties include stability, flow, air void in mix (VIM), void filled bitumen (VFB) and stiffness.

3 Result and Discussion

3.1 Penetration Test

A penetration test is the most widely used method to evaluate the consistency of a bituminous material at a certain temperature. Table 1 shows the penetration test result for bitumen used in this study. The average penetration value of the bitumen for the three samples was taken, which is 66.16 mm. In order to produce the mix design and achieve the optimum binder content, different value of the bitumen content required must follow the standard specification for road works. According to JKR Standard Specification for Road Works, the penetration value 66.16 mm was lies between 60 and 70 grades of bitumen which indicates 60/70 grade bitumen, so they can be used in this research.

Table 1 Result for penetration test

Standard penetration test: ASTM D5 Type of bitumen: 60/70 grade penetration, Load: 100 g, Time: 5 s, Temperature: 25 °C									
Sample No	A			B			C		
Point	1	2	3	1	2	3	1	2	3
Penetration (mm)	65.65	60.78	68.80	65.50	62.80	68.50	67.58	62.80	69.50
Average Penetration (mm)	66.16								

Table 2 Result for softening point tests

Softening point test: ASTM D36
Type of bitumen: 60/70 grade penetration

Sample No	A		B		C	
Ball	1	2	1	2	1	2
Softening point (°C)	50	51	49	50	49	50
Average Softening Point (°C)	50.5		49.5		49.5	

3.2 *Softening Point Test*

The softening point test was tested based on ASTM D36 using the Ring-and-Ball method at temperatures between 30 to 157 °C. Three samples of bitumen were tested to determine the mean value of the softening point. Table 2 displays the results of the softening point test.

From the result obtained in Table 2, the mean value of the three samples is 49.8 °C, where the bitumen began to show fluidity until it touched the bottom plate. Based on the JKR Standard Specification for Road Works, the softening point of 60–70 grade bitumen must be more than 48 °C and less than 56 °C, as well as the difference of temperature between Ring 1 and Ring 2 as steel ball passes through the bitumen until touches the bottom plates does not exceed 2 °C. Thus, the softening values for samples A, B and C are within the required range, so they can be utilized in this research.

3.3 *Bulk Specific Gravity*

The value of bulk specific gravity of the sample was calculated by weighting the sample in water and in air. The effect of the utilization of different types of filler in different percentages of binder content on the bulk specific gravity of the bituminous mixture was shown in Fig. 3. The bulk specific gravity for each filler showed an increase as the binder content increased from 4 to 6%. This is due to the bitumen filling in the void space of the aggregate particles. The JKR standard specification pavement design does not have a precise amount of bulk-specific gravity value. However, the maximum value for bulk specific gravity was observed to be 6% bitumen content.

3.4 *Stability*

Figure 4 demonstrates the stability graphs for the effect of various binder content and the filler types. The stability value was recorded directly from the Marshall

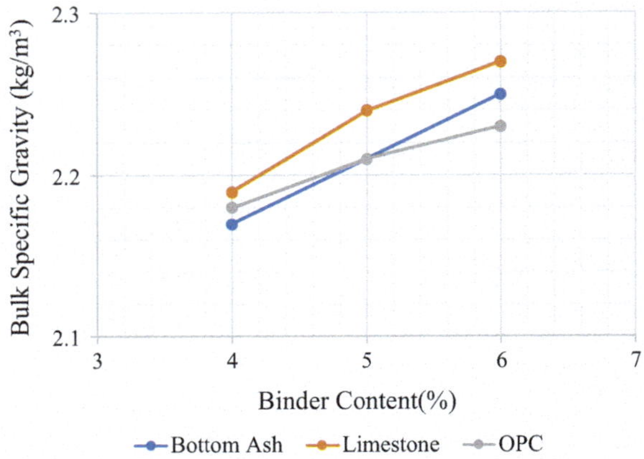

Fig. 3 Bulk Specific Gravity versus Binder Content

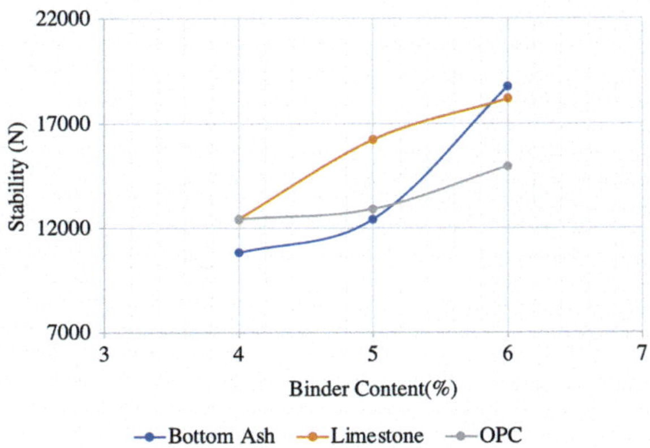

Fig. 4 Stability versus binder content

Testing Machine. The stability values indicate the maximum load that the sample can withstand before failure. By referring to Fig. 4, it can be observed that according to the bitumen content in the mixture, the impact of all fillers on stability follows the same trend, whereas the binder content in the mixture increase, stability values also increase as well. This is because when the binder content increases to the maximum value, the binder has the ability to fills the gap between the aggregate particles, which results in increasing the stability. Thus, due to the presence of binder content in the void between the aggregate particles, the strength of the binder mixture contact between the aggregates will continue to be built [12].

Based on the standard specification for the road work that has been set up by JKR, the stability value required for wearing course should be more than 8000 N. Thus, bottom ash, limestone and OPC as mineral filler in the bituminous mixture were approved for use in pavement construction based on their stability parameters value that meets the standard specification for road works as stated by JKR.

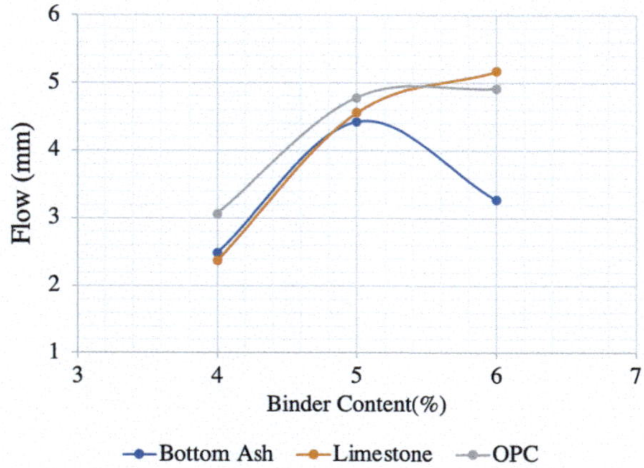

Fig. 5 Flow versus binder content

3.5 Flow

The flow value represents the vertical deformation when the maximum load is attained. As shown in Fig. 5, when binder content used in the mixture was increased from 4 to 6% for both limestone and OPC, the flow value obtained also increased except for the mixture with bottom ash, where the flow value increased by 4% and 5% of binder content then decrease at 6% of binder content. JKR indicated that the flow value for standard specifications for road improvements was in the range of 2.0–4.0 mm. Therefore, the mineral fillers that meet the JKR standard specification for road works are at 4% binder content for bottom ash, limestone and OPC, while for 6% binder content, only bottom ash fillers meet the requirement.

3.6 Air Void in Mix (VIM)

Figure 6 shows the percentage of air void in mix (VIM) against the binder content. It can be seen that the mixture with bottom ash at 4% of binder content shows the highest percentage of VIM compared to the other mineral fillers, while the lowest percentage of VIM is OPC at 6% of binder content. From the figure, it can be observed that the percentage of VIM for bottom ash, limestone and OPC are decreasing as the binder content increasing. This is because binder content in the mixture acts as a lubricant which can lower the VIM value. The presence of air void in the mixture reflects its porosity. If the quantity of the air voids in the mixture exceeds the acceptable value, the mixture may crack since there is an insufficient binder to coat all of the aggregate particles. However, there is not enough area to retain the binder if the air voids in mix present too low where can cause bleeding or plastic flow of binder known as abrasion. The proportion of air void in the mixture must be between 3.0 and 5.0%, as

Fig. 6 VIM versus binder content

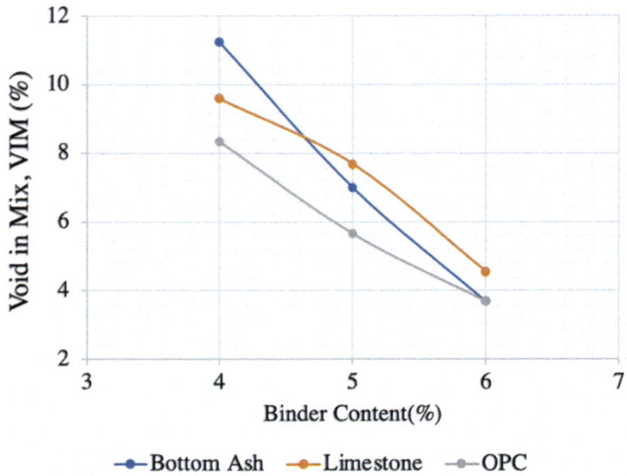

mentioned in the JKR standard specification of road word. Bottom ash, limestone and OPC in the mixture of 6% binder content are within the range of JKR requirement.

3.7 Void Filled Bitumen (VFB)

The impact of various types of fillers on the void filled bitumen (VFB) performance of a mixture is indicated in Fig. 7. As stated in the standard specification of road works from JKR, the value of void filled with bitumen is between 70 and 80%. Thus, the void filled with bitumen values for bottom ash, limestone and OPC as mineral filler in 6% bitumen content are within the range. This parameter has to do with the amount of workable bitumen content in the mixture. There will be less

Fig. 7 VFB versus Binder Content

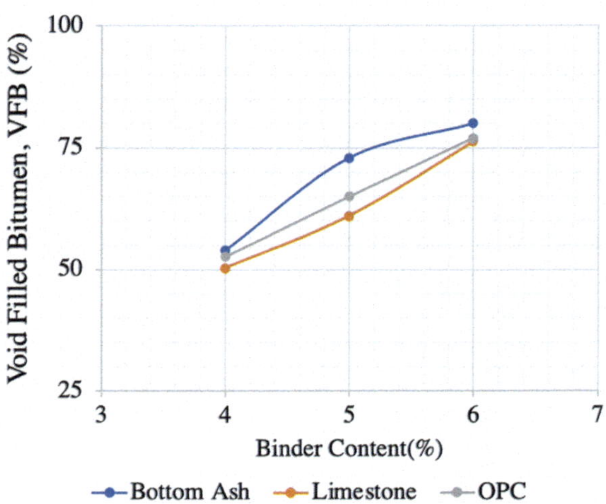

asphalt coating around the mineral aggregates if the number of voids filled with bitumen is less than the standard range. Therefore, lower bitumen coatings are more susceptible to moisture and weather influences, which can cause them to separate from the mineral aggregate, resulting in reduced performance.

3.8 Optimum Binder Content (OBC)

In the bituminous mixes, the percentage of binder content should be not too high or too low because it can affect the mixture. By adding enough binder content in the bituminous mixture until the aggregate is fully coated with binder and void within the bituminous material is the objective of optimum binder content design. Thus, the impermeability achieved can improve the bituminous pavement durability. Additionally, to maintain the aggregate from being pulled by the rough action of driving vehicles on the carriageway, the minimum binder content in the bituminous mixture is needed. However, the bituminous pavement will become brittle if the binder content is too high. This is because the use of excessive binder content reduces the resilience of bituminous pavement to deformation under traffic stress.

All data were evaluated and the value of the optimum binder content for different types of filler from the Marshal mix design test was obtained using all graphs presented (Fig. 3, 4, 5, 6 and 7). The average of five optimum binder content has been used to determine the mean optimum binder content as shown in Table 3.

Based on Table 3, the mean value optimum binder content for each mineral fillers for bottom ash, limestone and OPC was 5.42%, 5.65%, 5.54%, respectively. According to the JKR standard, the range of binder content for wearing course is within 4.0–6.0%. Thus, the optimum binder content for all mineral fillers meets the JKR requirements. All mineral fillers can be used for wearing course in performance analysis of hot mix asphalt. From the result achieved, the bottom ash has the lower optimum binder content value compared to limestone and OPC. When the lower binder content is used in the bituminous mixture, the cost for pavement construction will be reduced. This result is in agreement with previous studies done by [13].

Table 3 Optimum binder content

Percentage of binder content	Bottom ash	Limestone	Ordinary portland cement (OPC)
Bulk specific gravity at the peak curve	6.00	6.00	6.00
Stability at the peak curve	6.00	6.00	6.00
Flow at 3 mm	4.25	4.35	4.00
4% of VIM	5.85	6.00	5.80
75% of VFB	5.00	5.90	5.90
Average optimum binder content	5.42	5.65	5.54

4 Conclusion

This research was conducted to investigate the effect of the utilization of different types of mineral filler on the optimum binder content of hot mix asphalt. The aim of this research has been accomplished by conducting several testings in the lab and comparing the results obtained with the criteria in the standard specification JKR for road construction. The method used in this researched was a Marshall mix design test to determine the optimum binder content of hot mix asphalt for different types of filler with 4.0%, 5.0% and 6.0% of binder content. From the finding of the Marshall test, the optimum binder content of hot mix asphalt for the bituminous mixture with a different type of mineral filler meets the standard specification requirements. The optimum binder content for bottom ash, limestone and Ordinary Portland Cement (OPC) was 5.42%, 5.65% and 5.54%, respectively. All values of mineral filler used within the range of JKR standard specification (4–6%). From the result achieved, the bottom ash has the lower optimum binder content value compared to limestone and OPC.

References

1. Al-rubaie MF, Jahad IY (2011) The effect of using glass powder filler on hot asphalt concrete mixtures properties. Eng Tech J 29, 1, 2011. 29(1), 44–57
2. Chen Y, Xu S, Tebaldi G, Romeo E (2020) Role of mineral filler in asphalt mixture. Road Materials and Pavement Design, 0(0), 1–40
3. Al-Qadi IL, Elseifi MA, Yoo PJ (2004) Pavement Damage Due to Different Tires and Vehicle Configurations. Final Report Michelin Americas Research and Development Corporation, 515. Blacksburg, VA: Virginia Tech Transportation Institute
4. Al-Qadi IL, Hernandez JA, Gamez A, Ziyadi M, Gungor OE, Kang S (2018) Impact of Wide-Base Tires on Pavements: A National Study. Transportation Research Record, 2672 (40). https://doi.org/10.1177/0361198118757969
5. Hamzah MO, Gungat L, Golchin B (2016) Estimation of optimum binder content of recycled asphalt incorporating a wax warm additive using response surface method. Int J Pavement Eng. 1–11. https://doi.org/10.1080/10298436.2015.1121779
6. Sutradhar D, Miah M, Chowdhury GJ, Sobhan MA (2015) Effect of using waste material as filler in bituminous mix design. Am J Civil Eng 3(3), 88–94. https://doi.org/10.11648/j.ajce.20150303.16
7. Aodah HH, Kareem YNA, Chandra S (2012) Performance of Bituminous Mixes with Different Aggregate Gradations and Binders. International Journal of Engineering and Technology 2(11):1802–1812
8. Standard Specification for Road Works (2008) Section 4: Flexible Pavement. Jabatan Kerja Raya. (JKR/SPJ/2008-S4)
9. Abubakar, A. U., & Baharudin, K. S. (2013). Tanjung bin coal bottom ash: From waste to concrete material. Advanced Materials Research, 705(March 2016), 163–168. https://doi.org/10.4028/www.scientific.net/AMR.705.163
10. Kula I, Olgun A, Erdogan Y, Sevinc V (2001) Effects of colemanite waste, cool bottom ash, and fly ash on the properties of cement. Cem Concr Res 31:491–494. https://doi.org/10.1016/S0008-8846(00)00486-5

11. American Society for Testing and Materials, ASTM D1559 (1989). Resistance to Plastic Flow of Bituminous Mixtures using Marshall Apparatus, ASTM International, West Conshohocken, PA, USA.
12. Vasudevan G (2017) Effect on Coal Bottom Ash in Hot Mix Asphalt (HMA) as Binder Course. Lecture Notes in Engineering and Computer Science 2228(6):1046–1050
13. Vasudevan, G. (2013). Performance on Coal Bottom Ash in Hot Mix Asphalt. Int J Res Eng Technol 02(08), 24–33. https://doi.org/10.15623/ijret.2013.0208004

Experimental Analysis of Napier Grass Waste Pre-treatment Process for Biogas Production

N. E. Suhaimi, H. Mohamed, N. Kamaruzaman, M. E. Mohd Roslan, and A. H. Shamsuddin

Abstract Biomass utilisation can benefit from a pre-treatment process because the process disintegrates the link between hemicellulose and lignin, potentially increasing biomethane generation. This work aims to assess available pre-treatments for Napier grass to increase biomethane generation. Important factors such as working volume, temperatures, testing time, and mixture ratios were studied in this work. Furthermore, a biochemical methane potential (BMP) test was conducted to measure the volume of biomethane generated from two Napier grass conditions: ground sun-dried and ground sun-dried with additional oven heating Napier grass. The highest average volume generated by ground sun-dried substrates was 47.67 Nml. On the other hand, about 45.87 Nml methane was generated from the Napier grass with an additional oven heating pre-treatment making the percentage difference in methane volume production between the two conditions of Napier grass is 4%. Based on the result, we found that applying the additional oven heating treatment on Napier grass may seem unnecessary as it only shows a small percentage difference compared to sun-dried Napier grass in terms of methane volume produced. In addition, 1.05 g of volatile solids (VS) of pre-treated samples can produce approximately 50.74 Nml biomethane.

Keywords Napier grass · Pre-treatment · BMP · Inoculum · POME · Methane · Biogas

---

N. E. Suhaimi · H. Mohamed (✉) · N. Kamaruzaman · M. E. Mohd Roslan · A. H. Shamsuddin
Institute of Sustainable Energy, Universiti Tenaga Nasional, 43000 Kajang, Selangor, Malaysia
e-mail: MHassan@uniten.edu.my

M. E. Mohd Roslan
e-mail: Eqwan@uniten.edu.my

A. H. Shamsuddin
e-mail: Abdhalim@uniten.edu.my

N. Mohamed Noor et al. (eds.), *Proceedings of the 3rd International Conference on Green Environmental Engineering and Technology*, Lecture Notes in Civil Engineering 214,
https://doi.org/10.1007/978-981-16-7920-9_31

1 Introduction

One of the harmful greenhouse gases (GHG) is methane. It contributes to the greenhouse effect, leading to global climate change. In addition, agricultural activities, fossil fuel combustion, and organic waste decomposition in landfills can cause gas release. Malaysia is one of the countries that are highly dependent on fossil fuels to generate electricity. Therefore, biogas generation from organic waste decomposition seems to be a viable option to reduce the dependency on fossil fuels. At the same time, this option can reduce GHG emissions to the atmosphere.

The study focuses on Napier grass as the main substrate due to its high yield. The annual yield of Napier grass ranges from 25 to 35 oven-dried tons per hectare and can be harvested up to four times per annum. Additionally, the grass can grow easily without needing extra nutrients, resulting in low establishment costs [1–3]. Therefore, Napier grass can be considered suitable for biomass energy resources where we can use the biomethane gas released for biogas production.

A pre-treatment process is required to disintegrate the adhesive linkage of hemicellulose and lignin. Furthermore, it can enhance biomethane production by preventing the delay of the anaerobic digestion (AD) process. Lignin is an outer structural component that prevents microbes from gaining access as fermentable sugars [4]. According to the current study [5], biomass that undergoes a pre-treatment will generate a higher methane yield. There are various types of pre-treatment done on Napier grass, such as physical, biological, chemical, thermal, and physiochemical pre-treatments.

A biochemical methane potential (BMP) experiment is a method to evaluate the methane potential of a substrate that can be taken from organic wastes, sewage sludge, and crops. The BMP experiment is carried out by combining a substrate with an anaerobic bacteria culture, known as inoculum, which is typically obtained from an active digester. The mix of substrate and inoculum is kept at a stable temperature, either 35 °C or 55 °C [6]. During the period, biomethane gas will be released because of the anaerobic degradation of organic contents in the substrate.

Therefore, this research aims to assess the available pre-treatment methods for Napier grass by observing the biomethane yield from pre-treated Napier grass samples using the biochemical methane potential (BMP) experiment.

2 Materials and Method

2.1 Material

Napier grass was collected and pre-treated physically through cutting, drying and grinding. About 2 kg of fresh Napier grass was used in this study. The inoculum chosen in this project was taken Palm Oil Mill Effluent (POME) obtained from a biogas plant in Dengkil, Selangor. Inoculum is a waste with high bacteria content which can accelerate anaerobic digestion activity in a BMP test. About 10 L of POME were taken from the biogas plant. The POME was kept in a refrigerator to preserve its concentration and prevent bacteria from being active when not in use.

2.2 Method

Experimental tests were carried out for two Napier grass' conditions: sun-dried Napier grass and sun-dried with an extra treatment of oven-drying Napier grass to compare which pre-treatment contributes to higher methane production through BMP. Both Napier grass samples were first cut manually to 20–25 cm length and dried to sunlight before grinding into a smaller particle size of 1–2 mm. According to a previous study by Kang et al. smaller particle sizes can produce a higher amount of methane [7].

Next, the substrate's Total Solid (TS) and Volatile Solid (VS) values were analysed. The TS and VS values are necessary parameters for the BMP experiment [8]. Three 5 g of sun-dried Napier grass samples (equivalent to 8.76 g of fresh grass) were heated in a separate individual crucible at 110 °C for 5 h and 30 min, and the weight was measured and recorded to determine the TS value. The VS process is the continuation of the TS determination process. Next, the substrate samples were heated up to 550 °C for two hours before their weight was measured and recorded. Finally, half the ground sun-dried Napier grass was heated in a furnace at 60 °C for six hours as an extra pre-treatment to ensure the grass was completely dried before going through a BMP test.

The BMP test was done using the Automated Methane Potential Test System (AMPTS), at which the device has the function of measuring the volume of gas from a chosen substrate and generate the data in the software. The device consists of bioreactors, a gas volume measuring device, CO_2 absorption unit containing NaOH solution to absorb gases other than biomethane. In this experiment, seven bioreactors were used. To determine the average volume of methane produced, each condition of Napier grass was replicated three times. Three bioreactors were prepared for the mixture of sun-dried Napier grass and inoculum, labeled N1, N2 and N3. Then, another three bioreactors were reserved for the mixture of Napier grass with the additional pre-treatment and inoculum labeled as N4, N5 and N6. Finally, the last bioreactor was filled with inoculum only. The inoculum to substrate ratio (ISR) used in the BMP experiment is 3, where 300 ml inoculum mixed with 1.05 g VS Napier substrate added 100 ml of deionised water, making it a working volume of 400 ml in all bioreactors. The BMP test usually was carried out using the VS content and working volume of 400 ml for high accuracy results [8]. The pH value of all substrates, inoculum and mixtures were measured and recorded before the BMP experiment. The experiment was carried out for three days, and the volume of methane produced was observed in each bioreactor.

Table 1 TS and VS values for Napier grass

Substrates		Weight of sample + crucible (g) after 5 h 30 min at 110 degree C	Weight of sample + crucible (g) after 2 h at 550 degree C	g VS	Total solids (%)	Volatile solids (%)	TS (%)	VS (%)
Napier Grass	N1	68.24	65.39	1.15	80.0	57.0	86.7	65.7
	N2	70.3	66.24	0.94	100.0	81.2		
	N3	68.8	65.85	1.05	80.0	59.0		

3 Result and Discussion

3.1 Volatile Solids (vS) and Total Solids (TS)

Results from the VS and TS analysis are presented in Table 1. The table shows that 5 g of Napier grass equals the average value of 1.05 g of volatile solids (VS).

3.2 Th pH Values

Table 2 shows the pH readings for all the bioreactors. The pH values of the substrate are almost the same as neutral. Pre-treated Napier grass is slightly lower in pH values than untreated Napier grass, with a percent difference of 1.12% before the experiment. The pH value for inoculum is also almost the same for all bioreactors before the experiment, with an average of 4.11. However, the pH values increased

Table 2 The pH values before and after the BMP experiment

Sample	Bioreactors	pH reading before BMP		pH of the mixture before BMP experiment	pH of the mixture after BMP experiment
		Substrate	Inoculum		
Pre-treated Napier grass	N1	6.47	4.11	4.21	4.17
	N2	6.48	4.14	4.21	4.17
	N3	6.56	4.11	4.19	4.17
Untreated Napier grass	N4	6.56	4.09	4.19	4.18
	N5	6.58	4.09	4.19	4.17
	N6	6.59	4.12	4.20	4.18
Inoculum	POME	–	4.12	–	4.14

to an average of 4.19 once the substrate was mixed with inoculum, probably due to the inoculum substrate ratio of 3.

3.3 Biomethane Yield

The results in Figs. 1 and 2 show the amount of methane (Nml) produced in each bioreactor for every fifteen minutes. Based on Fig. 1, N1, N2 and N3 are bioreactors containing a mixture of Napier grass with additional oven heating treatment and POME. Meanwhile, N4, N5 and N6 are bioreactors containing a mixture of sun-dried Napier grass and POME. Figure 1 illustrates the increasing methane yield in each bioreactor throughout the entire experiment period. Mixtures in all seven bioreactors stopped producing gases at various times. The N6 mixture production has the steepest slope of the graph (highest production rate). The mixture gas production was also the last one to stop after three days.

Based on the bar graph shown in Fig. 2, N6 generated the highest amount of methane compared to others, which is 57.1 Nml. The N6 mixture most likely has the higher amount of bacteria in the inoculum since the sample was the last bioreactor filled with inoculum. Hence, it might have much more sedimentation and bacteria compared to other mixtures. In addition, the 400 ml of POME bioreactor generated an almost similar amount of methane with other bioreactors that contain 300 ml of POME, which is 50.9 Nml.

Based on the bar graph shown in Fig. 3, from 1.05 g VS of sun-dried Napier grass, 47.67 Nml of methane was produced on average. Meanwhile, Napier grass produced 45.87 Nml of methane after an additional treatment of oven drying. It is found that

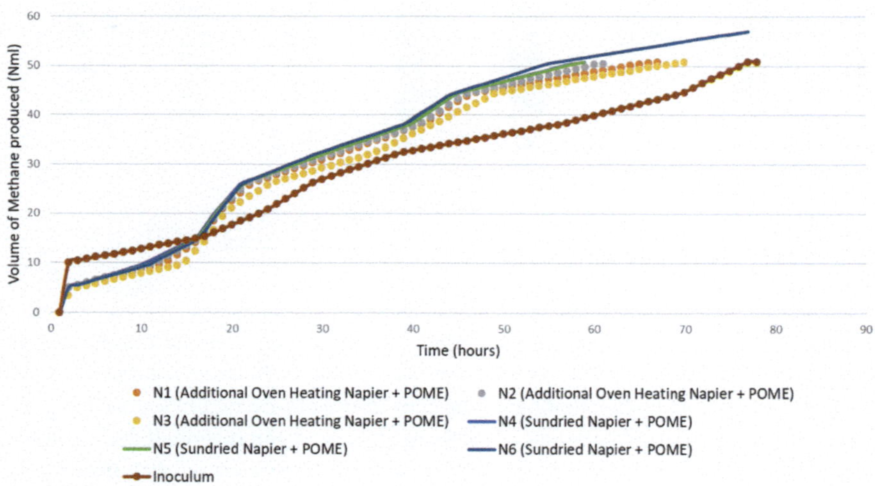

Fig. 1 Methane values produced by various mixtures of Napier grass and POME (Nml) vs time (hours)

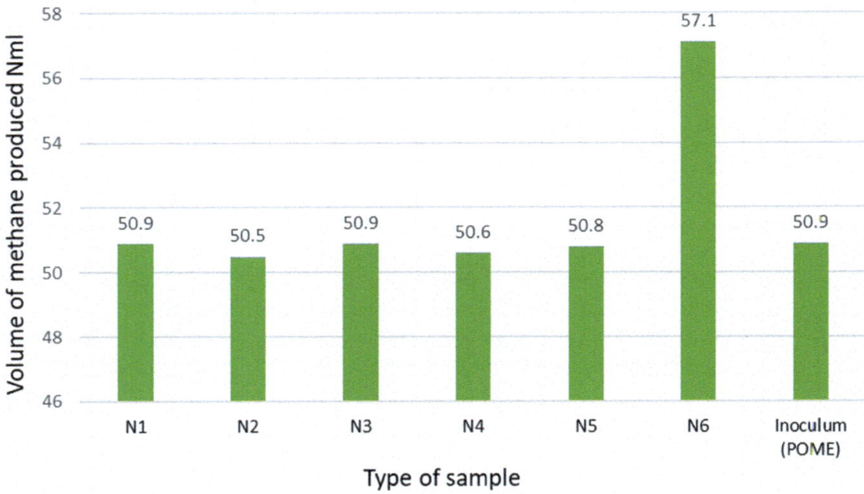

Fig. 2 Accumulation volume of methane produced in each bioreactor after three days

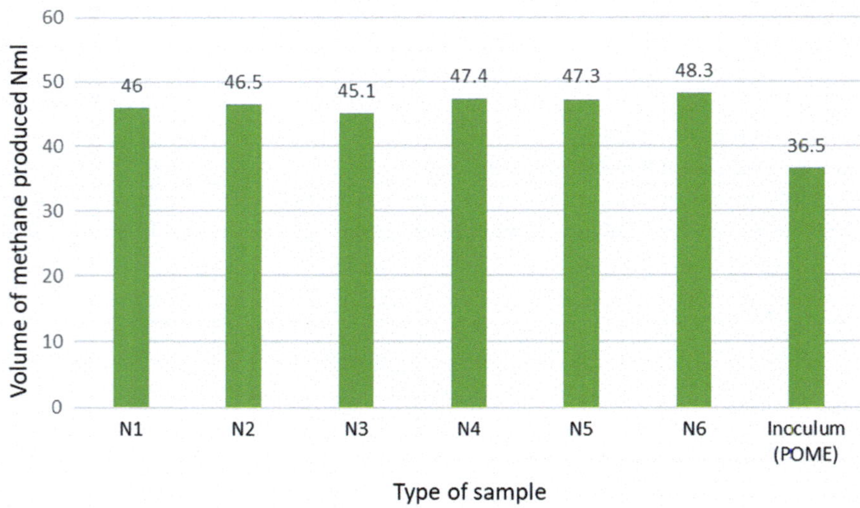

Fig. 3 Volume of methane produced at 50 h

the extra pre-treatment of oven heating after the sun drying may seem unnecessary because it makes a small volume change, with a 4% difference compared to the sun-dried substrate. On the other hand, the sun-drying pre-treatment seems adequate to generate a considerable amount of methane. Lastly, it can be concluded from the results that 1.05 g VS, which equals 8.76 g of fresh Napier grass, may potentially produce about 50.74 Nml of biomethane.

4 Conclusion

This experimental study identifies grinding and sun drying processes as the suitable and practical pre-treatment process for Napier grass. Based on the BMP results, the sun-dried Napier grass and Napier grass with the extra oven heating treatment produced an average methane volume of 47.67 Nml and 45.87 Nml, respectively. The percentage difference is small, approximately 4%. Hence, having the additional pre-treatment of oven drying might not be significant because the sun-drying process is adequate for the effective production of biomethane. In addition, the average 1.05 g VS of sun-dried Napier grass (equivalent to 8.76 g of fresh Napier grass) can generate approximately 50.74 Nml of biomethane. Furthermore, biomethane production can be enhanced by 20% when mixing Napier grass and POME as the inoculum.

Acknowledgements This study was supported by a grant from the AAIBE Chair of Renewable Energy, Institute of Sustainable Energy, Universiti Tenaga Nasional (UNITEN).

References

1. Mohammed IY, Abakr YA, Mokaya R (2019) Biofuel and valuable products recovery from Napier grass pre-processing: Process design and economic analysis. J Environ Chem Eng 7. https://doi.org/10.1016/j.jece.2019.10296
2. Song W, Peng L, Bakhshyar D, He L, Zhang J (2021) Bioresource Technology Mild O 2 -aided alkaline pretreatment effectively improves fractionated efficiency and enzymatic digestibility of Napier grass stem towards a sustainable biorefinery. Bioresour Technol 319:124162 . https://doi.org/10.1016/j.biortech.2020.124162
3. Mohammed IY, Abakr YA, Kazi FK, Yusup S, Alshareef I, Chin SA (2015) Comprehensive characterisation of Napier grass as a feedstock for thermochemical conversion. Energies 8:3403–3417. https://doi.org/10.3390/en8053403
4. Chandrasekhar K, Cayetano RDA, Ikram M, Kumar G, Kim S-H (2020) Evaluation of the biochemical methane potential of different sorts of Algerian date biomass. Environ Technol Innov 20:101180. https://doi.org/10.1016/j.eti.2020.101180
5. Phitsuwan P, Sakka K, Ratanakhanokchai K (2016) Structural changes and enzymatic response of Napier grass (Pennisetum purpureum) stem induced by alkaline pretreatment. Bioresour Technol 218:247–256. https://doi.org/10.1016/j.biortech.2016.06.089
6. Filer J, Ding HH, Chang S (2019) Biochemical methane potential (BMP) assay method for anaerobic digestion research. Water (Switzerland) 11
7. Kang X, Zhang Y, Song B, Sun Y, Li L, He Y, Kong X, Luo X, Yuan Z (2019) The effect of mechanical pretreatment on the anaerobic digestion of Hybrid Pennisetum. Fuel 252:469–474. https://doi.org/10.1016/j.fuel.2019.04.134
8. Bioprocess Control (2016) AMPTS II & AMPTS II Light. 1–95

Green Smart System Based on AI for Ammonia and Hydrogen Eco-Friendly Use in Naval Transport from Protected Wetlands

György Deák, Tudor Georgescu, Cosmin-Karl Bănică, Iasmina-Florina Burlacu, Irina Urloiu, and Irnis Azura Zakarya

Abstract Shipping is one of the major pressures that have a negative impact on conservation efforts in protected wetlands in Romania. Based mainly on the use of fossil fuels, shipping activities contribute significantly to water and air pollution. Reducing the carbon footprint by implementing ammonia and hydrogen on board has gained increasing interest in recent years, but innovative solutions are still needed to be competitive with traditional fuels. This paper is presented a green smart system based on Artificial Intelligence (AI) for ammonia and hydrogen eco-friendly use in naval transport. The preliminary test results showed that the proposed engine efficiency is almost over a half, thus the proposed solution is presenting a great potential of application in naval transport as a green solution towards zero CO_2 emissions economy, but it requires further improvements by refining the command and control software and by introducing a control of the hydrogen gas injection.

Keywords Ammonia/hydrogen fuel cells · Artificial Intelligence · Electric propulsion/thermal engines · Carbon-free naval transport · Protected wetlands areas

G. Deák · I.-F. Burlacu (✉) · I. Urloiu
National Institute for Research and Development in Environmental Protection Bucharest (INCDPM), 294, Splaiul Independentei, 6th District, 060031 Bucharest, Romania

G. Deák
e-mail: dkrcontrol@yahoo.com

T. Georgescu
Hypertech SRL, 080013 Giurgiu, Romania

C.-K. Bănică
Wing Computer Group SRL, 060782 Bucharest, Romania
e-mail: cosmin@wing.ro

I. A. Zakarya
Sustainable Environment Research Group (SERG), Centre of Excellence Geopolymer and Green Technology (CEGeoGTech), Universiti Malaysia Perlis, 01000 Kangar, Perlis, Malaysia
e-mail: irnis@unimap.edu.my

N. Mohamed Noor et al. (eds.), *Proceedings of the 3rd International Conference on Green Environmental Engineering and Technology*, Lecture Notes in Civil Engineering 214,
https://doi.org/10.1007/978-981-16-7920-9_32

1 Introduction

In Romania, the uncertain situation regarding the conservation status of protected wetlands [1, 2] in the context of climate change and the intensification of anthropogenic pressures, highlights the necessity to develop effective solutions for the decarbonization of the transport sector. Noise, air, and water pollution due to shipping activities are the main factors that generate pressures on aquatic ecosystems, interfering with biodiversity conservation objectives [3]. Currently, shipping activities are mostly based on fossil fuels, contributing to global carbon dioxide emissions by about 2.6% [4]. The decarbonization of naval transportation sector can be approached from two perspectives. First, the reduction of carbon dioxide emissions can be achieved by implementing solutions aimed at electrifying transport-using energy from renewable sources [5, 6]. The second way refers to reducing the carbon footprint of fuels by replacing fossil fuels with "green" alternatives, such as e-fuels [7]. Ammonia and hydrogen have emerged as potential candidates for achieving the global decarbonization targets of shipping sector.

Past research projects have proposed hydrogen as a solution for vessels propulsion systems, but the problems associated with its high-cost storage and transportation are a step backwards in its exclusive implementation [4]. On the other hand, the adoption of ammonia-based solutions is facilitated by the existence of mature infrastructure and technologies [8, 9]. Liquid ammonia can be stored at −34 °C at atmospheric pressure or pressurized at 20°C and have a volumetric energy density higher than that of liquid hydrogen, which makes it an attractive for on-board storage systems. Although addressed in several studies [10–12], the successful application of ammonia as a fuel for internal combustion engines requires overcoming technological and operational challenges due to its physico-chemical characteristics [12]. In this regard, previous research projects have been conducted on integration of computational intelligence for detection of the operating conditions of internal combustion engines powered by ammonia in dual fuel systems [13], combustion system optimization and control [14]. The major distinctive point of the Green Smart System proposed in this paper is its holistic point of view for the development in parallel, in two directions of fossil fuel substitution solutions. In the proposed Green Smart System based on an AI control and command component, ammonia is capitalized for both decarbonization pathways—as a source in fuel cells for electric propulsion and as an alternative fuel for boats combustion engines. Thus, AI component provides opportunities for costs reduction and for both optimize electric energy production process in fuel cells and the operating parameters of the boat engines. Therefore, the proposed smart technology aims to reduce the CO_2 emissions and the impact of tourist activities in protected natural areas by overcome barriers related to the use of ammonia and hydrogen for light naval transport.

2 Materials and Methods

The proposed Green Smart System includes two interconnected major sub-unities through common hardware accelerated AI core for control and command and automatic decision-making on (A) fuel cell operating parameters and for powering boat electric propulsion and (B) control of physico-chemical processes of combustion of ammonia and hydrogen to increase the total thermal engine propulsion efficiency. The core of the green smart system consisting in an AI-ML (Machine Learning) datacenter installed in a specialized container and an "AI at edge" type system will be part of each command and control system for fuel cell and thermal engine combustion processes. Operational parameters used: gas—E95 gasoline without additives; the engine has been installed on a test bench that can measure the response curve with a "normal" suction that can be modulated by software for certain parameters adjustment. The initial test version used Huawei's IA platform—Kirin 970 (HiKey 970 board) and an ML (Machine Learning) solution based on Huawei Atlas 200 and Kirin 990 was to be implemented as an interface unit. After Huawei came under embargo, the technical team reoriented itself to another processor manufacturer with NPU—RK 3399PRO which has a specialized processor for ML and interface. The SBC type system reaches 3TFOPS for neural computing operations and is a technical standby solution for advanced Arm-N77 platforms. The use of a decision-making system based on AI takes into account the multicriteria complexity and the strong nonlinearity of the answers to the variations of parameters that make the PID type loop control systems to be overcome. In this case, the system uses an offline analysis of engine performance and improves its performance through post-phenomenon self-analysis (ML) and the use of results in real-time settings (interface) so that over time the performance of this type of engine becomes comparable to fossil fuel engines. This latter approach, involving a closed-loop platform, will allow a secure and resilient environment for data collection management and use, ensuring support of the AI decision-making component of the system.

3 Results and Discussions

The proposed solutions for ammonia and hydrogen eco-friendly use in naval transport, schematically illustrated in Fig. 1, are developed both for the reduction of the pollution in protected wetlands areas and CO_2 emissions of the traditional internal combustion engines used in transport by small boats and ships in protected wetlands. The solution for converting heat engines from ships in operation to use part or even only e-fuel is attractive by reducing the total social costs associated with the transition to zero CO_2-free economy, however, the application of the proposed solution may be possible if there are financial incentives in this regard.

The preliminary test results of the proposed system presented in Fig. 2, showed that the power diagram it is not extremely favorable, the efficiency being almost over

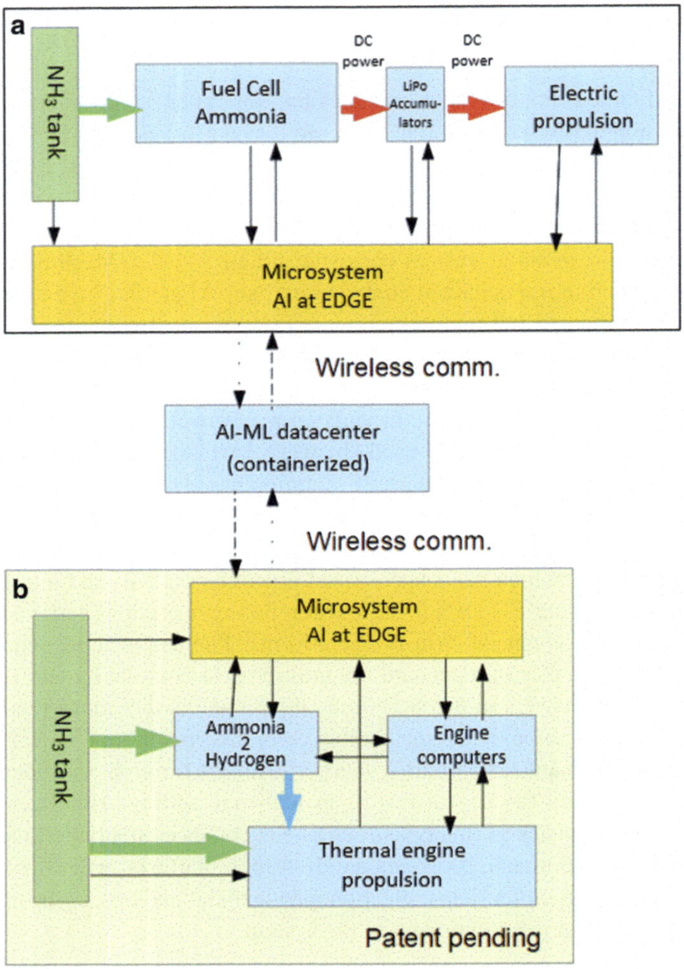

Fig. 1 Schematic representation of the Green Smart System based on AI for ammonia and hydrogen eco-friendly use in naval transport sector (patent application)

a half. Further researches will be conducted in order to increase the engine efficiency, through improvements of the command and control software and by introducing a control of the hydrogen gas injection.

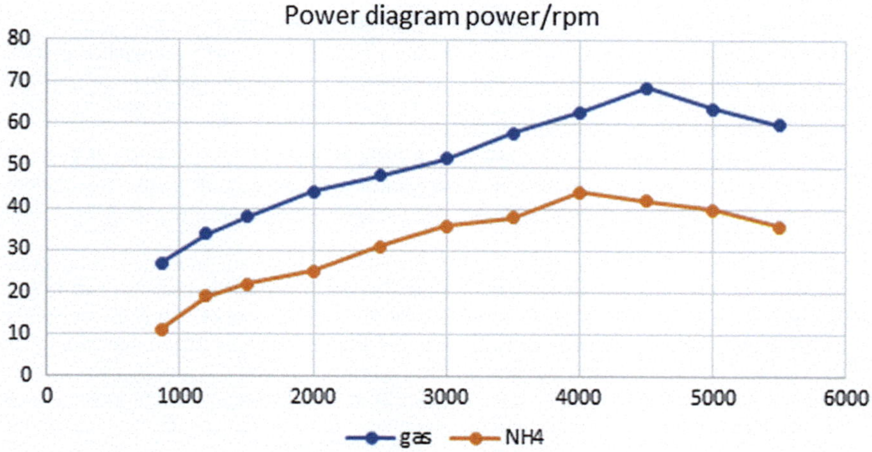

Fig. 2 Power diagram for the proposed engine

4 Conclusions

For successful deploying of carbon-free alternatives to fossil fuels in naval transportation systems, it is of great importance to maintain a balance between fuel and operational costs and optimization strategies of onboard installation performances. The pathway towards decarbonization of naval mobility sector may be achieved by implementation of green ammonia and hydrogen as a substitute of fossil fuels. The new Green Smart System based on AI for ammonia and hydrogen eco-friendly use presented in this paper proposes the parallel development of two solutions for fossil fuel substitution in naval transport in protected wetlands to reduce the impact of anthropogenic pressures and towards carbon dioxide zero emissions. The preliminary test results indicates that proposed solution needs further engine efficiency improvements as the power diagram is not extremely favorable, being almost over a half, but presents interesting perspectives for further researches.

References

1. Cîşlariu A G, Ichim P, Manzu C C (2020) Long term changes of wetlands in the context of anthropic influences: the case of ROSCI0222 (North-Eastern Romania). Carpath J Earth Env 15(1), 27–36. https://doi.org/10.26471/cjees/2020/015/105
2. Tanislav D (2014) Protected wetlands in Romania. In 2nd International Conference-Water resources and wetlands, Tulcea-Romania.
3. Jägerbrand A K, Brutemark A, Sveden J B, Gren M (2019) A review on the environmental impacts of shipping on aquatic and nearshore ecosystems. Sci Total Environ 695, 133637. https://doi.org/10.1016/j.scitotenv.2019.133637
4. Ayvalı T, Tsang S C, Van Vrijaldenhoven T (2020) The Position of Ammonia In Decarbonising Maritime Industry: An Overview And Perspectives. Johnson Matthey Technol Rev 65 (2),

275–290(16). https://doi.org/10.1595/205651321X16043240667033
5. Poteras G, Deak G, Burlacu I, Raischi N, Radu M (2021) Use of the bio-engineering technologies in the construction and upgrade of complex installations for obtaining energy from three renewable sources. Complex installations for flowing waters. E3S Web Conf 239, 00009. https://doi.org/10.1051/e3sconf/202123900009
6. Poteras G, Deak G, Baraitaru A, Olteanu M, Raischi N, Suriyani D, Che Halin D S (2020) Bioengineering technologies used for the development and equipment of complex installations to obtain energy from three renewable sources. Complex installations for coastal areas. IOP Conf Ser: Earth Environ Sci
7. Bicer Y, Dincer I (2018) Clean fuel options with hydrogen for sea transportation: A life cycle approach. Int J Hydrog Energy 1179-1193. https://doi.org/10.1016/j.ijhydene.2017.10.157
8. Van Biert L, Godjevac M, Visser K, Aravind PV (2016) A review of fuel cell systems for maritime applications. J Power Sources 327:345–364. https://doi.org/10.1016/j.jpowsour.2016.07.007
9. Hansson J, Brynolf S, Fridell E, Lehtveer M (2020) The Potential Role of Ammonia as Marine Fuel—Based on Energy Systems Modeling and Multi-Criteria Decision Analysis. Sustainability 12(8):3265. https://doi.org/10.3390/su12083265
10. Pochet M, Truedsson I, Foucher F, Jeanmart H, Contino F (2017) Ammonia-Hydrogen Blends in Homogeneous-Charge Compression-Ignition Engine. SAE Tech Pap
11. Mørch CS, Bjerre A, Gøttrup MP, Sorenson SC, Schramm J (2011) Ammonia/hydrogen mixtures in an SI-engine: Engine performance and analysis of a proposed fuel system. Fuel 90(2):854–864. https://doi.org/10.1016/j.fuel.2010.09.042
12. Lasocki J, Bednarski M, Sikora M (2019) Simulation of ammonia combustion in dual-fuel compression-ignition engine. In IOP Conf Ser: Earth Environ Sci 214 (1), 012081. IOP Publishing. https://doi.org/10.1088/1755-1315/214/1/012081
13. Aliramezani M, Norouzi A, Koch C R (2020) A grey-box machine learning based model of an electrochemical gas sensor. Sensor Actuat B-Chem 128414. https://doi.org/10.1016/j.snb.2020.128414
14. Castiglione T, Morrone P, Falbo L, Perrone D, Bova S (2020) Application of a Model-Based Controller for Improving Internal Combustion Engines Fuel Economy. Energies 13(5):1148. https://doi.org/10.3390/en13051148

Influence of Cement Paste Containing Municipal Solid Waste Bottom Ash on the Strength Behavior of Concrete

Roshazita Che Amat, Norlia Mohamad Ibrahim, Nur Liza Rahim, Khairul Nizar Ismail, Ainaa Syamimi Abdul Hamid, and Madalina Boboc

Abstract Cement in construction has become a vital requirement to build up the buildings, which may increase the expenses in construction. Materials that have the potential to replace cement would be proposing. This study used municipal solid waste incineration bottom ash (MSWIBA) as a partial replacement for cement. MSWIBA used in this study was a by-product from the incineration process and had compound content that was almost the same as cement. The treated bottom ash in the range of 0 to 30% and 10% of untreated bottom ash mixture use in this study. Mechanical and physical properties of concrete analysed with a few tests such as slump test, water absorption test, compressive strength test, heat exposure test and residual strength test after heating has proceeded. The workability of fresh concrete was measured by performing a slump test. Based on the compressive strength result, the 10% substitution of treated bottom ash was achieved the highest strength in testing in 7 and 28 days. Meantime, the control concrete obtained the best thermal insulator because of a smaller number of cracks on the surface of the concrete than that bottom ash concrete surface. After heated, the concrete was tested on compressive strength again to investigate the residual compressive strength. The highest residual surpasses gained by 10% bottom ash (treated) as a partial substitution in cement. Based on the

R. C. Amat (✉) · N. M. Ibrahim · N. L. Rahim · K. N. Ismail · A. S. A. Hamid
Faculty of Civil Engineering Technology, Universiti Malaysia Perlis, 02600 Arau, Perlis, Malaysia
e-mail: roshazita@unimap.edu.my

N. M. Ibrahim
e-mail: norlia@unimap.edu.my

N. L. Rahim
e-mail: nurliza@unimap.edu.my

K. N. Ismail
e-mail: nizar@unimap.edu.my

Sustainable Environment Research Group (SERG), Centre of Excellence Geopolymer and Green Technology (CEGeoGTech), Universiti Malaysia Perlis, 01000 Kangar, Perlis, Malaysia

M. Boboc
National Institute for Research and Development in Environmental Protection Bucharest (INCDPM), Splaiul Independentei Street, 6th District, 294 Bucharest, Romania

N. Mohamed Noor et al. (eds.), *Proceedings of the 3rd International Conference on Green Environmental Engineering and Technology*, Lecture Notes in Civil Engineering 214,
https://doi.org/10.1007/978-981-16-7920-9_33

overall test carried out, 10% of bottom ash replacement as cement is the optimum amount of bottom ash required to surpass the strength of the control sample.

Keywords Municipal solid waste incineration bottom ash · OPC · Density · Compressive strength · Elevated temperature

1 Introduction

As Malaysia is a developing country, then more construction activities have to develop. Especially residential buildings, public buildings, bridges and lots more. Furthermore, a large amount of cement demand severely increases to fulfil the request from developers. In Malaysia, cement production reached up to 1866 thousand tonnes by July 2020 [1]. Hence cement usages in construction have become a vital requirement to build up the edifice, which may increase the expenses in building construction, the new material proposed to replace cement. The use of bottom ash that relatively inexpensive materials are more environmentally and cheaper. So, Malaysia should find another alternative in reducing the use of cement by using waste products to enhance the workability of a building to ward off any challenges that occur in an uncertain situation.

Economic growth has brought prosperity, population increase, accelerated urbanisation and industrialisation. However, with the rise in population, there has also been a substantial rise in generating solid waste. The amount of waste produced each day in Malaysia is expecting to rise near to 41,035 tonnes by the year 2026 [2]. Besides, without proper management of waste would lead to serious environmental problems. Due to enhancement population, country development and high standard of living. It would affect the rate of municipal solid waste (MSW) in all municipalities in the country. As a result, Malaysia eventually led to environmental problems [3]. Incineration is the most effective in reducing the waste volume by controlled burning of waste at high temperature, sterilises and stabilises and may use as a landfill disposal option [4].

Municipal solid waste incineration will produce two types of ash which are fly ash and bottom ash. Bottom ash is the non-combustible residue of combustion inside a furnace or incinerator. Incinerator ash from the burning of solid waste consists of bottom ash and fly ash which bottom ash is heterogeneous consists of slag, metals, ceramics, glasses and partially other non-combustibles and unburned organic matter [5]. Bottom ash was generally coarse sandy with a range diameter between 0.1 and 100 mm. The incinerator residue consists of 865,944 metric tonnes of bottom ash (about 84.4%) and 160,056 metric tonnes of fly ash (about 15.6%). Bottom ash was classifying as the Si-Al-CA material where it was easy to produce vitrified slag [6]. Thus, this research used bottom ash from MSW as part of substitution in cement as an alternative to reuse the waste products and reduce the cement consumption.

2 Materials and Method

2.1 Material

Typically, bottom ash accounts for 80% of the whole amount of by-products in the incinerator plants. It generally contains coarse particles (0.1–100 mm) of slag, glass, rocks, metals and unburnt organic matters shows below in Fig. 1. The bottom ash was collecting from a municipal solid waste incinerator in Pangkor Island, Perak. MSWIBA then was oven-dried in an oven at a temperature range of 80–110 °C for 24 h until reaching a constant weight. Then MSWIBA was ground to smaller size particles in a size range below 63 microns as fine as cement. After sieved, the MSWIBA was treated by heating the MSWIBA into the furnace for 2 h at 500 °C temperature continuously to remove carbon content. MSWIBA then undergo a few laboratory tests to characterize the properties by carried out the X-ray fluorescence test (XRF) [7].

2.2 Method

Treated MSWIBA replaced cement in four mix designs 0, 10, 20 and 30% by volume. To test its potential as partial replacement in cement, specimens are required to achieve 20 N/mm^2 target characteristics strength. Similar size of mould was used to undergo this experiment which is 100 × 100 × 100 mm cube mould. Standard and calculation mixing had been referred by [8] specification. The mixture was prepared at water-ratio of 0.6. Table 1 shows the proportion concrete mixture design. Each specimen was prepared according to the proportion calculated. From Table 1 below shows the increasing proportion of bottom ash (BA) substitute as a cement calculated

Fig. 1 Samples of Municipal Solid Waste Incineration Bottom Ash (MSWIBA)

Table 1 Mix proportion incorporated with MSWIBA and cement

Material (Kg)	MIX 1	MIX 2	MIX 3	MIX 4	MIX 5
Water	0.23	0.23	0.23	0.23	0.23
MSWIBA	0.00	0.04	0.04	0.08	0.11
Cement	0.38	0.34	0.34	0.30	0.89
Fine aggregate	0.76	0.76	0.76	0.76	0.76
Coarse aggregate	1.52	1.52	1.52	1.52	1.52

in m^3 per kg. Fine aggregate and coarse aggregate undergo particle size analysis according to standard. sand with 4.75 mm maximum size was used as fine aggregate, fulfilling the requirements of [9] along with crushed stone of 25 mm maximum size used as coarse aggregate.

Slump test conducted in order of assessing the consistency and workability of the fresh concrete. The test procedures were done based on [10] standard. This test is commonly used in all construction works all over the world. It is useful to detect variation in the uniformity of nominal proportions and specifies procedures to determine the consistency of concrete in which the nominal maximum size of aggregate does not exceed 40 mm. Concrete casted around 24 h before replacing it into the water tank for 7 and 28 days curing period. Water absorption test used to determine the amount of water absorbed into the capillary pores of cement and bottom ash and [11] specification was used to determine the water absorption. Compressive strength is the most important test because it determines the durability of each ratio and compare it with control cube without any replacement of MSWIBA to see the durability differences. Heat exposure tests are being carried out to see how high temperatures affect the behavior of concrete materials and structures. Before heating, the specimens are preload, and the preload is maintained during the heating period. The preload might range from 20 to 40% of the concrete's compressive strength at ambient temperature. Heat is normally increased at a consistent pace until the target temperature is reached. Then the compression test after being exposed to elevated temperature at 800 °C was tested [12].

3 Result and Discussion

3.1 Chemical Composition of Cement and MSWIBA

Table 2 show concentration of chemical composition that present in cement and BA classified through XRF Test. The compound present in both materials is quite similar which is Alumina, Silica, and Iron as major compound while Calcium, Magnesium and Sulphate as minor compound. However, both compound that present contain different concentration unit. The presence of metallic substances Al is one of the

Table 2 Chemical Composition of Cement and MSWIBA

Compound	Concentration Unit for cement (%)	Concentration Unit for Bottom Ash (%)
Al_2O_3	2.60	5.20
SiO_2	10.80	27.30
SO_3	3.01	1.95
K_2O	1.08	1.94
Ca0	77.96	50.89
TiO_2	0.35	1.19
MnO	0.16	0.02
Fe_2O_3	3.52	0.18
CuO	0.03	0.11

largest barrier BA consumption cement caused by the release of hydrogen gas derived from reaction of metal. From the table data below show that BA has higher concentration of Al_2O_3 than that cement. BA was spotted having 3.6% of Al while cement have only 1.7% of Al in XRF test result. Higher concentration of metallic can gives large development of hydration gas which react at the early hydration process. The strength of concrete will reduce cause of the bubble that was blocked inside the concrete.

3.2 Workability of Fresh Concrete

The workability of fresh concrete was identified by carried out slump test during concrete casting. Compacting become decreased as the percentage displacement decreased. Factors that contributed on this circumstance is water content. Testing mix concrete was done before proceeding the exact amount of water cement ratio to determine the appropriate ratio that satisfy the criteria of true slump. The result of slump test was presented in Fig. 2.

The result shows the workability of concrete obtained by 0% of BA substitution (control sample) which is 30 mm after measurement. Followed by 10% substitution of untreated MSWIBA replacement in concrete, the reading reduces about 2 mm. For the last three sample also presented same reading where the result obtained was 30 mm same as control sample.

3.3 Water Absorption

Figure 3 show that the average percentage of water absorption of bottom ash concrete with increasing percentage of bottom ash as partial substitution in concrete. Addition

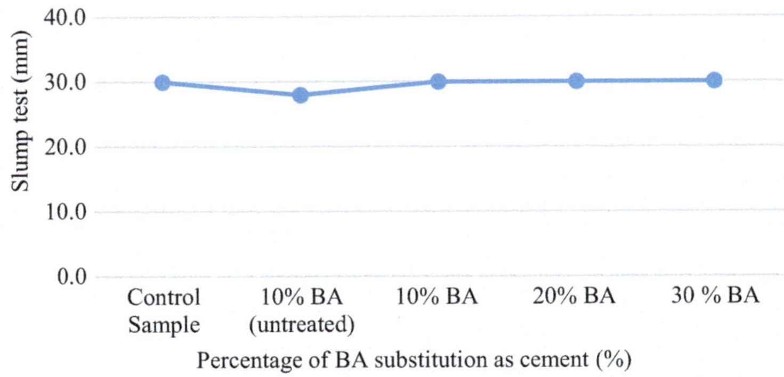

Fig. 2 Slump test of concrete

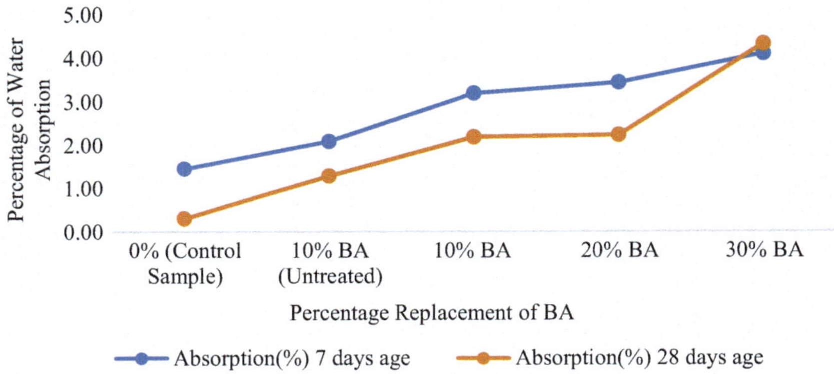

Fig. 3 Water absorption of concrete

of bottom ash in cement has led to the higher demand of water absorbed by the concrete. Higher porosity in concrete may result of higher water absorption because MSWIBA absorb unnecessary water for hydraulic bond.

3.4 Compressive Strength of Concrete

The results show (Fig. 4) that the strength development patterns of bottom ash concrete at all the levels of cement replacement with bottom ash. At the curing age of 7 days, concrete mixtures containing 10% (untreated) and 10% treated bottom ash as cement has gained 23.40 MPa and 23.88 MPa respectively exceed the compressive strength that achieved 22.84 MPa by the control sample mixtures. However, compressive strength of 20% bottom ash cement has gained 21.33 MPa fail to exceed control sample strength but still surpass the target characteristic strength. At curing age of

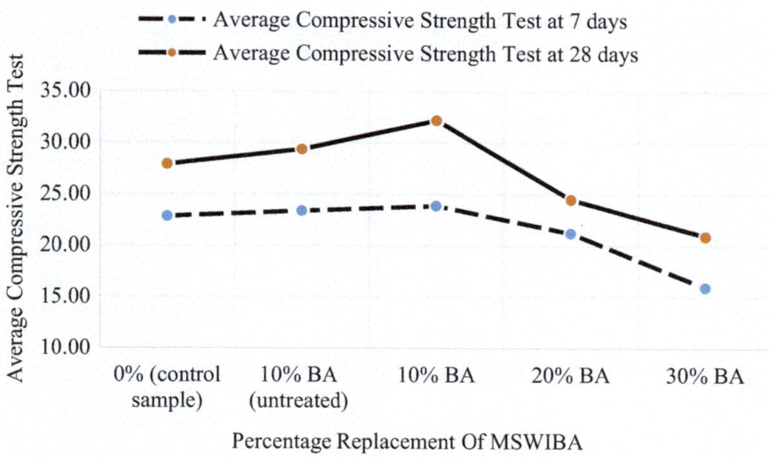

Fig. 4 Compressive strength of concrete

28 days, the compressive strength of concrete mixtures containing 10% (untreated) and 10% treated bottom ash as partial replacement of cement has surpassed that of control concrete by 29.4 MPa and 32.21 MPa.

The factors responsible for the decrease in compressive strength of bottom ash concrete mixtures for both curing age of 7 days and 28 days was due to the replacement of the stronger material (cement) with the weaker material (bottom ash) and absence of pozzolanic activity by the bottom ash. There is inadequate calcium hydroxide to react with pozzolanic. Other than that, reduction of strength of concrete can happen because pore spaces present in the matrix and because of higher concentration of metallic aluminum in bottom ash. Bubbles that are entrapped inside concrete has result on pore space thus weaken the specimen. Other than that, BA has higher silica content which can increase the hydration product by react with lime to from a better matrix. However, the silica will not be fully used during this process and left in concrete as unusable substances [13].

3.5 *Heat Expose at 800 °C Temperature*

Heat exposure test aim is to test the sample by heating with predetermined time. The heat exposure test aims to test the sample by heating with a predetermined time. The heat is fixing to have an accurate reading for every time. The heat used is up to 800°C for every specimen. Then after two hours of heat exposer, the bottom ash concrete shows lots of cracks compared to the control sample. The cracking that occurs on the concrete surface presented in Fig. 5 were scan using light microscopy. The higher the percentage of partial substitution of bottom ash, the more spalling observed on the surface.

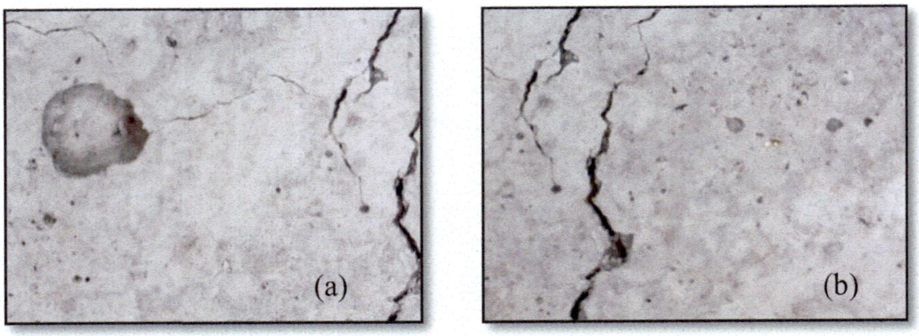

Fig. 5 Samples of **a** 20% MSWIBA concrete **b** 30% MSWIBA concrete

3.6 Residue Compressive Strength

The concrete cube burned in the furnace previously did not impact extreme cracking inside but only shows cracking at the outer side of the concrete. So, compressive strength then was tested again to determine the residual strength on the specimen. The residual strength was present in Fig. 6.

Based on the test conducted on every concrete, only 10% bottom ash (treated) concrete exceeding the control sample. The result show $Ca(OH)_2$ was decomposing during the heating range of 400 to 500 °C results in weakening the strength of concrete. The bottom ash concrete also shows decreasing strength with the increasing percentage of partial substitution of bottom ash.

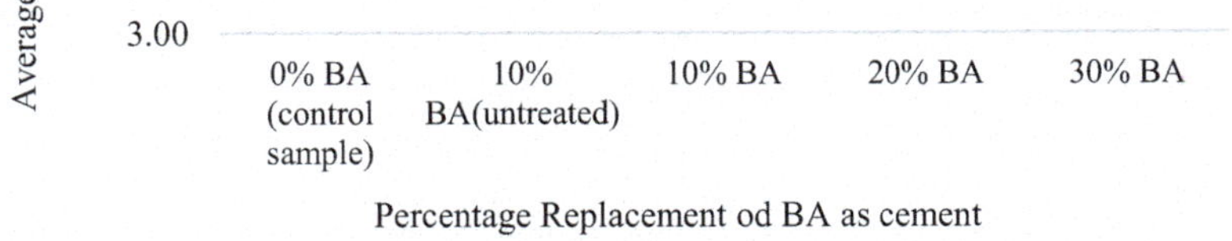

Fig. 6 Average residual strength of MSWIBA concrete

4 Conclusion

In conclusion, this study shows the utilizing municipal solid waste incineration bottom ash is suitable as a partial replacement of cement in concrete. The result indicates that MSWIBA concrete from Pangkor Island could use as the partial substitution in cement with the appropriate amount. Based on the results, 10% of bottom ash substitution in cement is the optimum amount to produce higher strength. Compressive strength decreased with the increasing percentage of bottom ash in concrete. Water absorption of bottom ash concrete increased approximately with the increase in cement replacement with bottom ash in concrete. Residual strength after heat exposer also increasing proportionately with percentage with replacement of 10% BA and then slightly decrease after 20% of BA replacement.

References

1. Trading Economics, Cement Production (2020) Retrieved from https://tradingeconomics.com/country-list/cement-production.
2. Chua HS, Bashir MJ, Tan KT, Chua HS (2019) A sustainable pyrolysis technology for the treatment of municipal solid waste in Malaysia. AIP Conference Proceedings, 2124(July). https://doi.org/10.1063/1.5117076
3. Mendes A, Sanjayan J, Collins F (2008) Phase transformations and mechanical strength of OPC/Slag pastes submitted to high temperatures. Mater Struct 41(2):345–350. https://doi.org/10.1617/s11527-007-9247-8
4. Badgie D, Samah MAA, Manaf LA, Muda AB (2012) Assessment of Municipal Solid Waste Composition in Malaysia: Management, Practice, and Challenges. Pol J Environ Stud 21(3):537–547
5. Li X, Bertos MF, Hills CD, Carey PJ, Simon S (2007) Accelerated carbonation of municipal solid waste incineration fly ashes. 27(July 2004), 1200–1206. https://doi.org/10.1016/j.wasman.2006.06.011
6. Chiou I, Wang K, Tsai C (2009) Enhancing performance and durability of slag made from incinerator bottom ash and fly ash. Waste Manage 29(2):501–505. https://doi.org/10.1016/j.wasman.2008.03.014
7. American Society of Testing and Materials 2013, Method for Determination of Elemental Content of Polyolefins by Wavelength Dispersive X-Ray Fluorescence Spectrometry, ASTM D6247–10 – 2013, retrieved from American Society of Testing and Materials Online.
8. British Standard 1997, Methods for Specifying Concrete Mixes, BS 55328: Part 2 - 1997, retrieved from British Standard Online
9. American Society of Testing and Materials 2006, Standard test method for sieve analysis of fine aggregate and coarse aggregate, ASTM C136 – 2006, retrieved from American Society of Testing and Materials Online.
10. British Standard 1983, Testing Concrete: Method for Determination of Slump, BS 1881: Part 102 - 1983, retrieved from British Standard Online.
11. British Standard 1983, Method for Determination of Water Absorption, BS 1881: Part 122 - 1983, retrieved from British Standard Online
12. British Standard Institution. (1983). Method for determination of compressive strength of concrete (BS 1881–116:1983). Retrieved from https://www.azobuild.com/article.aspx?ArticleID=4777
13. Remya Raju MMP, Aboobacker KA (2014) Strength performance of concrete using bottom ash as fine aggregate. International Journal of Research in Engineering & Technology 2(9):111–122

Investigating Palm Oil Mill Effluent Furnace Heating Pre-treatment for Biogas Production

Nurhamizah Kamaruzaman, Hassan Mohamed, Nurmi Ezzatul Suhaimi, Mohd Eqwan Mohd Roslan, and Abd Halim Shamsuddin

Abstract This study evaluates the furnace heating pre-treatment of palm oil mill effluent (POME) for biogas production by measuring the amount of biomethane generated with and without pre-treatment using a biochemical methane potential (BMP) test. A pre-treated POME and an untreated POME were tested in a biochemical methane potential (BMP) experiment to assess the biomethane production capability. The methods for performing the biochemical methane potential test and the pre-treatment process of POME using a furnace heating method are also explained in this paper. The findings show that the biomethane volume produced by furnace heated POME is comparable to that produced by untreated POME.

Keywords Pre-treatment · Heating · POME · Wastewater sludge · Biogas · BMP

1 Introduction

Biomass, a renewable energy resource, can be derived from plants, animals, food, or solid wastes. For example, empty fruit bunch (EFB) and palm oil mill effluent (POME) are two common biomass wastes produced by palm oil plantations. Since Malaysia has more than 7 million hectares of palm oil plantation by 2020, this paper focuses on palm oil mill effluent (POME). Based on previous studies [1, 2], it is projected that one hectare of palm oil plantations produces 10,725 L of POME. In addition, POME can release a colourless methane gas through the anaerobic digestion (AD) process of its biodegradable contents [3]. This gas can later be used to generate electricity and heat.

N. Kamaruzaman · H. Mohamed (✉) · N. E. Suhaimi · M. E. Mohd Roslan · A. H. Shamsuddin
Institute of Sustainable Energy, Universiti Tenaga Nasional, 43000 Kajang, Selangor, Malaysia
e-mail: mhassan@uniten.edu.my

M. E. Mohd Roslan
e-mail: Eqwan@uniten.edu.my

A. H. Shamsuddin
e-mail: Abdhalim@uniten.edu.my

N. Mohamed Noor et al. (eds.), *Proceedings of the 3rd International Conference on Green Environmental Engineering and Technology*, Lecture Notes in Civil Engineering 214,
https://doi.org/10.1007/978-981-16-7920-9_34

Palm oil mill effluent contains a significant amount of chemical and biological oxygen demands, contributing to pollution in the environment. Raw POME is currently handled by dumping it into a pond before undergoing anaerobic digestion without pre-treatment [4]. Hence, the main objective of this work is to investigate the potential of furnace heating pre-treatment for enhancing biomethane production during the anaerobic digestion (AD) process. According to previous research, the significant effect of heat pre-treatment of POME is that it produces a higher volume of biogas and positively impacts methane production even though the process is considered simple [4].

2 Materials and Method

2.1 Materials

The thick brownish wastewater produced during palm oil processing is known as palm oil mill effluent (POME). A previous study reported [5] that this wastewater contains approximately 25,000 mg/L biological oxygen demand (BOD), 51,000 mg/L chemical oxygen demand (COD), 18,000 mg/L suspended solids, and 34,000 mg/L total volatile solids, and 4,000–6,000 mg/L oil and grease with a pH value of 4.2. The sludge from the POME pond acts as an inoculum containing bacteria that can stimulate biomethane production during the anaerobic digestion process in the BMP experiment. The inoculum contains 15,180 mg/L BOD and 40,563 mg/L COD with an almost neutral pH value (7.4 to 7.6). Moreover, it has high composition of phosphorus (1.2 mg/L), potassium (5.1 mg/L), calcium (2.5 mg/L) and magnesium (1.4 mg/L) [6]. In this study, POME and wastewater sludge were obtained from a biogas plant in Dengkil, Selangor and immediately stored in a refrigerator (below 4 °C) to inhibit any microbial activities (Fig. 1).

2.2 Method

In this study, thermal pre-treatment, which is furnace pre-heating, has been chosen as pre-treatment. The furnace heating temperature fluctuated from 75 to 80 °C for an hour. Prior to and after the pre-treatment process, the weight of the substrate was measured. The goal is to see how much of a difference there is in weight after an hour of heating.

The Automatic Methane Potential Test System (AMPTS®) II was then used to conduct a biochemical methane potential (BMP) test to assess the potential of biomethane produced in anaerobic conditions [7]. Before starting the experiment, all the substrate and inoculum pH values were measured before setting up the automatic methane potential test system and after the experiment. This step is necessary to

(a) (b)

Fig. 1 Samples of **a** Palm Oil mill Effluent (substrate), **b** wastewater sludge (inoculum)

ensure that the mixture does not become too acidic, which could inhibit the biodegradation process. We used 400 mL of total substrate and inoculum in 500 mL bioreactor bottles as suggested in the manual [8]. Eight bioreactor bottles were used. Three of them were filled with 300 ml of untreated POME and 100 ml of inoculum. Another three reactors contained 300 ml of treated POME and 100 ml of inoculum. The other two bioreactors were filled with treated and untreated POME only, without inoculum. The experiment was conducted for three days. The setup of the BMP test is shown in Fig. 2, with a water bath temperature maintained at 35 °C.

Fig. 2 Biochemical methane potential (BMP) test setup

3 Result and Discussion

Figure 3 shows that the biogas production of the P1 mixture was stopped too soon, at 15 h. It halted because the connecting tube between the CO_2 absorption and gas volume measuring devices became clogged. However, the production of biogas from the P2 sample seemed to continue even after the three days. The P2 and P3 samples generated about 99.3 and 91.9 Nml of biogas, respectively. Furthermore, compared to other mixtures of untreated POME and inoculum bioreactors, such as P5 and P6, P4 produced a small amount of methane. It is most likely due to the fact that microbial activity had already ceased at 85 h. As mentioned in the previous study, methane released is produced exclusively due to microbial activity [9].

P2, P3, P5, and P6 are acceptable data from this experiment. The volume of methane produced at 71 h was used to calculate the average biomethane yield. Thus, the average volume between P2 and P3 is 91.25 Nml, while the average volume between P5 and P6 is 91.0 Nml. Results also show that the biomethane yield by POME and inoculum mixtures is larger than that of POME without inoculum (P7 and P8 samples). Furthermore, because inoculum contains more bacteria, adding the inoculum to the bioreactor container is recommended to promote the production of biomethane gas [10]. The presence of bacteria aids in the production of methane (Fig. 4).

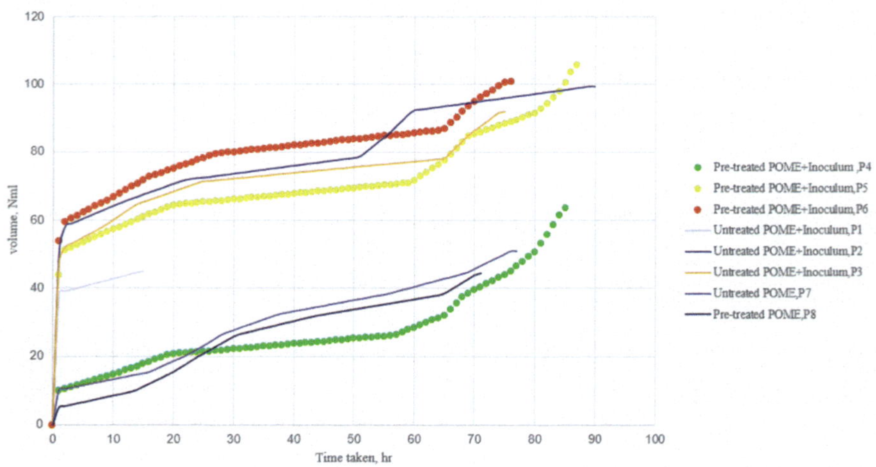

Fig. 3 Results of methane volume against the time (hour)

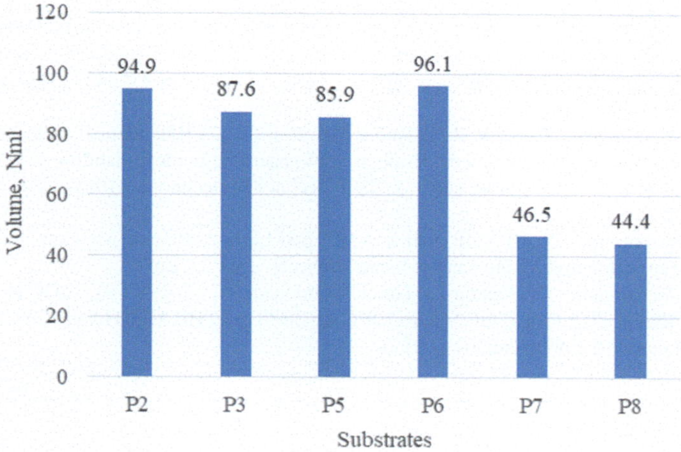

Fig. 4 Results of methane volume produced at 71 h

4 Conclusion

The purpose of this experiment was to investigate the potential of furnace heating pre-treatment for enhancing biomethane production during the anaerobic digestion (AD) process. Temperature and time were taken into consideration during the pre-treatment process. Based on the average data at 71 h, it was concluded that the biomethane produced from the furnace heated POME is comparable to the untreated POME.

Acknowledgements This study was supported by a grant from the AAIBE Chair of Renewable Energy, Institute of Sustainable Energy, Universiti Tenaga Nasional (UNITEN).

References

1. Joshua Levin W-U (2012) Profitability and Sustainability in Palm Oil Production. Analysis of Incremental Financial Costs and Benefits of RSPO Compliance.
2. GreenPalm GreenPalm: Palm oil is the most popular vegetable oil on the planet. https://www.greenpalm.org/about-palm-oil/what-is-palm-oil. Accessed 2 Nov 2020
3. Sawyerr N, Trois C, Workneh T, Okudoh V (2019) An overview of biogas production: Fundamentals, applications and future research. Int J Energy Econ Policy 9, 105–116. https://doi.org/10.32479/ijeep.7375
4. Khadaroo SNBA, Grassia P, Gouwanda D, Poh PE (2020) The impact of thermal pretreatment on various solid-liquid ratios of palm oil mill effluent (POME) for enhanced thermophilic anaerobic digestion performance. J Clean Prod 261:. https://doi.org/10.1016/j.jclepro.2020.121159
5. Zainal NH, Jalani NF, Mamat R, Astimar AA (2017) A review on the development of palm oil mill effluent (POME) final discharge polishing treatments. J Oil Palm Res 29, 528–540. https://doi.org/10.21894/jopr.2017.00012

6. Baharuddin AS, Hock LS, Yusof MZM et al (2010) Effects of palm oil mill effluent (POME) anaerobic sludge from 500 m3 of closed anaerobic methane digested tank on pressed-shredded empty fruit bunch (EFB) composting process. African J Biotechnol 9:2427–2436. https://doi.org/10.4314/ajb.v9i16
7. Raposo F, Fernández-Cegrí V, de la Rubia MA et al (2011) Biochemical methane potential (BMP) of solid organic substrates: Evaluation of anaerobic biodegradability using data from an international interlaboratory study. J Chem Technol Biotechnol 86:1088–1098. https://doi.org/10.1002/jctb.2622
8. Bioprocess Control (2020) AMPTS II & AMPTS II Light Automatic Methane Potential Test System Operation and Maintenance Manual. Scheelevägen 22, Lund
9. Cadena S, Cervantes FJ, Falcón LI, García-Maldonado JQ (2019) THE EARTH AND ITS RESOURCES THE ROLE OF MICROORGANISMS IN THE METHANE CYCLE. 7:1–8. https://doi.org/10.3389/frym.2019.00133
10. What bacteria can teach us about efficient methane production? https://www.biooekonomie-bw.de/en/articles/news/what-bacteria-can-teach-us-about-efficient-methane-production#sitetop. Accessed 5 Aug 2021

Iron Removal Efficiency in Synthetic Acid Mine Drainage (AMD) Treatment Using Peat Soil

Mohd Syazwan Mohd Halim, Abdul Haqi Ibrahim, Tengku Nuraiti Tengku Izhar, Suhaina Ismail, Ku Esyra Hani Ku Ishak, and Andreea Moncea

Abstract Acid mine drainage (AMD) formation is due to the sulfide minerals reaction either chemically or biologically when exposed to atmospheric conditions. The AMD formation often occurred in the region involved with anthropogenic activities, including mining, agricultural plantation, urban development and logging. Treatment of AMD is a challenging part of most mining operations around the world. Selection of method treatment is crucial depending on the area's geological, mineralogical, topography and AMD characteristic. There are two types of method treatment; active and passive treatment method. In this study, passive treatment method was adopted; which is successive alkalinity producing system (SAPS). The study aims to analyze effect of variable parameters on iron (Fe) concentration and propose optimum operating condition for AMD treatment. Peat soil and limestone aggregate was used as treatment media in treatment tank. Synthetic AMD was formulated using sulphuric acid (H_2SO_4) and iron sulfate ($FeSO_4$) to represent actual AMD. Once the synthetic AMD was introduced, water samples were collected and analyzed using UV–Vis test

M. S. M. Halim (✉) · A. H. Ibrahim · T. N. T. Izhar
Faculty of Civil Engineering Technology, Universiti Malaysia Perlis, 02600 Perlis, Malaysia

A. H. Ibrahim
e-mail: abdulhaqi@unimap.edu.my

T. N. T. Izhar
e-mail: nuraiti@unimap.edu.my

Centre of Excellence, Water Research and Environmental Sustainability Growth, Universiti Malaysia Perlis, 02600 Perlis, Malaysia

S. Ismail · K. E. H. K. Ishak
School of Materials and Mineral Resources Engineering, Universiti Sains Malaysia, Engineering Campus, 14300 Nibong Tebal, Pulau Pinang, Malaysia
e-mail: suhaina@usm.my

K. E. H. K. Ishak
e-mail: kuesyrahani@usm.my

A. Moncea
National Institute for Research and Development in Environmental Protection Buchares (INCDPM), 294, Splaiul Independentei Street, 6th District, 060031 Bucharest, Romania

N. Mohamed Noor et al. (eds.), *Proceedings of the 3rd International Conference on Green Environmental Engineering and Technology*, Lecture Notes in Civil Engineering 214,
https://doi.org/10.1007/978-981-16-7920-9_35

after 6 to 48 h' retention time. Based on the analysis, the proposed methodology has successfully reduced more than 85% iron content only after 6 h of retention time. The maximum Fe removal percentage recorded was 95%, using the higher peat soil depth configuration. The statistical analysis results show that the optimum operating condition for SAPS with high Fe removal is using high peat soil depth. Experiments with higher peat soil depth provide satisfactory results in treating the high initial Fe concentration regardless of the retention time for the AMD treatment.

Keywords Passive treatment · AMD · Successive alkalinity producing system (SAPS) · Peat soil · Heavy metal

1 Introduction

Acid mine drainage (AMD) formation is due to the sulfide minerals reaction either chemically or biologically when exposed to atmospheric conditions. The AMD formation often occurred in the region involved with anthropogenic activities, including mining, agricultural plantation, urban development and logging. Production of the AMD is a conversion of solid-phase activity (sulfide minerals) to solution-phase acidity with high heavy metals content. A simple explanation of AMD formation can be represented by pyrite (FeS_2) mineral. The process is initiate with pyrite oxidation and release of ferrous iron (Fe^{2+}), sulfate ($SO_4{}^{2-}$) and hydrogen (H^+). Emission of H^+ results in low pH of the AMD solution. The Fe^{2+} reacts with dissolved oxygen to produce iron oxide (Fe_2O_3) precipitates, commonly called "yellow boy" and embedding on stream or ocean bed.

Treatment of AMD is a challenging part of most mining operations around the world. Selection of method treatment is crucial depending on the area's geological, mineralogical, topography and AMD characteristic. There are two types of method treatment: active and passive treatment method. The active treatment is also known as a chemical treatment where chemical agents such as hydrated lime ($Ca(OH)_2$), calcium oxide (CaO), caustic soda (NaOH) and soda ash (Na_2CO_3) are introduced into AMD water. The active treatment methods are effective in neutralizing pH levels and suitable for active mining operations. However, the operation and maintenance cost is expensive as the system requires an electrical supply, continuous chemical usage, comprehensive surveillance by the skilled worker, and proper waste management [1].

Alternatively, passive treatment methods are developed to treat AMD water and improve the water quality through natural flow [2]. There are several passive treatment methods adopted to treat AMD include aerobic wetlands, anaerobic wetlands, anoxic limestone drain (ALD), open limestone channel (OLC) and successive alkalinity producing system (SAPS). The mechanism of passive treatment methods is to take advantage of natural chemical and biological processes that do not require any continuous chemical supply to treat contaminated AMD water [3]. The passive treatment methods have two categories, which are oxidizing and reducing approaches. First is an oxidizing system, where the Fe is sequestered by precipitation as iron hydroxide ($Fe(OH)_3$). Second, is a reducing system is vice versa to the oxidizing system where the dissolved Fe and SO_4^{2-} in the AMD experienced reduction process and forming the compounds such as FeS, FeS_2 and H_2S [4].

In the passive treatment method, the selection of organic substrates with high carbon content is essential. The carbon content produces sulfate reduction bacteria (SRB) as an agent to sequester heavy metals in the AMD. The SRB reduces sulfate (SO_4^{2-}) to hydrogen sulfate (H_2S), which reacts with dissolved metals to form solid-phase metal sulfate. There are many types of organic substrate used in previous studies, such as compost mushroom spent [5, 6], animal manure [7–9] and sewage sludge [8]. Recent studies of passive treatment indicate that iron (Fe) has the most research number compared to other metals. The iron removal mechanism in the passive treatment system is due to the redox reaction, where the oxidation–reduction reaction occurs by electrons exchange from the reducing agent to the oxidizing agent [10]. As a result, the iron tends to oxides to ferrous ions by acting as a reducing agent.

In this study, SAPS was adopted to treat a formulated synthetic AMD using peat soil as an organic substrate. The synthetic AMD contained a high Fe concentration and low pH value. Iron was the only heavy metal element in the synthetic AMD because it is the most abundant element in the AMD among the dissolved heavy metals. It is often the principal contaminant of AMD concern from the mining site and the most challenging metal to be removed from AMD by using passive treatment method due to coating or armoring of limestone. The peat soil was selected as an alternative treatment media for the common organic substrate used by researchers. This study aims to identify the efficiency of SAPS in Fe removal and correlation between variable parameters to achieve optimum operating conditions was presented.

2 Materials and Method

2.1 Material

The material used for SAPS laboratory-scale setup is peat soil, limestone aggregate and synthetic AMD. The peat soil sample was obtained from a local nursery located in Nibong Tebal, Penang, Malaysia. The limestone aggregate (10–20 mm) was collected from a limestone quarry in Baling, Kedah, Malaysia. The synthetic AMD was formulated using distilled water, sulphuric acid (H_2SO_4) and iron sulfate ($FeSO_4$).

2.2 Design of Experiment

Design of Experiment (DOE) was a method that has been used to investigate the optimum operating conditions to reduce Fe concentration in synthetic acid mine

Table 1 Experimental run of 2^3 with four center points factorial design

Run order	Variable		
	Peat soil depth (cm)	Initial Fe concentration (ppm)	Retention time (hour)
1	5.0	50	12
2	5.0	50	48
3	5.0	100	12
4	5.0	100	48
5	10.0	50	12
6	10.0	50	48
7	10.0	100	12
8	10.0	100	48
9	7.5	75	30
10	7.5	75	30
11	7.5	75	30
12	7.5	75	30

drainage (AMD) solution. The effect of the peat soil depth, Fe concentration, and retention time was studied using Minitab 19 (Minitab Inc, USA) software. A factorial model of two power of three (2^3) designs with four center points implemented using this software. There are twelve experiments configuration in total were conducted for this study as shown in Table 1. Effluent water samples were collected every 6 h at the effluent point and preserved before measure Fe concentration via UV–Vis spectrophotometric for the respective hour.

2.3 *Laboratory Set-Up*

The experiment was set up using transparent polypropylene tanks. First, synthetic AMD was prepared in a 60 L solution tank then supplied into the treatment tanks by gravitational flow. Then, limestone aggregate and peat soil were placed in the treatment tanks, respectively, as shown in Fig. 1.

The synthetic AMD thickness throughout this study was constant; 5 cm. The synthetic AMD solution was supplied from the solution tank through the influent pipe at the treatment tank top. Then the synthetic AMD was flown downward through the peat soil before contact with limestone. Eventually, the treated solution was collected at effluent pipe within 6 h' interval of retention time. During effluent water sample collection, pH and temperature were measured in-situ and 15 ml of the sample was collected to measure the Fe concentration by using UV–Vis spectrophotometry.

Fig. 1 Actual treatment tank configuration

3 Result and Discussion

Fe concentration removal efficiency was calculated based on the concentration difference between influent and effluent samples to determine the effectiveness of the SAPS toward the experiment's variables. Referring to Fig. 2, not much difference between

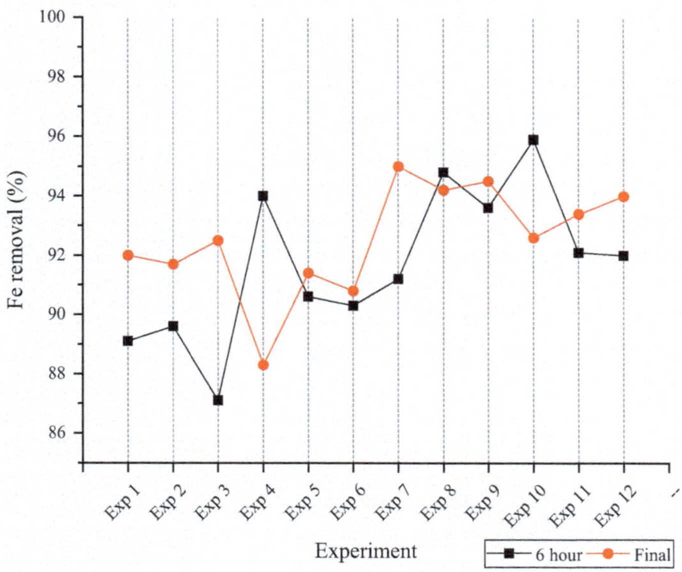

Fig. 2 Comparison of percentage Fe removal between initial 6 h and final retention time

Experiment 1 and Experiment 2 at the initial 6 h and final retention time. Even though Experiment 2 was conducted longer retention time. The results indicate that the SAPS in both experiments experience insufficient carbon sources which generate H_2S to further reacts with Fe ions to form insoluble metal sulfide precipitates.

Experiment 3 and Experiment 4 have the same variables setup; 5.0 cm of peat soil depth and 100 ppm of initial Fe concentration, but different total retention times; 12 h and 48 h, respectively. At the initial 6 h of retention time, percentage Fe removal in Experiment 3 was lower than Experiment 4 significantly. However, Experiment 3 recorded a higher percentage of Fe removal at the end of the experiment compared to Experiment 4. Comparison between both experiments shows that the efficiency of the SAPS is decreased over time when treating a high concentration of Fe with a lower depth of peat soil layer. The difference in the percentage of Fe removal due to a decrease in dissolved organic carbon content in the SAPS over time contributed to decreased biological sulfate reduction. Hence, less activity of SRB occurred, which supposed to react with dissolved Fe to form FeS in the peat soil layer.

Overall, the treatment tanks containing higher than 5.0 cm of the peat soil depth have higher removal rates at 6 h of retention time until the final hour, which is more than 90%. It can be concluded that a tank that contains higher peat soil depth could sustain the efficiency of percentage Fe removal for a long time. The result suggested that Fe removal's effectiveness in SAPS depends on the amount or depth and organic carbon content of the organic substrate. Microbial activity of SRB more responsive to high metal loading of AMD at initial few hours. The efficiency of Fe removal rates should then decrease over time since the available surface for the adsorption on organic substrates becomes smaller and limited [12].

4 Conclusion

A characterization study of the AMD quality is essential before designing the SAPS using peat soil as an organic substrate. The results suggest that the optimum operating condition for SAPS with high Fe removal is using high peat soil depth. Experiments with higher peat soil depth provide satisfactory results in treating the high initial Fe concentration regardless of the retention time for the AMD treatment. It is shown that higher peat soil depth could provide a sufficient organic carbon source for bacteria activity to reduce Fe concentration in the synthetic AMD.

However, it does not mean that lower peat soil depth cannot be used in SAPS. For instance, the lower peat soil depth could treat a lower Fe concentration of AMD. Based on experimental results in Experiment 1 (low peat soil depth; 5.0 cm, low initial Fe concentration; 50 ppm and low retention time; 12 h), the percentage of Fe removal efficiency is high, which is 92.0%. Therefore, suggest that a lower peat soil depth configuration can be constructed as a final AMD treatment stage before the water is discharged from the mining tenement area.

References

1. Torres E, Lozano A, Macías F, Gomez-Arias A, Castillo J, Ayora C (2018) Passive elimination of sulfate and metals from acid mine drainage using combined limestone and barium carbonate systems. J Clean Prod. https://doi.org/10.1016/j.jclepro.2018.01.224
2. Zipper, C, Skousen J, Jage C (2011) Passive treatment of acid mine drainage. Passive Treatment of Acid-Mine Drainage, 9780470487(April 2014), 339–353. https://doi.org/10.1002/9781118749197.ch30
3. Skousen JG, Ziemkiewicz PF, McDonald LM (2019) Acid mine drainage formation, control and treatment: Approaches and strategies. Extractive Industries and Society 6(1):241–249. https://doi.org/10.1016/j.exis.2018.09.008
4. Trumm D (2010) Selection of active and passive treatment systems for AMD - Flow charts for New Zealand conditions. NZ J Geol Geophys 53(2–3):195–210. https://doi.org/10.1080/00288306.2010.500715
5. Yim G, Ji S, Cheong Y, Neculita CM, Song H (2015) The influences of the amount of organic substrate on the performance of pilot-scale passive bioreactors for acid mine drainage treatment. Environmental Earth Sciences 73(8):4717–4727. https://doi.org/10.1007/s12665-014-3757-9
6. Muhammad SN, Kusin FM, Zahar MSM, Halimoon N, Yusuf FM (2015) Passive Treatment of Acid Mine Drainage Using Mixed Substrates: Batch Experiments. Procedia Environ Sci 30:157–161. https://doi.org/10.1016/j.proenv.2015.10.028
7. Singh S, Chakraborty S (2020) Performance of organic substrate amended constructed wetland treating acid mine drainage (AMD) of North-Eastern India. J Hazardous Mater 397, 122719. https://doi.org/10.1016/j.jhazmat.2020.122719
8. Mashalane TB, Novhe ON, Coetzee H, Wolkersdorfer C (2018) Metal removal from mine water using integrated passive treatment system at Witkrantz discharge point in the Ermelo Coalfields 11th ICARD/IMWA/MWD Conference: Risk to Opportunity 146–150
9. Zhang M, Wang H (2014) Organic wastes as carbon sources to promote sulfate reducing bacterial activity for biological remediation of acid mine drainage. Miner Eng 69:81–90. https://doi.org/10.1016/j.mineng.2014.07.010
10. Limper D, Fellinger GP, Ekolu SO (2018) Evaluation and microanalytical study of ZVI/scoria zeolite mixtures for treating acid mine drainage using reactive barriers - Removal mechanisms. J Environ Chem Eng 6(5):6184–6193. https://doi.org/10.1016/j.jece.2018.08.064
11. Halim MS, Ibrahim AH, Izhar TT, Ismail S, Jaafar ZM (2020) Peat soil for synthetic acid mine drainage treatment: Characteristic study. IOP Conference Series: Earth and Environmental Science, 616(1). https://doi.org/10.1088/1755-1315/616/1/012069
12. Ji S, Kim S, Ko J (2008) The status of the passive treatment systems for acid mine drainage in South Korea. Environ Geol. https://doi.org/10.1007/s00254-007-1064-4

Performance of Two Phase Anaerobic Digestion on Food Waste for Biogas Production

Irnis Azura Zakarya, Tengku Nuraiti Tengku Izhar, Siti Khadijah Zaaba, Nur Adlina Mohd Hilmi, Mohammad Rizam Che Beson, and Monica Matei

Abstract Every year, the total amount of solid waste generated in Peninsular Malaysia grew, according to the Malaysia government. The typical solid waste management system practiced in developing country brings many problems that can cause risks and hazards for living things and the environment in Malaysia if there are not managed properly. For example, illegal dumping cause groundwater and soil pollution. The methane gas produced from the landfill causes greenhouse effect. Food waste is categorized under household waste, where it is produced in residential areas, restaurants, cafeteria, markets and commercial areas. Therefore, anaerobic digestion process is introduced for organic waste with higher solids contents such as food waste as an alternative method. The effectiveness of anaerobic digestion process can be investigated through this process. In this study, two phase of anaerobic digesters was proposed. The reactor was operated at a temperature 35 °C, analyzed for biogas

I. A. Zakarya (✉) · T. N. Tengku Izhar · N. A. M. Hilmi
Faculty of Civil Engineering Technology, Universiti Malaysia Perlis, 02600 Arau, Perlis, Malaysia
e-mail: irnis@unimap.edu.my

T. N. Tengku Izhar
e-mail: nuraiti@unimap.edu.my

N. A. M. Hilmi
e-mail: dynahilmi@studentmail.unimap.edu.my

Sustainable Environment Research Group (SERG), Centre of Excellence Geopolymer and Green Technology (CEGeoGTech), Universiti Malaysia Perlis, 01000 Kangar, Perlis, Malaysia

S. K. Zaaba
Center for Energy Management & Sustainable Campus (COSCEM), Universiti Malaysia Perlis, 02600 Arau, Perlis, Malaysia
e-mail: khadijah@unimap.edu.my

M. R. C. Beson
E-Idaman Sdn.Bhd. Wisma Idaman, Alor Setar, Kedah, Malaysia
e-mail: mohdrizam@e-idaman.com

M. Matei
National Institute for Research and Development in Environmental Protection Bucharest (INCDPM), Splaiul Independentei Street, 6th District, 294 Bucharest, Romania

N. Mohamed Noor et al. (eds.), *Proceedings of the 3rd International Conference on Green Environmental Engineering and Technology*, Lecture Notes in Civil Engineering 214,
https://doi.org/10.1007/978-981-16-7920-9_36

production, pH values and C/N ratio. The highest biogas production in this study is 50.4%. For a co-digestion containing more sludge can increase the pH value and biogas production.

Keywords Anaerobic digestion · Two-phase reactor · Food waste · Biogas

1 Introduction

Municipal solid waste is primarily consists of waste produced by domestic households, even though it includes some commercial and industrial waste. This waste is made up of waste, organics and recyclable materials with the municipality overseeing its disposal. Municipal solid waste is collected then separated to send to landfill or recycling centre for recycling. The waste that is sent to landfill is dump into designated cell, compacted and covered [1].

Due to the demand of Malaysian for quality of life increase, the solid waste generation rate also increase. In year 2001, the solid wastes were generated from 16,200 tonnes/day to 19,100 tonnes/ day in year 2005 in Peninsular Malaysia. Then, total amount of solid waste that were generated in Malaysia has increased to 1.3 kg/capita/day in year 2006 and expected to reach 1.5 kg/capita/day in most cities [2].

Developing country can causes many problems when practiced the conventional solid waste management system such as open dumping without regard for air and water pollution, as well as vermin and fly breeding. These problems can cause risks and hazards for living things and the environment if there are not managed properly [3]. Besides, approximately 95–97 percent of collected waste transferred to landfill for disposal, which necessitates the construction of additional landfills for dumping and only the remaining of waste is taken for intermediate treatment. They are sent for treatment at small incineration plants; diverted to recyclers and reprocessors or dumped illegally [4]. Illegal dumping cause groundwater and soil pollution. The methane gas produced from the landfill causes greenhouse effect.

One alternate option for tackling the concerns mentioned above is anaerobic digestion. This is a biological treatment that can help to reduce the air and water pollution, landfill usage, control groundwater and soil pollution and harness biogas as a sustainable energy source such as cooking gas, electricity and fuel. Anaerobic bacteria compost organic materials in the absence of oxygen and produces leachate, carbon dioxide and methane gas [5].

Food waste is classified as household waste because it is generated in residential areas, restaurants, cafeterias, markets and commercial areas. It typically has a total solids content of 18.1 percent to 30.9 percent, and the anaerobic digestion system requires a large amount of water to homogenize it. These require more energy for pumping, mixing and heating. Therefore, for organic waste with larger solids contents, such as food waste, dry anaerobic digestion procedure has been introduced [6, 7].

In order to facilitate solid waste degradation, two phase anaerobic digestion, which separates acidogenesis and methanogenesis, is an appealing enhancement over single phase digestion for solid waste decomposition. Each phase which tuned separately, resulting in high rate of degradation and energy recovery [8].

Two reactors make up a two phase anaerobic digestion system. The first rector performs hydrolysis, acidogenesis and acetogenic reactions, while the second reactor uses those acids for methanogenesis. Anaerobic digestion by microbial digestion of the organic part of wastes. The organic matter component of food waste is turned into biogas during the anaerobic degradation process in a chain process that consists primarily of four phases. Hydrolysis, acidogenesis, acetogenesis, and methanogenesis are the four reactions involved [5].

The application of anaerobic digestion as a process in a solid waste management system was proposed in this research to reduce the impact of waste on the environment and human health, as well as landfill utilization in Malaysia. The method used in this study is also discussed.

2 Materials and Method

2.1 Material

For this anaerobic digestion, the food wastes used were collected from the food stall area of campus, which is located at Campus Jejawi, Perlis. This is represented kitchen waste from food stores area. It contained residues of chicken, fish and meat, as well as spoiled rice. The bones and shells from food waste were taken out before fed into the reactor.

Two phase anaerobic digestion was made up from two reactors, which are acidogenic reactor for first reactor and methanogenic reactor for second reactor. In this study, two identical glass fabricated reactors measuring 20 cm in diameter and 30 cm in height with capacity of 7L were used. The inoculum, which are sludges from Taman Tunku Sarina (TTS) and Kulim were used for this anaerobic digestion process.

2.2 Method

The first reactor was fed with 1 kg of food waste and 1 kg of inoculum while second reactor was fed with 2 kg inoculum and left in mesophilic condition (35 °C) and acclimatized for 2 weeks. Feeding and loading were carried out manually by opening the cap of valve and pump out the sample. After 2 weeks of acclimatization, 400 ml of leachate from acidogenic reactor is transferred to methanogenic reactor and then 400 ml of food waste is fed into acidogenic reactor to maintain the weight of the reactor. The sludge from TTS was used in both reactor and sludge from Kulim was used together during day 40 in the second reactor. Sample was taken from acidogenic reactor and methanogenic reactor for every 3 days and 7 days respectively to do the tests [5].

Anaerobic digestion needs a suitable sludge to perform well. Parameters such as pH determination after sampling, carbon and total nitrogen (HACH method) were carried out after the process. The inoculum and Calcium Carbonate ($CaCO_3$) was added to increase the pH in the reactor [9]. The amount of leachate that is transferred from acidogenic reactor to methanogenic reactor should be maintain with same amount of feeding. Gas production was measured and analyzed before the samples were collected for biogas analysis using the portable gas analyser GA5000. A tedlar gas sample bag attached to the reactor was used to collect gas.

3 Result and Discussion

Anaerobic digestion is one of the methods that apply in the Malaysia in the waste management. Sludge and food waste co-digestion show positive effect in the methane gas production. To determine the methane gas production that can be achieved through co-digestion, two types of reactors can be used: single phase and two phase reactors. In this study, two-phase reactor was used to determine the methane gas production. The gas was measure in the both reactors. Before being fed into the reactor, the pH of food waste was measured, and it range from 5.88 to 6.62.

This research focused on methane gas production and the characteristics of sludge that affect gas production. Before food wastes were feeding in the reactor 1, the inoculum was acclimatized and the main composition for both inoculum from TTS and Kulim were shown in the Table 1.

Figure 1 shows that the pH and methane gas production the reactor 1. The food wastes were feeding on the 11th day and the pH become acidic on the 14th day, which the value is below 4.3. The pH was drop from 7.7 to 4.1 on that day. The values of pH decrease on day 15 and slightly increase on the 27th day. This condition happened due to acidogenesis phase occurs and turn to acidic condition.

During the stage of hydrolysis and fermentation, non-methanogenic microorganisms are playing role and adapting to low pH [10]. The high concentration of volatile acid in the early phase causes the pH values to drop as well and caused unsuitable condition for the growth of methanogenic [11]. To increase the pH in the reactor 1, the Calcium Carbonate ($CaCO_3$) was added on the 32th day, however the pH still

Table 1 Main composition of inoculum from TTS and Kulim

Test parameter	Unit	TTS	Kulim
Organic matter	%	5.1	29.6
Nitrogen	%	1.6	2.6
C/N ratio	%	0.35: 1	5.69: 1
pH	–	7.5	7.7

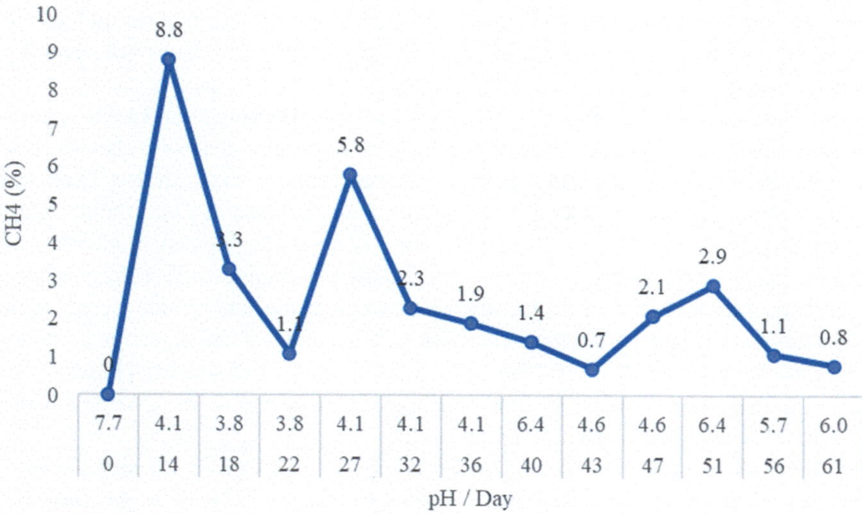

Fig. 1 Results of pH and CH_4 for Reactor 1

maintains at pH 4.1, which is acidic. Then, the dosage of $CaCO_3$ was used increase from 10 to 40 g on day 36. From the result, it showed that the pH was increase on day 40 to 6.1. $CaCO_3$ was used for controlling the pH in the reactor 1.

Figure 2 shows the pH values of reactor 2 of inoculum slightly increase during acclimatized from 7.7 to 8.2 on the 14th day. However, the pH value dropped back to 7.7 on day 18 after feeding time on day 14 and the pH value slowly dropped after feeding time. The pH value increase from 6.7 to 7.2 because on day 40, 2 kg of

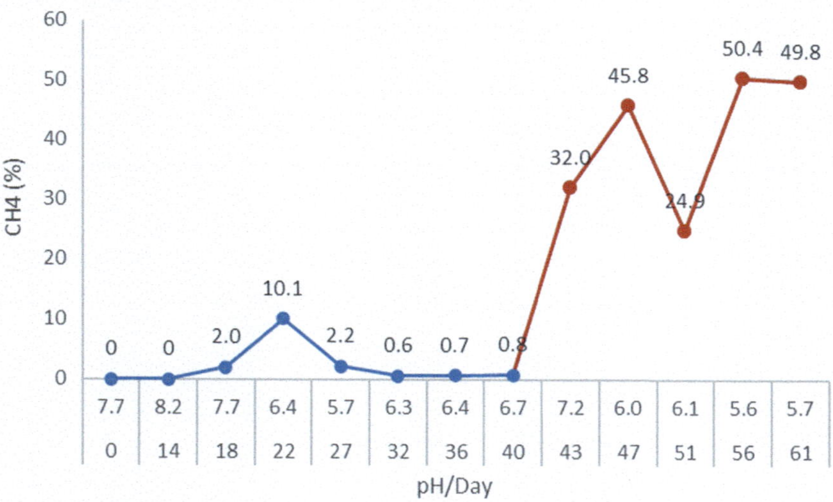

Fig. 2 Results of pH and CH_4 for Reactor 2

inoculum from Kulim was added. Inoculum was added to maintain the pH and to increase the methane production. However, the pH also slowly drops after feeding on the next day.

The methane gas production low in reactor 1 because hydrolysis and acidogenesis occurs in this phase. However, the methane gas production in reactor 2 also low and the higher percentage of methane gas produced only 10% on the 22th day. Then, the inoculum from Kulim was added in the reactor 2 on the 40th day and the methane gas production increase on the 43th day with 32.0%. The highest production of methane gas was on 56th day with 50.4%. Small pH changes affect the methane formers (methanogens) in anaerobic digestion, whereas the acid producers can work adequately over a large pH range. The buffering capacity of the digestive contents affects digestion stability [12]. The acid and methane fermentation phases were well separated, as evidenced by low and high methane yields of 15% and 60%, respectively, in the first and second phases [13].

Figure 3 shows the C/N ratio for both reactors. The C/N ratios for inoculum for both reactors are 5.69. The range of C/N ratio for reactor 1 is 18.78 on the 18th day and 40.96 on the 51th day while for reactor 2 are 5.83 on the 18th day and 14.5 on the 56th day. When the inoculum from Kulim was added on the 40th day in the reactor 2, the C/N ratio was drop from 14.84 to 7.65. This is occurred because of the C/N ratio for the inoculum from Kulim is 0.35 and it is give effect to the C/N ratio in the reactor 2. The C/N ratio for reactor 1 and reactor 2 is 37.48 and 14.25 respectively that produce highest methane gas on day 56. The optimal condition for biogasifcation is 29.6 for C/N ratio in mesophilic condition [14].

Bacteria need a suitable ratio of carbon to nitrogen (C:N) for their metabolic processes. If the C/N ratio is less than 1.0, the bacterial community will digest the substrate more quickly, using carbon 25–30 times faster than nitrogen. A excess of nitrogen will result in the formation of ammonia, which will inhibit digestion [10]. According to Cheerawit [15] a high nitrogen content results in a high C/N ratio, which

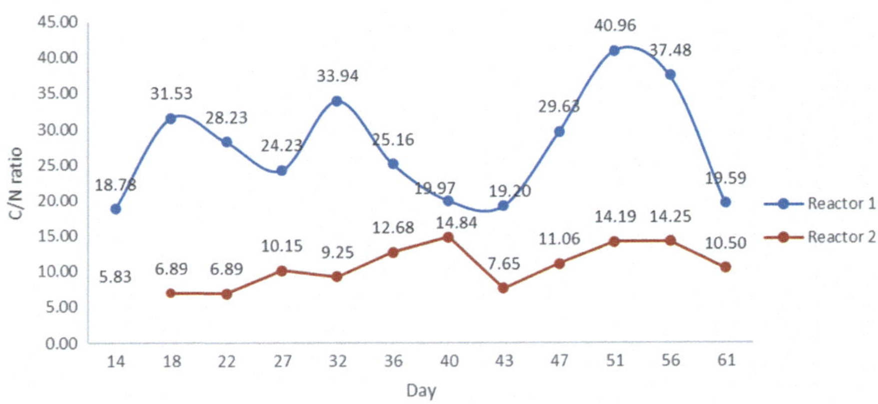

Fig. 3 Results of C/N ratio for Reactor 1 and Reactor 2

increases nitrogen consumption during the starting process, reducing lignocellulosic biomass and led to poor anaerobic digestibility.

The pH values in the reactor 1 can give effect to the pH values in the reactor 2. $CaCO_3$ can be used to maintain the pH in the reactor 1 and need to be added when the feeding occurs. The pH value in the reactor 2 was affected by the sludge from reactor 1. Therefore, for pH value in reactor need to maintain in range 6.5–5.5. For methane gas production, the amount of inoculum is important. The inoculum were added during the feeding also can increase the pH values and the gas production in the reactor 2.

4 Conclusion

In conclusion, the pH value and C/N ratio of co-digestion of food waste and sludge give impact to the production of methane gas. The effectiveness of anaerobic digestion process in two phase anaerobic digester was shown from the methane production, which increased from 32.0% to 50.4%. The amount of sludge used also influenced on the methane gas production for mesophilic condition. In this condition, it can contribute to the microorganism in the sludge that needed for anaerobic digestion, which the process for acidogenesis, hydrolysis and methanogenesis also depends on the microbe from the sludge.

An anaerobic digestion system can become one of the ways to reduce the organic municipal waste and reduce the impact of waste on environment and human health in Malaysia. Waste management system in the Malaysia can be more develop by using this method.

References

1. Jouhara H, Czajczyńska D, Ghazal H, Krzyżyńska R, Anguilano L, Reynolds AJ, Spencer N (2017) Municipal waste management systems for domestic use. Energy 139:485–506. https://doi.org/10.1016/j.energy.2017.07.162
2. Periathamby A, Shahul F (2009) Evolution of solid waste management in Malaysia: impacts and implications of the solid waste bill, 2007. J Mater Cycles Waste Manag 11:96–103. https://doi.org/10.1007/s10163-008-0231-3
3. Samsudin MDM, Don MM (2013) Municipal solid waste management in Malaysia: current practices, challenges and prospect. J Teknol Sci Eng 62:95–101. https://doi.org/10.11113/jt.v62.1293
4. Afroz R, Masud MM (2011) Using a contingent valuation approach for improved solid waste management facility: evidence from Kuala Lumpur, Malaysia. Waste Manag 31:800–808. https://doi.org/10.1016/j.wasman.2010.10.028
5. Zakarya IA, Abustan I, Ismail N, Yusoff MS (2013) Production of methane gas from organic fraction municipal solid waste (OFMSW) via anaerobic process: application methodology for the Malaysian condition. Int J Environ Waste Manag 12:121–129. https://doi.org/10.1504/IJEWM.2013.055588

6. Abas MA, Wee ST (2016) Municipal solid waste management in Malaysia: an insight towards sustainability. SSRN J. https://doi.org/10.2139/ssrn.2714755
7. Fullerton D, Kinnaman TC (1995) Garbage, recycling, and illicit burning or dumping. J Environ Econ Manag 29:78–91. https://doi.org/10.1006/jeem.1995.1032
8. Chow WL, Chong S, Lim JW, Chan YJ, Chong MF, Tiong TJ, Chin JK, Pan GT (2020) anaerobic co-digestion of wastewater sludge: a review of potential co-substrates and operating factors for improved Methane yield. Processes 8(1):39. https://doi.org/10.3390/pr8010039
9. Wisconsin Department of Natural Resources Wastewater Operator Certification (1992) Advanced anaerobic digestion study guide. https://dnr.wisconsin.gov/. Accessed 21 May 2021
10. Paritosh K, Kushwaha SK, Yadav M, Pareek N, Chawade A, Vivekanand V (2017) Food waste to energy: an overview of sustainable approaches for food waste management and nutrient recycling. Biomed Res Int. https://doi.org/10.1155/2017/2370927
11. Venkiteshwaran K, Bocher B, Maki J, Zitomer D (2015) Relating anaerobic digestion microbial community and process function: supplementary issue: water microbiology. Microbiol Insights 8s2:MBI.S33593. https://doi.org/10.4137/mbi.s33593
12. Schnaars K (2012) What every operator should know about anaerobic digestion. https://www.wef.org/MAGAZINE. Accessed 21 May 2021
13. Ince O (1998) Performance of a two-phase anaerobic digestion system when treating dairy wastewater. Water Res 32:2707–2713. https://doi.org/10.1016/S0043-1354(98)00036-0
14. Yan Z, Song Z, Li D, Yuan Y, Liu X, Zheng T (2015) The effects of initial substrate concentration, C/N ratio, and temperature on solid-state anaerobic digestion from composting rice straw. Bioresour Technol 177:266–273. https://doi.org/10.1016/J.BIORTECH.2014.11.089
15. Rattanapan C, Sinchai L, Suksaroj TT, Kantachote D, Ounsaneha W (2019) Biogas production by co-digestion of canteen food waste and domestic wastewater under organic loading rate and temperature optimization. Environment 6(2):16. https://doi.org/10.3390/environments6020016

Removal of Oil from Oil-Contaminated Sand Using Chemical-Free Bubbles in Turbulent and Laminar Flows: A Conceptual Paper

Aliff Radzuan Mohamad Radzi, Sharifah Nadra Sofia Syed Hassan, Junaizad Jamil, Nor Shahirah Mohamad Nasir, and Amin Safwan Alikasturi

Abstract Regardless of the type of oil spill in our environment, numerous studies have demonstrated the ecological and economic consequences of our activities. There are many techniques available that can be used to treat oily-contaminated sand. However, significant improvements in efficiency, energy consumption, and treatment costs are required to make the technologies more environmentally friendly. This paper demonstrates the concept of micro and macro bubbles in the turbulent and laminar flows to remove the oil from the contaminated sand. The parameters tested in this study are macro and microbubbles, laminar and turbulent flows, saline and tap water, flowrate, retention time, and oil-sand ratio. The analysis proposed in this study is the chemical oxygen demand, total organic compound, thermogravimetric analysis, and Fourier Transform Infrared Spectroscopy. The preliminary results indicate that when microbubbles are introduced at the appropriate pressure, flowrate, salinity level, and temperature, the removal efficiency of oil from the oil-in-water mixture was 95.8%.

Keywords Remediation · Bubbles · Laminar · Turbulent · Oil

A. R. M. Radzi (✉) · S. N. S. S. Hassan · J. Jamil · N. S. M. Nasir · A. S. Alikasturi
Universiti Kuala Lumpur Branch Campus Malaysian Institute of Chemical and Bio-Engineering Technology, Kawasan Perindustrian Bandar Vendor, Taboh Naning, Lot 1988, 78000 Alor Gajah, Melaka, Malaysia
e-mail: aliffradzuan@unikl.edu.my

S. N. S. S. Hassan
e-mail: snadra.hassan@s.unikl.edu.my

J. Jamil
e-mail: junaizad.jamil@s.unikl.edu.my

N. S. M. Nasir
e-mail: norshahirah@unikl.edu.my

A. S. Alikasturi
e-mail: aminsafwan@unikl.edu.my

N. Mohamed Noor et al. (eds.), *Proceedings of the 3rd International Conference on Green Environmental Engineering and Technology*, Lecture Notes in Civil Engineering 214,
https://doi.org/10.1007/978-981-16-7920-9_37

1 Introduction

In August 2017, 9000 tons of palm oil spill happened when the Global Apollon ship, carrying raw palm oil, collided with another ship in Hong Kong. The situation deteriorated further was when the Hong Kong government discovered the spill, the palm oil already washing up on the beaches [1]. It caused 11 beaches to be closed temporarily to carry out the cleaning process of the beach. This spillage creates a slew of problems, as the surrounding beaches begin to smell foul and dead fish begin to cover the shores. There was solid palm oil floating around the ocean, and it was found that there were four inches of palm oil layer already seeping through the sand at the shore. The cleaned-up process was done by scooping up the oil waste, and some volunteers helped collect the Styrofoam-like solid on the shoreline [2] (Fig. 1).

The University of Hong Kong conducted a study to determine the palm spillage's degradation, bioaccumulation, and toxicity after it occurred, which took approximately 18 months [1]. After a year, only 10% of palm oil was disintegrated and degraded. The study also discovered that a few marine species have abnormally high fatty acid levels in their bodies. This study demonstrates how critical it is to remove palm oil from our shoreline immediately. A prompt response can help mitigate the environmental impact of palm oil spills [3].

Currently, a variety of methods are used commercially to clean up oil spills. One of the most frequently used techniques is oil booms, which contain the oil and prevent it from spreading further into the ocean. The second type is skimmers, which scoop oil

Fig. 1 Solidified palm oil stranded in Victoria Harbour [1]

from the water. It was designed to absorb the oil in the same way that a vacuum cleaner does. The third technique is sorbent, which involves introducing certain substances to absorb or adsorb the floating oil. Fourth is in-situ oil combustion, which can remove up to 98 percent of an oil spill [4]. It cannot, however, be conducted if the oil layer is not sufficiently thick. Fifth is with dispersant, which sprays and disintegrates the oil. It will create a chemical bond between oil and water by increasing the surface area. Sixth is the hot water and high-pressure washing, where it operates by spraying water up to 170 °C using high-pressure nozzles. Seventh employs manual labour to remove oil and debris from the shoreline using hand-held tools such as rakes and shovels. The eighth method is bioremediation, in which microbes such as bacteria, fungi, and algae are used to metabolise and break down the toxic molecules [4].

One of the well-known methods is to incorporate the dissolved air flotation (DAF) produced with microbubble. DAF technique is a viable technique of remediation used to separate and recover oil from polluted sand/soil by aiding the adhesion of the contaminant particles [5]. Previously, many studies have been conducted to study the usage of microbubbles flotation of oil from contaminated soil/sand [6–8].

Research is being conducted on the chemical-free separation of DAF mineral oil and water from sand [9]. However, no study examines the effects of various flow and bubble types on the separation process. The turbulent flow of macrobubbles is proposed in this study as a means of separating the oil molecule from the sand surface. This study has the potential to introduce a more efficient method of sand washing separation. A few other parameters were covered in this study: type of oil, salinity, flow type, experimental duration, and oil-sand ratio. If this approach can be successfully proven, we might have techniques with qualities such as being chemical-free, low energy consumption, and high removal efficiency.

The motivations of the study are to design a novel in-situ sand washing system with laminar and turbulent flow functions and investigate the performance of the parameters. In addition, this study is carried out to forecast the removal efficiency of oil from oily contaminated sand. There are three research questions from this study: What is the impact of bubble size on the removal of palm oil from sand? Secondly, how does the parameter tested will impact the process efficiency? Lastly, how does the type of flow will impact the removal of palm oil from sand?

2 Materials and Method

Two palm oil types will be used: Crude Palm Oil (CPO) and cooking oil. CPO would be obtained from the local palm oil mill. As for the cooking oil, a local cooking oil brand will be bought from a nearby supermarket. Sand beach samples and seawater used are from Pengkalan Balak, Melaka beach. As for the tap water, it is supplied by Syarikat Air Melaka Berhad.

2.1 Methodology

The sand samples will be prepared by air-dried the sand overnight to remove any moisture content and then sieve to remove debris such as wood chips, gravel, and stones. Sieving will be carried out to separate the sand into specific particle sizes (1000 μm, 500 μm, 125 μm). Initially, the physical and chemical properties such as density, viscosity, and API gravity will be identified for both palm oil used. The sand mix with the 200 g of palm oil in a beaker containing 500 g of sand dried and will be left to mix homogeneously overnight before using it for the experiments. The ratios of oil and sand are manipulated in this research work. Microbubbles will be generated along with laminar flow by the DAF rig that has been installed in UniKL-MICET. In contrast, macrobubbles will be generated by a pump with the highest capacity of 16 L/min of air and are intended to create a turbulent flow inside.

2.2 Floatation Experiments

The floatation experiments will be conducted in a floatation tank made from glass with a dimension of 0.4 m length, 0.3 m width, and 0.3 m height. The micro and macro bubbles generated will be introduced from the bottom of the contact zone of the washing tank. A steel strainer (0.25 m height, 0.09 m diameter) will hold the oily-contaminated sand and locate it inside the washing zone. During these experiments, a few parameters will be tested: type of bubble (macro/micro), type of flow introduced (laminar/turbulent), saline/tap water, flowrate, retention time, and oil-sand ratio.

2.3 Analysis

Several analyses will be conducted, such as the Fourier Transform Infrared Spectroscopy (FTIR), Chemical Oxygen Demand (COD), and Thermogravimetric Analysis (TGA). Once the FTIR evaluates all elements in the samples, the samples will be fed into TGA with known temperature and period. These analyses will observe the initial and final oil in the oily contaminated sand and determine the compounds present. The efficiency was then calculated using the following equation:

$$\frac{X_{initial} - X_{final}}{X_{initial}} \times 100\% \qquad (1)$$

$X_{initial}$ is the initial reading for all the analysis, and the X_{final} are the final reading for all the analyses. COD will be used to check the water quality using HACH Spectrophotometer. It is assumed that most of the oil floats to the water's surface to be easily skimmed off.

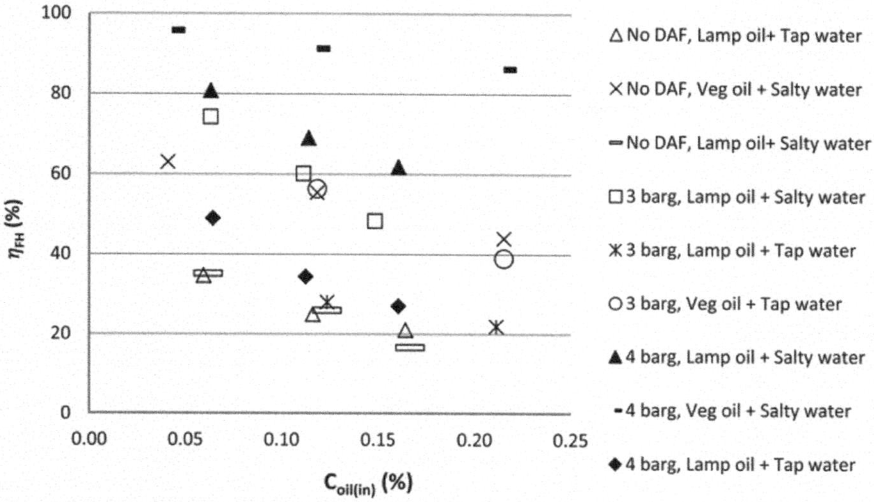

Fig. 2 Performance of oil droplets removal from the oil-in-water mixture by using FastHEX method

3 Hypotheses and Preliminary Result

This paper illustrates the effects of macrobubbles and microbubbles in a turbulent and laminar flow to remove the oil from oil-contaminated sand. Eight parameters such as oil to the sand ratio (w/w%), type of oil, type of flow, experimental duration, type of continuous phase, DAF pressure level, salinity, and size of sand particles to remove oil will be observed. It is a new idea to utilise macrobubbles in turbulent flow to remove oil. It is predicted that macrobubbles can help in increasing the overall rise velocity so that sufficient lighting force is supplied and increase the floatability of oil contaminants [6]. On the other hand, a mathematical model needs to be developed to understand better the effect of macrobubbles and turbulent flow towards the efficiency of oil-sand removal.

The investigations have been done to utilise the microbubbles in a laminar flow to remove the oil from the oil-in-water mixture [9]. The results show more than 90% of oil can be separated with the help of optimised parameters, as illustrated in Fig. 2. The previous research will be the benchmark in continuing the utilization of air bubbles in treating oil-contaminated sand.

4 Conclusion

Previous work has shown promising evidence on this new technique of sand washing. Although there is a lack of study to observe the effect of macro and micro bubbles combined with the turbulent flow in the treatment of oil-contaminated sand, the preceding work demonstrates high chances for this study to be successful. In this conceptual paper, new approaches have been introduced in the separation technology world.

Acknowledgements We would like to express our gratitude to Universiti Kuala Lumpur for providing the Short-Term Research Grant (STRG). On the other hand, we would like to acknowledge Majlis Amanah Rakyat (MARA) for the Graduate Excellence Programme (GrEP-MARA).

References

1. Kao E (2017) Questions over two-day delay on notice of palm oil spill that left 11 Hong Kong beaches closed. South China Morning Post
2. Chow L (2017) 11 beaches closed in Hong Kong After 9,000-Ton Palm Oil Spill. EcoWatch
3. Zhou GJ, Lai RWS, Sham RCT, Lam CS, Yeung KWY, Astudillo JC, Ho KKY, Yung MMN, Yau JKC, Leung KMY (2019) Accidental spill of palm stearin poses relatively short-term ecological risks to a tropical coastal marine ecosystem. Environ Sci Technol 53(21):12269–12277. https://doi.org/10.1021/acs.est.9b04636
4. Mi News Network (2021) 10 methods for oil spill cleanup at sea. In: Mar. Insight. https://www.marineinsight.com/environment/10-methods-for-oil-spill-cleanup-at-sea/. Accessed 7 May 2021
5. Silva EJ, Silva IA, Brasileiro PPF, Correa PF, Almeida DG, Rufino RD, Luna JM, Santos VA, Sarubbo LA (2019) Treatment of oily effluent using a low-cost biosurfactant in a flotation system. Biodegradation 30:335–350. https://doi.org/10.1007/s10532-019-09881-y
6. Lim MW, Lau EV, Poh PE (2018) Micro-macrobubbles interactions and its application in flotation technology for the recovery of high density oil from contaminated sands. J Pet Sci Eng 161:29–37. https://doi.org/10.1016/j.petrol.2017.11.064
7. Abousnina RM, Manalo A, Shiau J, Lokuge W (2016) An overview on oil contaminated sand and its engineering applications. Int J GEOMATE 10:1615–1622. https://doi.org/10.21660/2016.19.150602
8. Mat-Shayuti MS, Tuan Ya TMYS, Abdullah MZ, Othman NH, Alias NH (2021) Exploring the effect of ultrasonic power, frequency, and load toward remediation of oil-contaminated beach and oilfield sands using ANOVA. Environ Sci Pollut Res. https://doi.org/10.1007/s11356-021-14776-8
9. AliffRadzuan MR (2017) Removal of oil droplets from oil-in-water mixtures by dissolved air flotation (DAF). University of Surrey
10. Aliff Radzuan MR, Abia-Biteo Belope MA, Thorpe RB (2016) Removal of fine oil droplets from oil-in-water mixtures by dissolved air flotation. Chem Eng Res Des 115, 19–33. https://doi.org/10.1016/j.cherd.2016.09.013

Study on the Potential Reutilization of Dyes from Batik Textile Effluent as Dye Sensitizers in Dye Sensitized Solar Cell (DSSC)

Weng Yi Lee, Ayu Wazira Azhari, Dewi Suriyani Che Halin, and Abdul Kareem Thottoli

Abstract An alternative method for waste dyes management by reutilization as dye sensitizer of DSSC is introduced in this study. The waste dye based DSSCs are fabricated by using wastewater that is collected from fixing process, dewaxing process and effluent of batik industry using the standard fabrication method. Absorption spectra, band gap, dyes concentration, pH value as well as functional group are the important characteristics of photosensitizer that is being studied in this research. Among the processes, the DSSC that was fabricated from waste dyes produced from dewaxing process had the highest energy conversion efficiency ($5.68 \times 10^{-8}\%$) due to its highest dye concentration. However, further study is required to improve the performance of waste dye based DSSC since its energy conversion efficiency is extremely poor compared to the current highest efficiency of 13% and undesirable in being applied in real life.

Keywords DSSC · Textile effluent · Dye sensitizers · Solar cells

W. Y. Lee · A. W. Azhari (✉)
Faculty of Civil Engineering Technology, Universiti Malaysia Perlis, 02600 Jejawi, Arau, Perlis, Malaysia
e-mail: ayuwazira@unimap.edu.my

Water Research and Environmental Sustainability Growth, Centre of Excellence (WAREG), Universiti Malaysia Perlis (UniMAP), Perlis, Malaysia

D. S. C. Halin
Center of Excellence Geopolymer and Green Technology (CEGeoGTech), Universiti Malaysia Perlis (UniMAP), Kangar, Perlis, Malaysia
e-mail: dewisuriyani@unimap.edu.my

A. K. Thottoli
Department of Physics, Pocker Sahib Memorial Orphanage College, Tirurangadi Malappuram 676306, Kerala, India
e-mail: abdulkareem@psmocollege.ac.in

N. Mohamed Noor et al. (eds.), *Proceedings of the 3rd International Conference on Green Environmental Engineering and Technology*, Lecture Notes in Civil Engineering 214,
https://doi.org/10.1007/978-981-16-7920-9_38

1 Introduction

A study carried out by Kelantan Department of Environment (DOE) between January and September in 2010 showed that the regulatory compliance level of batik industry in Kelantan was very low, only 5% when compared with other manufacturing industries [1]. Not many of the batik entrepreneurs have carried out wastewater treatment to treat the effluent before discharging it into water bodies as this industry is still dominated by the small and medium enterprises and cannot afford to have a wastewater treatment system [2]. As the result, the surrounding water bodies of batik industry have been highly polluted by dyes. To stop the receiving stream from being polluted, management on waste dye is required to remove the dyes from the effluent of batik industry before discharging it into surrounding water bodies.

According to [3], the removal of dye from wastewater by using physicochemical methods can create a secondary pollution due to the generation of waste with high concentrated dyes. Proper disposal or further treatment is required for these wastes. On the other hand, the complex chemical structure of dye has result in the degradation of dye becoming more difficult and complex than removal of dye. For example, bioremediation and photochemical technique are the existing method which can degrade the dyes. The efficiency of bioremediation in degrading dye has reduced since the microorganism will be killed by the toxic by-product that formed during biodegradation of dye. While photochemical technique consumed a long time to break down the complex synthetic dyes into simple chemical structure [3].

Compared to waste treatment, waste reutilization is more preferred in managing the waste as the resources can be used maximally before disposing it [4]. This had brought out the interest on finding a better way in managing the batik industrial waste dyes as the waste dyes have potential in being reused as dye sensitizer for DSSC. However, there are several uncertain factors that will affect the suitability of wastewater from batik industry in being the dye sensitizer of DSSC. Those factors are the various colours of dyes in wastewater, concentration of waste dyes in wastewater, the attachment of waste dyes on oxide semiconductor, the band gap of waste dyes as well as the impurities in wastewater.

The colour of the waste dyes is one of the problems that need to be consider in this study. Every batik will be dyed with various colours; some are using synthetic while others are using natural dyes. Besides that, the amount of colour used to dye the batik is also varied. This resulted in the inconsistence colours of waste dyes. On the other hand, the concentration of waste dyes is also another problem which cannot be ignored. The study by [5, 6] stated that, the higher the concentration of dye sensitizer, the better the energy conversion efficiency of DSSC. Unfortunately, the concentration of waste dyes in the effluent of batik industry is very diluted since it has been mixed with huge amount of wastewater. Therefore, the energy conversion efficiency of waste dye based DSSC may be low.

Another important factor to be considered is the attachment of waste dyes on oxide semiconductor. Oxide semiconductor is one of the DSSC's components which is used to provide an area for dyes to anchorage on it. The more dyes that are attached to the oxide semiconductor, the higher the energy conversion efficiency of DSSC [7]. The attachment of dye on the oxide semiconductor is mainly affected by the anchoring group of dyes, such as phosphonic acids and carboxylic acids. To form a bond with oxide semiconductor, the anchoring group of dye should be able to react with surface hydroxyl group of the oxide semiconductor. The pH value of waste dyes must not be more than 9 as the basic condition will make the dyes desorbed from oxide semiconductor easily [8]. The waste dyes must have pH value smaller than 9 while having anchoring functional group to attach on the oxide semiconductor [9].

Finally, the impurities in waste dyes are also one of the problems in this research. Other than waste dyes, the wax, soda ash and sodium silicate are the common impurities that can be found in untreated effluent of batik industry in this case of study. Currently there are no finding stated that these impurities will give effect on the dyes' potential in being photosensitizer of DSSC.

2 Materials and Method

Batik effluents were collected from three collection tank which were the fixing process tank, dewaxing process tank and final tank of the effluent. Standard 1060 (B)—collection of samples, as stated in standard methods for the examination of water and wastewater (1915), were used to collect the samples. The collected batik effluents were filtered with 0.45 μm filter paper prior to analysis. Before fabrication of the DSSCs, the absorption spectra, functional group, band gap, dye concentration and pH value of the effluent were determined. For comparison purposes, methyl orange based DSSC was also fabricated.

2.1 Preparation of Dye Sensitizers

The photosensitizer solution was prepared by filtering the effluent with a 0.45 μm filter paper to remove wax and suspended solid. After filtration, 20 ml of potassium permanganate solution was added into 100 ml of the samples and heated in an oven at 45 °C for 24 h. The solution was dried in an oven at 35 °C until all the water was removed. The residual was then dissolved in 20 ml ethanol and was left for two days to ensure complete reaction between the solvent and the effluent. The solution was then filtered to remove any undissolved particles. The methyl orange-based sensitizer was prepared by mixing 4 mg of methyl orange powder with 20 ml ethanol and was left overnight for complete reaction. Hydrochloric acid (HCl) was added to adjust the pH of the solution to pH 8.5.

2.2 Wastewater Characteristics Analysis

The collected waste dye samples were filtered prior to analysis to determine the wastewater characteristics. The UV/Vis spectrophotometer (Perkin-Elmer, Lambda 950) was used to measure the absorption spectra, band gap and dyes concentration of the samples. Since the prerequisite for dye to function as photosensitizer of DSSC is the absorption in the visible region of solar spectrum, the absorption wavelength of waste dye was measured within 400 to 700 nm (Hug et al. 2014). On the other hand, Fourier transform infrared (FTIR) spectrometer was used to analyse the functional group of the samples. This FTIR studies were conducted by using Perkin-Elmer Fourier Transformer Spectrophotometer (Version 10.03.08) in the wavelength range 650–4000 cm^{-1}.

2.3 Device Fabrication

Fluorine-doped (FTO) conductive glasses cut into 2.5 cm × 2 cm were cleaned with acetone, ethanol, and distilled water in ultrasonic cleaner. After that, the active area was clean with acetone prior to coating with titanium dioxide (TiO_2) paste by using doctor blading technique. The TiO_2 electrode were then sintered at 450°C in muffle furnace for 30 min and let to cool down.

To prepare the counter electrode, the FTO glasses were cleaned with acetone prior to coating with carbon. A graphite pencil was used to coat the active area of the plates while any loose graphite particles were gently removed. The carbon layer serves as a catalyst for the triiodide to iodide regeneration reaction. Iodide-triiodide electrolyte solution was prepared by mixing 63.45 mg iodine prills and 830 mg potassium iodide with 10 ml ethylene glycol.

To prepare the DSSC, the photoanode was immersed into the sensitizer solution for 24 h and then washed gently with ethanol to remove any excess dye. A few drops of the electrolyte solution were put onto the TiO_2 film. The photoanode was then sandwiched with the graphite electrode while ensuring that the active area of both electrodes was facing each other, and no bubble exist within the electrodes and electrolyte solution. To determine the efficiency of the DSSCs, the energy conversion efficiency (η) and incident photon-to-current conversion efficiency (IPCE) were measured.

3 Result and Discussion

3.1 Absorption Spectra, Band Gap and Dyes Concentration Analysis

UV–Vis spectrometer was used to measure the absorption spectra of the batik effluent. Figure 1 shows the results obtained from UV–Vis spectrometer in measuring the absorption spectra of the batik effluent samples. From the graph, the absorbance value of the different stages of the batik effluents were higher than that of the blank plastic cuvette. However, no absorption peak is present that may be cause by the overlapping of absorption spectra between various colours of dyes that are mixed in the wastewater [10].

The absorption spectra in Fig. 1 shows that effluent from the dewaxing stage has the highest absorption value followed by the final effluent, while the effluent from the fixing process shows the lowest absorption. Since the absorption value is directly proportional to the concentration of the absorbing material it is anticipated that the performance of DSSCs produced from the dewaxing effluent will show the highest efficiency compared to the others. This is because the provision of dyes from dewaxing effluent is most sufficient due to its highest dye concentration and result in full anchoring of dyes on the pores of titanium (IV) oxide, TiO_2 film [5].

3.2 pH Value Analysis

Previous study by [11] show that the Remazol dyes used in dyeing of the batik textile will form slight acidic solution after being dissolved into the water. However, study on the pH value at the various stages of the batik effluent shows that the pH is relatively

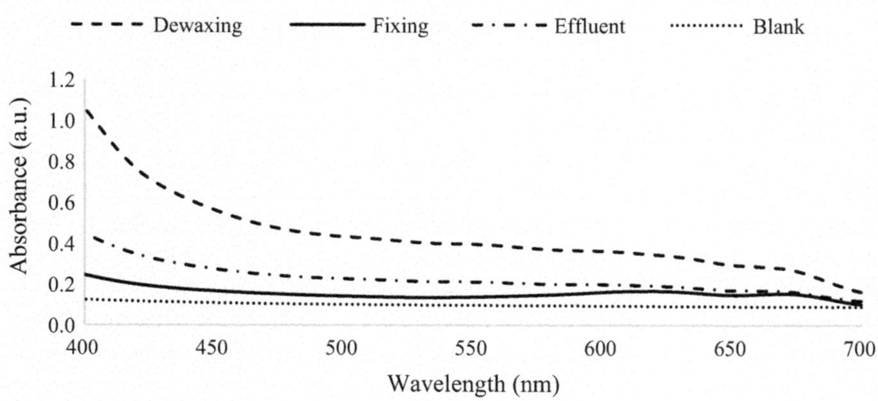

Fig. 1 Absorption spectra of waste dye samples and blank plastic cuvette

Table 1 pH value of waste dyes from different processes of batik industry

Process	pH Value			
	Sample 1	Sample 2	Sample 3	Average
Fixing	11.186	11.178	11.182	11.182
Dewaxing	10.053	10.044	10.044	10.047
Effluent of industry	10.369	10.362	10.364	10.365

high as shown in Table 1. It is believed that this is resulted from the addition of sodium silicate during fixing process and soda ash during dewaxing process. From Table 1, it is apparent that the effluent produced from fixing process shows the highest pH value at about 11.2. This followed by the final effluent with the pH of 10.4 which is slightly higher than effluent from the dewaxing process as both additives are accumulated in the final effluent.

Since desorption of dyes from TiO_2 film occur when the pH of sensitizer is higher than 9 [8], lower pH is required in attaching the dye onto TiO_2 film as the hydroxide ion will attract the hydrogen ion of hydroxyl group located on the surface of the TiO_2 film result in bond formation between remained oxide radical ion and anchoring group of dye [12]. Therefore, 0.1 M of hydrochloric acid is required in adjusting the pH value of waste dye samples to pH 8.5 prior to dyeing the TiO_2 film. Since pH value of the waste dye-based photosensitizer is controlled at pH 8.5, pH value is not a significant factor that will affect the performance of DSSCs in this study.

3.3 Functional Group Analysis

Figure 2 shows that there was not much difference between the wastewater produced from dewaxing process and fixing process in term of the functional group analysis. Both had alkenes group (C = C group) and alcohol group (O–H group). The relative IR spectra for C = C group is 1620 to 1680 cm^{-1}, and for O–H group is 3200 to 3550 cm^{-1}. Since the effluent was a mixture of wastewater produced from fixing and dewaxing processes, hence it can be considered that the effluent of batik industry also had similar functional group.

Remazol dye is the most common used dye batik industry [10]. It originally possesses sulfonic acid and sulfone as its functional group. The formation of alkene group in chemical structure of Remazol dye may cause by the reaction between sulfonic acid and hydroxide ion during fixing process. Since the sulfonic acid was removed from Remazol dye during fixing process, the waste dyes were lack of anchoring group and will result in poor attachment of waste dye on TiO_2 film. Hence a new anchoring group should be formed by changing the alkene group to sulfonic acid or carboxylic acid group. Compared to the formation of sulfonic acid, formation of carboxylic acid group from alkene group is easier as it can be prepared directly by heating the waste dye with solutions of potassium permanganate ($KMnO_4$) [13].

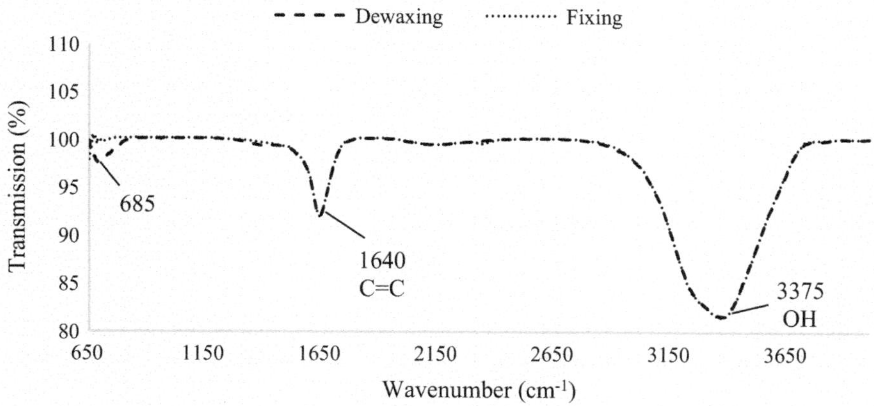

Fig. 2 FTIR spectra of waste dyes from dewaxing and fixing processes

On the other hand, the existence of alcohol group in the samples may be due to the soda ash and sodium silicate that formed hydroxide ions when dissolved in water. During the UV–Vis analysis, the existence of alcohol group will shift the absorption spectra of the samples towards longer wavelength. This restriction from alcohol group had increase the difficulties in analysing peak absorption of the waste dye samples that was carried out by using UV–Vis spectrometer [14].

3.4 Energy Conversion Efficiency Analysis

Open circuit voltage (V_{oc}) and short circuit current (I_{sc}) are the important parameters for calculating energy conversion efficiency of DSSC. They can be found in I-V characteristic curve as shown in Fig. 3. From this figure, the efficiency of the DSSCs which was fabricated by using waste dye produced from dewaxing process shows the highest one among the waste dye based DSSCs.

As shown in Fig. 4, the power output of the DSSC which was fabricated by using waste dye produced from dewaxing process was higher than other waste dye-based DSSCs but was less than methyl orange-based DSSC. The higher the power output, the higher the energy conversion efficiency of DSSC. Therefore, the methyl orange based-DSSC has the highest energy conversion efficiency ($88.974 \times 10^{-9}\%$) then followed by dewaxing process DSSC ($56.815 \times 10^{-9}\%$). The DSSC which fabricated by using waste dye produced from fixing process has the worst energy conversion efficiency that is $1.991 \times 10^{-9}\%$. This prove that the hypothesis which had been made before that stated that the performance of DSSC which fabricated by using waste dye produced from dewaxing process as its photosensitizer will be the best among the waste dye based DSSCs, was acceptable. The V_{oc}, I_{sc}, fill factor (FF) as well as η of DSSCs were summarized in Table 2.

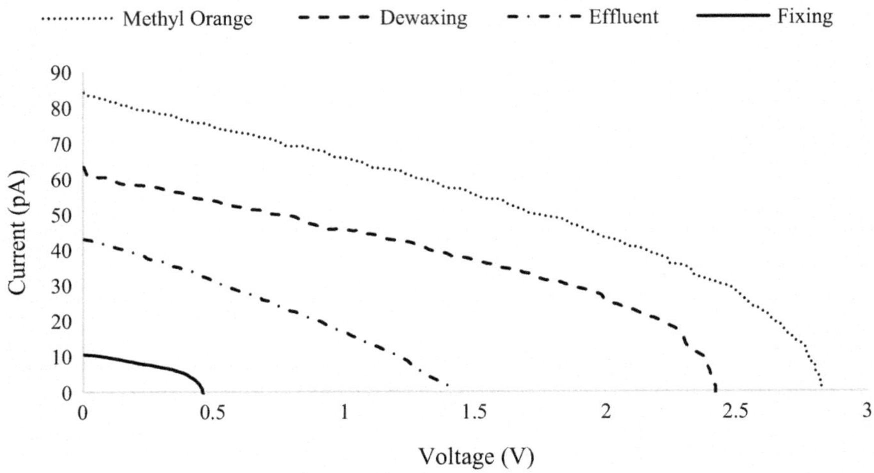

Fig. 3 I-V characteristic curve of waste dye-based DSSCs and methyl orange-based DSSC that shows Voc and Isc of the DSSCs

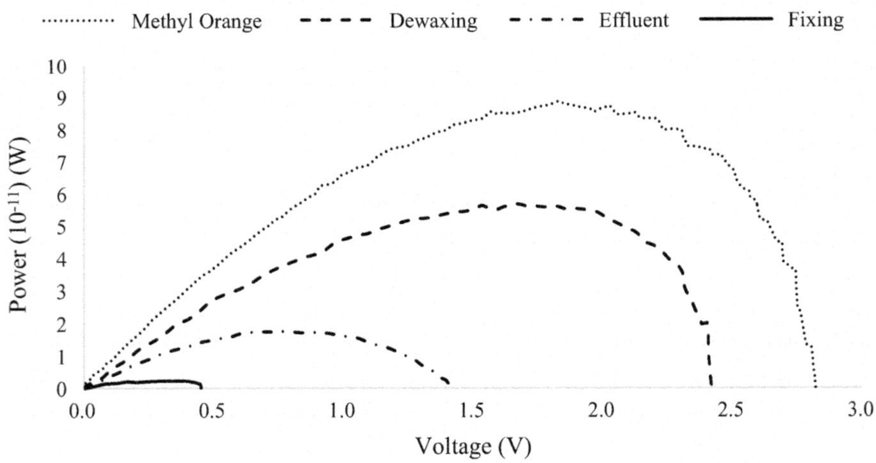

Fig. 4 Output power of DSSCs as the voltage changing

Table 2 Summary of energy conversion efficiency of DSSCs

Photosensitizer	V_{oc} (V)	J_{sc} (10^{-7} A/m^2)	FF	η (10^{-9}%)
Waste dye from fixing process	0.450	1.0442	0.423	1.991
Waste dye from effluent of industry	1.405	4.2503	0.300	17.924
Waste dye from dewaxing process	2.420	6.2704	0.374	56.815
Methyl orange	2.825	8.4156	0.373	88.794

4 Conclusion

The absorption spectra, band gap, pH value as well as functional group were not the significant characteristics of waste dyes in affecting the performance of waste dye based DSSCs, except the dye's concentration. This was because the absorption spectra, pH value and functional group of the waste dye samples were almost similar with each other. On the other hand, the concentration of dyes in wastewater was the most important parameter that will affect the performance of waste dye based DSSCs. Among the waste dye based DSSCs, the DSSC which was fabricated by using waste dye produced from dewaxing process as its photosensitizer had best performance in term of energy conversion efficiency and IPCE. This is because it has the highest concentration of dyes. It can be stated that the photosensitizer which has higher concentration of dyes results in higher absorbance in visible region of solar spectrum will produce DSSC with better energy conversion efficiency. Unfortunately, the performance of waste dye-based DSSC is lower than methyl orange-based DSSC. Figure 4 shows that the energy conversion efficiency of methyl orange-based DSSC is around 36% much higher than dewaxing's waste dye-based DSSC. Improvement is required for waste dyes in order to enhance the performance of waste dye-based DSSC while increase the economic potential of waste dye-based DSSC.

The energy conversion efficiency of waste dye based DSSCs, which were fabricated in this research, were quite low. With such low energy conversion efficiency, the waste dye-based sensitizers are currently not favourable for fabrication of the DSSCs. It is believed that this extremely low energy conversion efficiency may be due to the various impurities that are available in the waste dyes. Hence, a detail study on the effect of different impurities on the dye sensitizer is important before a suitable pre-treatment method can be used to improve the quality of the waste dyes before reutilization as photosensitizer for the DSSCs.

References

1. Subki NS, Rohasliney H (2011) A preliminary study on batik effluent in Kelantan state: A water quality perspective. In: International conference on chemical, biological and environment sciences, Bangkok, pp 274–276
2. Nuzul Z, Talib SN, Wan Johari WL (2020) Water quality of effluent treatment systems from local batik industries. IOP Conf Ser: Earth Environ Sci 476:012097. https://doi.org/10.1088/1755-1315/476/1/012097
3. Pereira L, Alves M (2011) Dyes—environmental impact and remediation. In Environmental protection strategies for sustainable development. Springer, Netherlands.
4. Gertsakis J, Lewis H (2003) Sustainability and the waste management hierarchy – a Discussion Paper, EcoRecycle Victoria
5. Mat Isa SS, Anhar NAM, Selamat NM, Adib NSC, Muda MR, Ramli MM, Ahmad Hambali NAM (2015) The effect of different remazol dye concentrations and soaking times in dye-sensitized solar cell. Appl Mech Mater 815:203–211. https://doi.org/10.4028/www.scientific.net/AMM.815.203

6. Sowmya S, Inbarajan K, Ruba N, Prakash P, Janarthanan B (2021) A novel idea of using dyes extracted from the leaves of Prosopis juliflora in dye – Sensitized solar cells. Opt Mater 120:111429. https://doi.org/10.1016/j.optmat.2021.111429
7. Baglio V, Girolamo M, Antonucci V, Aricò AS (2011) Influence of TiO_2 Film thickness on the electrochemical behaviour of dye-sensitized solar cells. Int J Electrochem Sci 6(8):3375–3384
8. Hug H, Bader M, Mair P, Glatzel T (2014) Biophotovoltaics: natural pigments in dye-sensitized solar cells. Appl Energ 115:216–225. https://doi.org/10.1016/j.apenergy.2013.10.055
9. Eop TS, Azhari AW, Halin DSC (2019) Natural dyes extracted from leaves, fruits, and roots of Piper Betel, Adonidia Merrillii and Morinda Citrifolia as photosynthesizers for dye sensitized solar cells. IOP Conf Ser: Mater Sci Eng 572(1). http://dx.doi.org/https://doi.org/10.1088/1757-899X/572/1/012113
10. Isah KU, Sadik AY, Jolayemi BJ (2017) Effect of natural dye co-sensitization on the performance of dye-sensitized solar cells (DSSCs) based on anthocyanin and betalain pigments sensitisation. Eur J Appl Sci 9(3)
11. Rashidi HR, Sulaiman NN, Hashim NA (2012) Batik industry synthetic wastewater treatment using nanofiltration membrane. Procedia Eng 44:2010–2012. https://doi.org/10.1016/j.proeng.2012.09.025
12. Syafinar R, Gomesh N, Irwanto M, Fareq M, Irwan YM (2006) Optical characterization using nature based dye extracted from hibiscus's flower. J Eng Appl Sci 10(15):6336–6340
13. Ho HY (2011) Organic chemistry. Petaling Jaya: Pearson Malaysia Sdn Bhd.
14. Che Halin DS, Azhari AW (2016) Synthesis of chlorophyll thin film from noni leaves via dip coating process. Mater Sci Forum 857:142–145

The Effectiveness of Non-woven Geotextiles as a Filter Media for Total Suspended Solid Removal

Suzana Ramli and Mohd Rizal Firdaus Wahhit

Abstract Water with high turbidity can significantly reduce the aesthetic quality of lakes and streams, having a harmful effect on recreation and tourism. Besides, it may harm fish and other aquatic life by reducing food supplies and affecting the function of fish's gill. Turbidity and Total Suspend Solids (TSS) are among the main parameters used in water quality indicators. The objectives of this study are to remove total suspended solid (TSS) concentration and turbidity (NTU) by using Non-Woven Geotextiles (NWG) as a filter media. Other parameters such as BOD, COD, DO, pH and NH3-N were also taken into account to determine the quality of Penchala River in Kuala Lumpur. Performance of NWG was also assessed with Water Quality Index (WQI) calculation. The baseline data showed the TSS value was 150 mg/L (Class III) and the value for turbidity was 154 NTU (highly turbid). NWG properties used are 100% polypropylene materials, with properties design of 190 μm (0.19 mm) AOS, 2.2 mm thickness and permittivity 2S-1. Results showed that the NWG was effectively removed TSS and decreased the NTU level by 70% reduction, therefore, significantly improved the river classification from Class III to Class II. Other parameters that showed improvement was COD, with average percentage improvement of 17.2%. Meanwhile data such as BOD, pH and NH3-N were remained unchanged throughout the study period. As overall, the WQI for Penchala River during prior time of study was in range 83.78–86.70 (Class II) and will be transformed to Class I after continuous treatment with NWG filtration media.

Keywords TSS · Turbidity · Non-woven geotextiles · Water quality

1 Introduction

Turbidity is a common problem in Malaysian surface waters that interfere with the aesthetic value of the river. It is due to alluvial soil conditions, organic and inorganic materials from natural causes and also from climate factors. High turbidity

S. Ramli (✉) · M. R. F. Wahhit
School of Civil Engineering, College of Engineering, Universiti Teknologi MARA, 40450 Shah Alam, Selangor, Malaysia

N. Mohamed Noor et al. (eds.), *Proceedings of the 3rd International Conference on Green Environmental Engineering and Technology*, Lecture Notes in Civil Engineering 214,
https://doi.org/10.1007/978-981-16-7920-9_39

can be caused by total suspended solids from silt, dirt, algae, plant pieces, ashes or chemicals in the water. The presence of these particles produces cloudy and dirty water. The best method to manage turbidity is to discourse its sources. This include reducing stormwater run-off, restoring eroding stream and lake shorelines, and applying industry-specific best management practices (BMPs) [1]. Settling ponds, re-vegetating steep slopes, maintaining a minimum of 25 feet of vegetation around streams, lakes, and other waterbodies, and maintaining all drainage systems are among mitigations steps taken to reduce turbidity issues in surface water. One of the new approaches that has been introduced is the application of Non-Woven Geotextiles filtration media. This study aims to determine the effectiveness of NWG as filtration media to remove TSS and turbidity. No river yet in Malaysia has been provided with non-woven geotextiles as filter media facilities for Total Suspended Solids (TSS) removal in river, therefore this study valuable to the researchers as well engineers, who are interested to explore more about the non-woven geotextiles material application in river cleaning program. Geotextiles make hard soil more workable, designed to allow building in areas where it would otherwise be impossible. Many infrastructure projects, such as highways, ports, landfills, drainage systems, and other civil projects apply the use of geotextiles. Zornberg and Thompson [2] also stated in their reports, that application geotextiles are widely used in geotechnical and structural applications such as roads and railways, earth structures, drainage and filtration systems, hydraulic works and many other applications. Geotextiles have favorable water conductivity, which are used as drainage channel. The water in the soil structure in the geotextiles can be collected and slowly discharged along the geotextiles [3]. In most filtering applications, geotextiles are used in combination with systems that have either horizontal or vertical water flow. Use of a drainage geotextile ensures ongoing drainage of fluids with minimum pressure loss [4]. In a drainage system, geotextile also preventing fine particles from migrating between soil layers. Many studies have proved that NWG can improve water quality [5, 6]. Improving the level of water quality depends on several factors. Among the factors are the Apparent Opening Size (AOS), thickness, permittivity/permeability and the ability of the NGW exposed to external resistance such as weather or chemicals for a certain period of time. Sarma et al. [7] in their study, proved that the effectiveness of nonwoven geotextiles in reducing the nutrients and suspended solids for lake water quality. In year 2005, One State One River (1S1R) has been launched by Malaysia's Department of Irrigation and Drainage (DID). The program aims to help state DID organizes river conservation programs and improve river water quality. Penchala River was selected by DID State of Selangor for this program. Based on Department of Environment Malaysia [8] in year 2005, Klang River is the most polluted river and Penchala River is one of tributaries contributed to polluted problems. Among the causes of pollution that occur in the Penchala River are the disposal of domestic waste from nearby housing, thick silt at the bottom of the river as well as restaurants that drain food waste without going through the filtering method first. Due to that, the WQI on that time was felt under Class IV which the range 31–51.9. In year 2018 the water quality dropped to Class III and stayed constant till 2019 [9]. Due to that,

Penchala River is the reason why this study was conducted. As the time goes by, the river is now getting worse as it loses their ability to make a self-purification [10].

2 Material and Method

2.1 Material

Geosynthetics, specialty nonwovens in the civil engineering field, form the backbone of numerous civil engineering projects. With nonwoven technology, fabrics can be designed to take on an array of characteristics that other materials simply cannot match; they are penetrable and are normally used for drainage applications and filtration applications such as highway and slope constructions. Rapid developments in infrastructure have led to huge demand for nonwovens in water conservancy, highway, railway, sea and airport projects as they satisfy the qualities needed for these demanding environments. Nonwoven fabrics are a dynamic, value-added alternative to more traditional materials. These geotextiles offer excellent filter properties and help to prevent the dirt and aggregate from clogging up drains thus extending the life and increasing the performance. Non-woven polypropylene geotextile provides excellent physical and hydraulic properties in addition to high tensile strength. The advantages of nonwoven filter over the woven filters are high filtration efficiency, high permeability, less blinding tendency, no yarn slippage as in woven media, there is no limitation for thickness, high production rate and last but not least continuous process line. The pore sizes are much smaller than the pore sizes of woven or knit fabrics. Due to the overlapping of yarns, the size of the pore, or the opening between the yarns, is quite large; which can be seen by holding up a piece of these filter fabrics into the light. Woven and knit fabrics are two-dimensional filter bag media in which yarns overlap.

Non-Woven Geotextiles (NWG) was selected for filter media for this study due to the effective filtration performance. It is also widely use in civil engineering projects as separators or filtrations namely in highway construction, slope rehabilitation/stability, landfill cover and drainage system. NWG in this study is developed and fabricated by manufacturing in Malaysia. The material used for NWG is 100% polypropylene and UV stabilized. The apparent opening size and thickness of the NGW are 0.09 mm and 2.2 mm (. The filters were cut in rectangular shape with dimension 700 mm × 600 mm and the total of 20 filters ($8.4m^2$) was carried out for this study. To ensure the effectiveness of this filter, it is arranged horizontally against the river flows and in three rows. These filter sheets are hung on 50 mm diameter iron rods across the river. To avoid the filter sheet to ripple according to the velocity, at the bottom of each filter sheet is placed a weight, so that it will support and give the stability of the filter. Each of these filter sheets is arranged with a distance of 750 mm from center to center. Figures 1 and 2 show the filters and their arrangements on site.

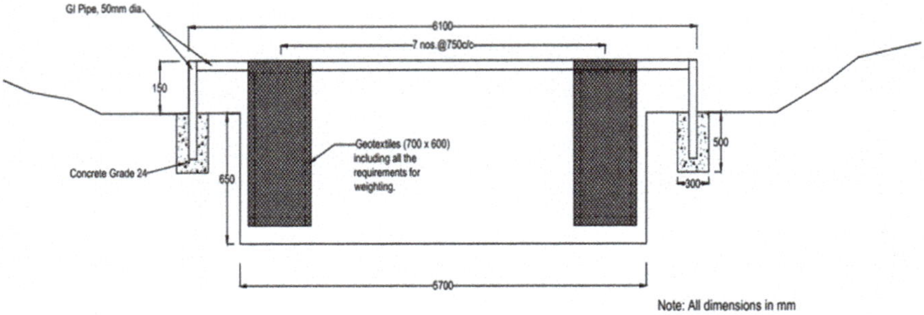

Fig. 1 Dimensions of the filters and their arrangement

Fig. 2 Material arrangement at the site

Based on this study, it was found that the lifespan of the filter when exposed in water directly is 3 months. After the 3 months, it can be seen that the filter filled with suspended matter on it. Therefore, it can be concluded that the maintenance the filters should be done by replacing them with new ones for every 3 months.

2.2 Method

This study focused at the upstream of Penchala River near to TPBKKL inlet pond. The aim is to determine the total suspended solids and turbidity impact on the water quality of Penchala River. This study involved both on-site and laboratory tests. For the baseline data, parameters NTU and TSS was determined by using Lamotte and Partech 7450 Model sensor respectively and utilized procedures from [11]. Baseline value of total suspended solids and turbidity was recorded before the application of the non-woven geotextiles filter media. Baseline data or benchmark is very important to ensure that its effectiveness by comparing events before and after implementation.

Other than total suspended solids and turbidity, another five (5) standard parameters for water quality were also being recorded such as BOD, COD, pH and Ammonia. All the experiments were conducted at the Department of Chemistry Malaysia except for Dissolved Oxygen (DO) which was conducted in situ. Sampling of river water is taken for a year and the frequency is once a month in clear weather and at normal flows of river. The data obtained were analysed and compared to the National Water Quality Standards for Malaysia (NWQSM). Analysis for this study was done by comparing the baseline data with thirteen (13) samples reading taken from December 2019 until December 2020. A guideline from National Water Quality Standard (NWQS), Water Quality Index Classification from DOE was used to classify the river in order to determine the effectiveness of the filters. Tables 1 and 2 show the Water Quality Index Classification (WQIC) and Water Quality Classification Based on Water Quality Index, DOE standard. DOE also established water classes and uses for classified the condition of water depending on WQI result.

Table 1 Water quality index classification

Parameters	Unit	Classes				
		I	II	III	IV	V
Ammoniacal Nitrogen	mg/l	<0.1	0.1–0.3	0.3–0.9	0.9–2.7	>2.7
BOD5	mg/l	<1	1–3	3–6	6–12	>12
COD	mg/l	<10	10–25	25–50	50–100	>100
DO	mg/l	>7	5–7	3–5	1–3	<1
pH		>7	6–7	5–6	< 5	>5
TSS	mg/l	<25	25–50	50 -150	150–300	>300
*Turbidity	NTU	5	50	–	–	–
WQI	mg/l	> 92.7	76.5–92.7	51.9–76.5	31.0–51.9	<31.0

Note/Source National Water Quality Standard, (DOE 2019)

Table 2 Water quality classification based on water quality index

Parameters	Index range		
	Clean	Slightly polluted	Polluted
SIBOD	91–100	80–90	0–79
SIAN	92–100	71–91	0–70
SISS	76–100	70–75	0–69
WQI	81–100	60–80	0–59

Note/Source National Water Quality Standard, (DOE 2019)

3 Results and Discussions

3.1 *Physical Characteristics*

Baseline data is the data taken as a benchmark before an experiment is performed. These benchmarks can be identified positive or negative based on the analyst itself. In this study, especially for TSS and Turbidity if the results after implementation exceeded the baseline data it will be considered as a negative (failed) or vice versa. The baseline for TSS and Turbidity is 150 mg/l and 154 NTU respectively. When referring to the NWQS that has been established by the DOE, this baseline data was in the Class III for TSS and SI_{ss} (53.60) was in the Polluted category. Turbidity, on the other hand, the baseline results showed that it was not either in class I or class II but in highly turbid. This is because, in the NWQS table, the value that can be measured are starting from 0–5 (Class I) and 6–50 (Class IIA or IIB). With the baseline data obtained, it showed that it has strengthen the evidence that the condition of Penchala River is contaminated with high content of TSS and Turbidity.

3.2 *Total Suspended Solids (TSS)*

All TSS levels from December 2019 onwards fluctuated from ranged between 42 and 48 mg/L. At the beginning of the result, the value of TSS was recorded as 46 mg/L and it was slightly dropped to 45 mg/L in January 2020. Afterwards, in February 2020 the value of TSS went up to 48 mg/L and in March and April 2020 the data decreased to 43 mg/L and remained steady afterwards. A month later in May 2020, the TSS value was declined to 42 mg/L which was the lowest value recorded throughout the study duration. In August 2020, the data was significantly increased to 48 mg/L. In September 2020 it showed that the result slightly decreased to 43 mg/L and increased back in October 2020 to 47 mg/L and remained stable until November 2020. Finally, at the end of prior of study in December 2020 the result was recorded 45 mg/L. The result is shown in Fig. 3.

3.3 *Turbidity*

Turbidity is a measure of the extent to which light is either absorbed or scattered by suspended material. Figure 4 shows the turbidity levels at the same point's location of TSS sampling. The baseline level also was established in November 2019 and the turbidity level was 154 NTU. The water samples were collected on a monthly basis to monitor the turbidity levels. All turbidity levels data from December 2019 fluctuated with range between 45 and 50 NTU. The highest turbidity value was 50 NTU recorded in February, July and August 2020 and the lowest turbidity value was

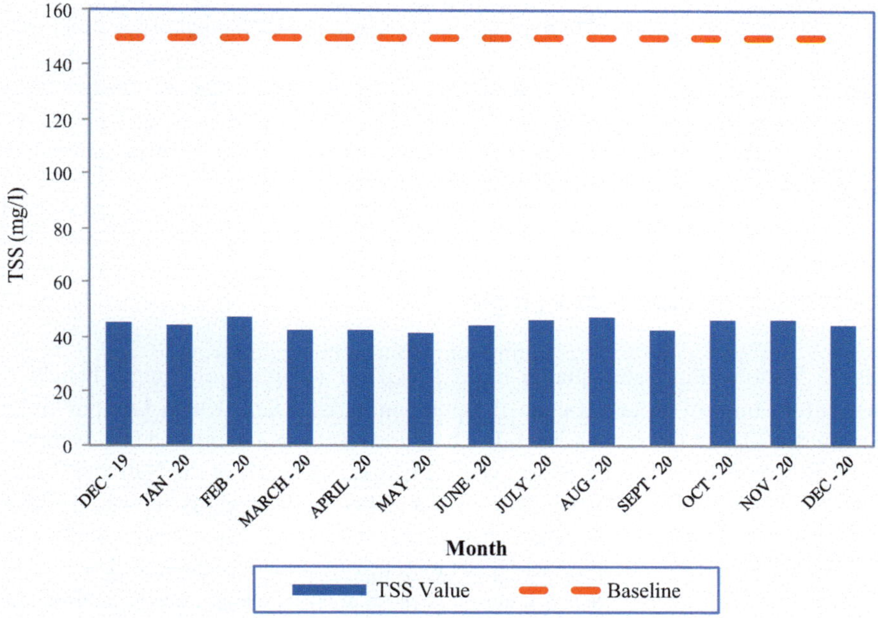

Fig. 3 Monthly results of TSS

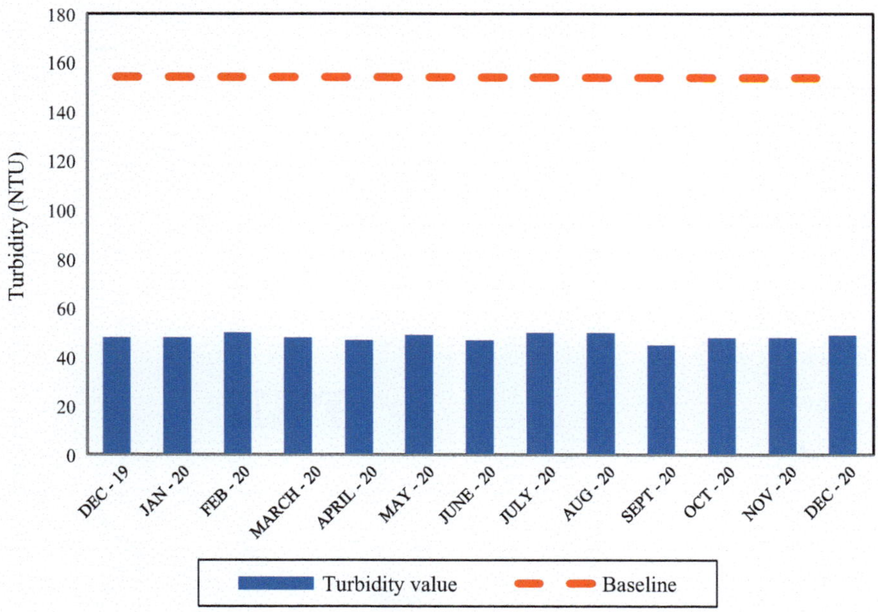

Fig. 4 Monthly results of TSS

45 NTU which was recorded in September 2020. Starting December 2019 to January 2020, the turbidity remained constant for 48 NTU. However, the value was slightly decreased to 48 NTU to 47 NTU in March 2020 and April 2020. In October 2020 to December 2020 the bar chart showed the values of turbidity remained stable. As a whole, the results obtained were significantly below the baseline level and within the National Water Quality Standard Class II limit.

3.4 *Water Quality Index (WQI)*

Figure 5 shows the water quality index (WQI) of upstream of Sungai Penchala just before inlet of Kolam Taman Persekutuan Bukit Kiara, Kuala Lumpur from November 2019 to December 2020. The baseline level was established in November 2019.The WQI value was 81.36 and was classified as Class II of the National Water Quality Standard. The WQI levels from December 2019 onwards showed a slight improvement in comparison with the baseline levels and ranged between 83.78 and 86.79 mg/L which are still Class II of the National Water Quality Standard. The results WQI generally showed improvement in comparison with baseline levels established in November 2019. Even though the highest value WQI is 86.79 and still in Class II was recorded in the end of prior of this study, but the increasing was significant

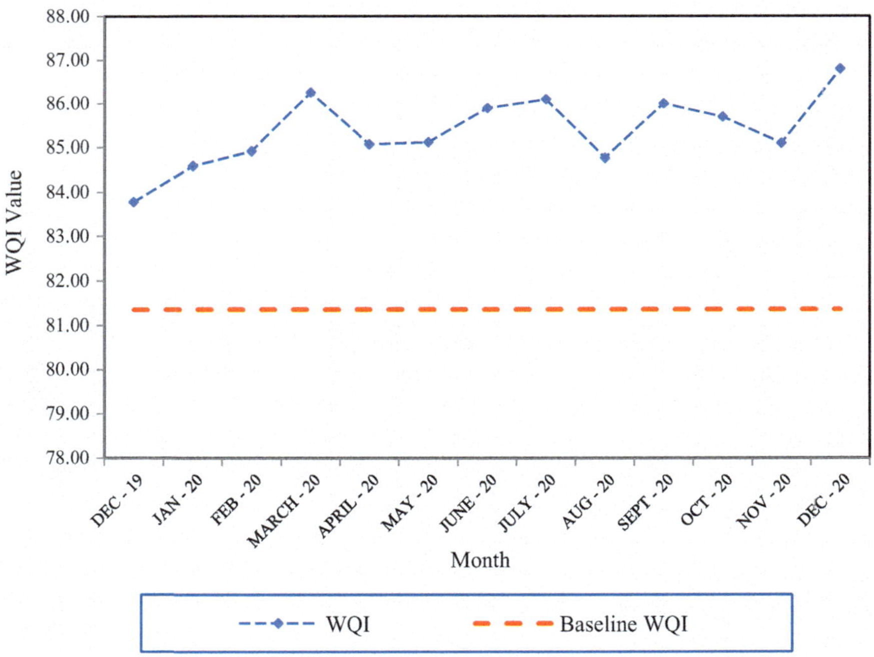

Fig. 5 Monthly WQI value

starting from the beginning and strongly believed the trends could be achieved standard Class 1, since no data was recorded below the baseline. Other than that, river conditions itself has natural elements characteristics such as and existence of aquatic plants plus cleanliness of river also contributing to increase of WQI.

4 Conclusion

Based on the findings, it shows that the reduction of TSS and NTU level were almost 70% after treatment with NWG filter media. It can be concluded that the high level of TSS content in that area had caused high turbidity level in the river. Meanwhile, it can also be seen that the quality of the river is increasing with an average of 4.94% almost reaching Class 1 (Class 1 needs an increase of 12% from the baseline data) based on WQI. If this study is extended for a few more months the likelihood of achieving class 1 is high due to the graph patents pushing towards the increase. As conclusions, this study has achieved the objectives study to design a filter by using NGW and determination of removal the TSS and Turbidity were also accomplished. Comparison with NWQS is correspondingly performed in order to classify the river quality which showed improvement in the WQI value.

References

1. Lloyd D, Koenings J, LaPerrire J (1987) Effects of turbidity in fresh waters of Alaska. J Fish Manag 7(1):18–33. https://doi.org/10.1577/1548-8659(1987)7%3c18:EOTIFW%3e2.0.CO;2
2. Zornberg JG, Thompson N (2016) Application guide and specifications for geotextiles in roadway applications. Tex Dep Transp 7:128. http://www.utexas.edu/research/ctr/pdf_reports/0_5812_1.pdf
3. Bezgovšek Š, Gorjanc DŠ, Pulko B, Lenart S (2020) Influence of structural parameters of nonwoven geotextiles on separation and filtration in road construction. Autex Res J 20(4):449–460. https://doi.org/10.2478/aut-2019-0038
4. Mitra A (2013) Geotextiles and its application in coastal protection and off-shore engineering. J Texte Assoc 74(1):5–11
5. Palakkeel VD, Mulligan CN, Eliaslara M, Castillo AE, Bhat, S. (2017). Geotextile filtration for improving surface water quality of eutrophic lakes.In: 15th international conference on environmental engineering, pp 201–210
6. Kutay ME, Aydilek AH (2012) Filtration performance of two-layer geotextile systems. Geotech Test J 28(1):79–91. https://doi.org/10.1520/gtj12580
7. Sarma S, Mulligan CN, Kim K, Veetil DP, Bhat S (2016) Use of nonwoven geotextiles for removing nutrients and suspended solids from a eutrophic lake. Proc Annu Conf-Can Soc Civ Eng 2:1061–1071
8. Department of Environment Malaysia (DOE) (2019) Water quality standard. https://www.doe.gov.my/portalv1/wp-content/uploads/2019/05/Standard-Kualiti-Air-Kebangsaan.pdf
9. Zainudin Z (2010) Benchmarking river water quality in Malaysia. Jurutera (February) 12–15

10. Wu H, Yao C, Li C, Miao M, Zhong Y, Lu Y, Liu T (2020) Review of application and innovation of geotextiles in geotechnical engineering. Materials 13(7):1–21. https://doi.org/10.3390/MA13071774
11. ASTM (1999) D4491 - standard test methods for water permeability of geotextiles by permittivity. ASTM D4491:1–6

Unravelling the Performance of Microbial Fuel Cell for Simultaneous Binary Dyes Remediation and Bioelectricity Generation

Sing-Mei Tan, Soon-An Ong, Yee-Shian Wong, Che Zulzikrami Azner Abidin, Li-Ngee Ho, Norazian Mohamed Noor, and Andreea Moncea

Abstract This study was to examine the bioremediation of wastewater containing a binary mixture of methyl orange (MO) and reactive black 5 (RB5), as well as bioelectricity generation in single chamber up-flow membrane-less microbial fuel cell (UFML-MFC). Decolourisation performance and power density of MFC were studied with different concentrations of binary dye (25, 50 and 75 mg/L) as influent. The colour removals evaluated from the standard curve of dye against optical density at its maximum absorption wavelength were 80.40%, 70.33% and 56.99%, respectively in the three phases of study. The decolourisation extent of RB5 is comparatively higher than MO in the binary mixture, due to the presence of more sulphonate groups in the molecular geometry of RB5. The reductive cleavage of azo bond was denoted by UV spectroscopic analysis.

S.-M. Tan (✉) · S.-A. Ong · Y.-S. Wong · C. Z. A. Abidin
Centre of Excellence Water Research and Environmental Sustainability Growth (WAREG), Universiti Malaysia Perlis, 02600 Arau, Perlis, Malaysia

Y.-S. Wong
e-mail: yswong@unimap.edu.my

C. Z. A. Abidin
e-mail: zulzikrami@unimap.edu.my

Faculty of Civil Engineering Technology, Universiti Malaysia Perlis (UniMAP), 02600 Arau, Perlis, Malaysia

L.-N. Ho
Faculty of Chemical Engineering Technology, Universiti Malaysia Perlis, 02600 Arau, Perlis, Malaysia
e-mail: lnho@unimap.edu.my

N. M. Noor
Sustainable Environment Research Group (SERG), Centre of Excellence Geopolymer and Green Technology (CeGeoGTech), Universiti Malaysia Perlis, 02600 Arau, Perlis, Malaysia
e-mail: norazian@unimap.edu.my

A. Moncea
National Institute for Research and Development in Environmental Protection Bucharest (INCDPM), 294, Splaiul Independentei Street, 6th District, 060031 Bucharest, Romania

N. Mohamed Noor et al. (eds.), *Proceedings of the 3rd International Conference on Green Environmental Engineering and Technology*, Lecture Notes in Civil Engineering 214,
https://doi.org/10.1007/978-981-16-7920-9_40

Keywords Microbial fuel cell · Binary dyes · Decolourisation · Bioelectricity generation

1 Introduction

Azo dyes are the most eminent group of colour additive that have been tremendously used in textile industry [1]. The discharge of dye-containing effluent wastewater into water body is undesirable due the presence of toxic, carcinogenic and teratogenic substances in the breakdown of dye products, which persist as environmental pollutants [2]. Physio-chemical approaches are generally used to treat dye effluents. Natheless, these approaches have drawbacks such as high cost, production of a massive amount of secondary pollutants, emanations of harmful substances, and generation of refractory substances [3]. On that account, biological wastewater treatment that *necessitates* the use of an active microbial biomass can be a good replacement for the conventional physiochemical wastewater treatment unit operations.

Microbial fuel cell (MFC) is a green sustainable technology that harnesses the interactions of respiring microorganisms in converting chemical energy into electricity. As a new type of biological approach, MFCs have received great attention for their innovative features and environmental benefits [4]. Over the past few years, MFCs have also been used for refractory wastewater treatment, where sufficiently good removal efficiency of pollutants and bioelectricity generation can be attained using an MFC treatment unit [5]. Previous MFC studies had attempted to utilize MFCs for degradation of single azo dye. Cao et al. [6] examined the electricity generation upon degradation of Congo Red in single chamber air–cathode MFC. Fang et al. [5] also treated reactive brilliant red X-3B (ABRX3) with MFC and highest decolourisation rate and power density of 95.6% and 0.852 W/m^3, respectively were achieved. Despite a great number of studies were performed on decolourisation of single azo dye, very few attempts were made for binary azo dyes degradation with concurrent bioelectricity generation in MFC system. In this study, the performance of single chamber up-flow membrane-less MFC for simultaneous binary dyes remediation and bioelectricity generation was evaluated by using azo dyes such as, Methyl Orange (MO) and Reactive Black 5 (RB5). The performances of MFC at different initial dye concentrations were discussed.

2 Materials and Method

2.1 Dye

MO and RB5 were purchased from Sigma-Aldrich, and were of analytical grade. The dye structures were presented in Table 1.

Table 1 Aspects of azo dyes used in this study

Azoic dye	Molecular geometry	Maximum Wavelength (λ_{max})
Methyl orange ($C_{14}H_{14}N_3NaO_3S$)	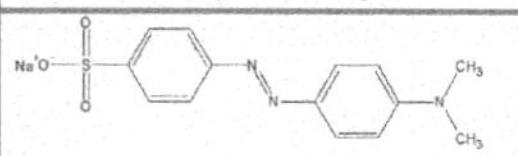	475
Reactive black 5 ($C_{26}H_{21}N_5Na_4O_{19}S_6$)	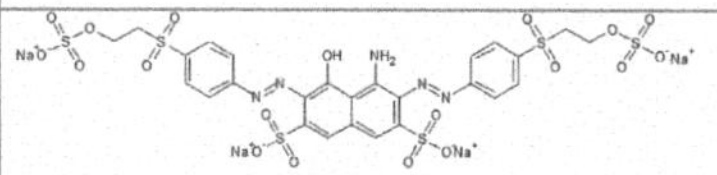	600

2.2 *Inoculation and Medium*

Activated sludge was acquired from a rubber glove fabricating industry's wastewater treatment plant. Synthetic wastewater used to feed the UFML-MFC bioreactors contained 0.81 g/L sodium acetate (CH_3COONa) as the main organic substrate and other constituents as buffer solutions.

2.3 *MFC Configuration and Operation*

Figure 1 shows the schematic set up of UFML-MFC. Single chambered UFML-MFC bioreactor (Height = 30 cm, Diameter = 6 cm) was constructed using poly-acrylic plastic column. They were operated under ambient temperature. Carbon felt was layered as anode. The sampling point was positioned along the column at 18 cm from the bottom of the reactor.

A peristaltic pump (BT100-100 M Longer Precision Pump) was used to transfer the synthetic wastewater prepared in the influent container to feed the bioreactors in continuity and hydraulic retention time (HRT) of 1 day. An aerator was positioned at the cathode compartment to provide supplementary aeration. Air flow meter (LZB-3, China) was used to keep up the aerator flow rate at 50 mL/min. An external 1kΩ resistor was connected to anode–cathode terminals by copper wires throughout the entire start-up and operations. It was disconnected only when polarization curves have to be done. The output voltages of UFML-MFC system were recorded by using a data logger.

3 Results and Discussion

3.1 *Decolourisation Performance of MFC*

The decolourisation performance of MFC in influent, anode and effluent was displayed in Fig. 2. The decolourisation efficiency slightly declined when 25 mg/L

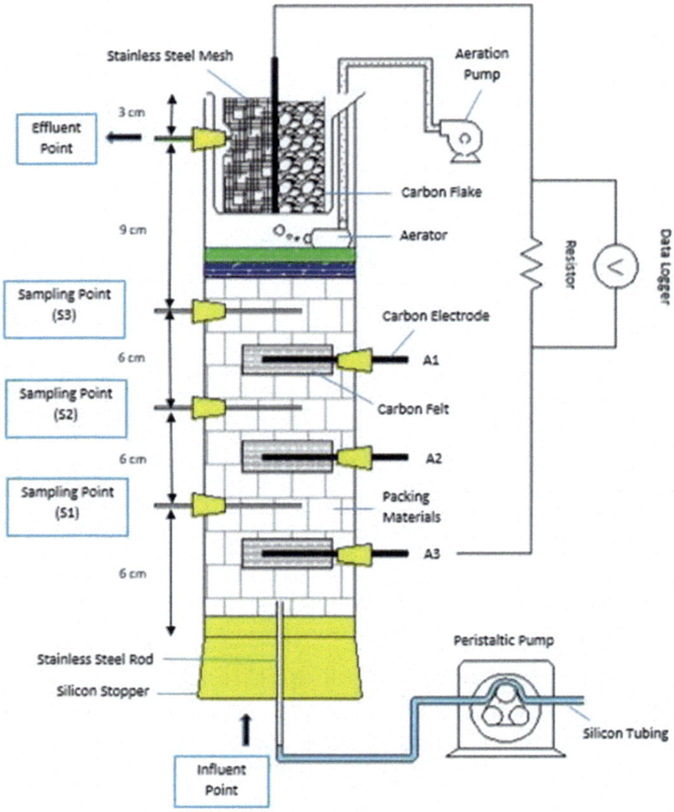

Fig. 1 Schematic set up of UFML-MFC

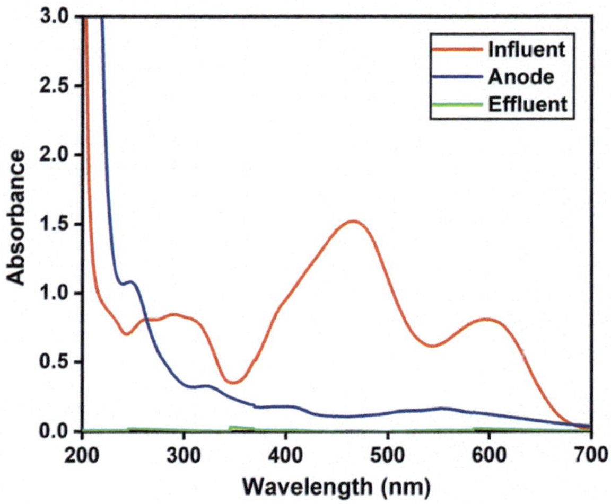

Fig. 2 Decolourisation performance of MFC

of binary dye was introduced into the system, by which the colour removal of MO and RB5 were 97.96% and 99.63%, respectively. However, the decolourisation efficiency of both MO and RB5 in the binary mixture of dyes was found to have further deteriorated from 65.3 to 76.5% to 48.61–52.3% when the initial concentration of dye increased from 50 mg/L to 75 mg/L. The decolorization efficiency decreased remarkably with an increase in dye concentration, as the carbon fuel was inadequate to support the increased number of azo bonds. These uncleaved azo bonds had caused the decolourisation performance of MFC to become progressively worse, especially when the concentration of dye increases. In the binary dyes, it was noticed that the decolourisation of RB5 is better than MO. There are two reasons for this tendency. First, this is due to the greater number of sulphonate groups presence in the molecular geometry of RB5. RB5 consists of 4 sulphonate groups, while MO has only 1 sulphonate group. Sulfonates are molecular fragments that departs with a pair of electrons in heterolytic bond cleavage and contingent upon nucleophilic substitution in the presence of the bases required to generate the carbanion. Thus, the greater the number of sulphonate groups would contribute to a greater colour removal [7]. Second, metabolic intermediates produced during degradation of RB5 possess electron-shuttling characteristics. These electrons-shuttling compounds would be a causative factor in promoting new avenues for electron transfer, and hence improving the decolourisation performance of RB5.

3.2 *Effect of Initial Dye Concentration on Bioelectricity Generation*

Figure 3 shows the voltages and power output of MFC using a binary mixture of MO and RB5 as electron acceptor at the anodic region. The same anode electrode (A3) was employed throughout the study. It can be observed that the power density obtained was 84.47 mW/m^3 (372.5 mV) when 25 mg/L of initial dye concentration was supplied to the system. The power density significantly reduced about 77% to 19.35 mW/m^3 (267.6 mV) as the initial concentration of the binary dye was increased by twofold (50 mg/L). Accordingly, the internal resistance of the MFC system was elevated to 1000 Ω, which doubled the internal resistance (500 Ω) of the system with 25 mg/L of binary mixture of dyes. Further deterioration of the bioelectricity performance of the MFC system was observed as 75 mg/L of binary dye was provided to the system, with the power density of 10.97 mW/m^3 (161.5 mV) recorded. The power density and voltage declined when a higher concentration of binary dye was introduced into the system. This phenomenon could be due to the competition of electrons between anode and azo dyes. The introduction of binary dyes had resulted in less electrons for bioelectricity generation, as the presence of azo bonds, which are electron acceptors, may compete with the anode for electrons. According to Sun et al. [8], azo dye was more preferable in receiving electron, as compared to anode.

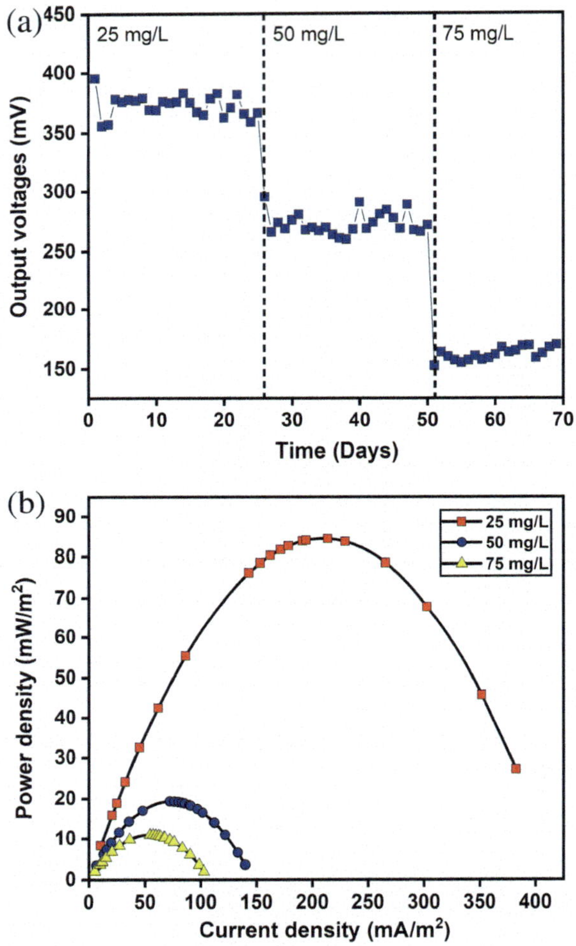

Fig. 3 Output voltages (**a**) and power density (**b**) at different initial dye concentration

4 Conclusion

In this study, the capability of the UFML-MFC in simultaneous treatment of binary mixture of dyes and bioelectricity generation has been examined. Mono azo MO and diazo dyes RB5 were used as model of azo dyes. The colour removal efficiency of binary dye declined remarkably when higher initial dye concentration was introduced into the system. The maximum power density and voltage were deteriorated as the dye concentration increases. A greater colour removal efficiency of RB5 (10%) was achieved, as compared to MO. This can also be concluded that different azo dyes, with different molecular geometries can affect the performance of MFC.

References

1. Solanki K, Subramanian S, Basu S (2013) Microbial fuel cells for azo dye treatment with electricity generation: a review. Bioresour Technol 131:564–571. https://doi.org/10.1016/j.biortech.2012.12.063
2. Sandhya S (2010) Biodegradation of azo dyes under anaerobic condition: role of azoreductase 39–57. https://doi.org/10.1007/698_2009_43
3. Abraham TE, Senan RC, Shaffiqu TS, Roy JJ, Poulose TP, Thomas PP (2003) Bioremediation of textile azo dyes by an aerobic bacterial consortium using a rotating biological contactor. Biotechnol Prog 19:1372–1376. https://doi.org/10.1021/bp034062f
4. Mu Y, Rabaey K, Rozendal RA, Yuan Z, Keller J (2009) Decolorization of azo dyes in bioelectrochemical systems. Environ Sci Technol 43:5137–5143. https://doi.org/10.1021/es900057f
5. Fang Z, Song HL, Cang N, Li XN (2015) Electricity production from Azo dye wastewater using a microbial fuel cell coupled constructed wetland operating under different operating conditions. Biosens Bioelectron 68:135–141. https://doi.org/10.1016/j.bios.2014.12.047
6. Cao Y, Hu Y, Sun J, Hou B (2010) Explore various co-substrates for simultaneous electricity generation and Congo red degradation in air-cathode single-chamber microbial fuel cell. Bioelectrochemistry 79:71–76. https://doi.org/10.1016/j.bioelechem.2009.12.001
7. Ong SA, Ohm MM, Ho LN, Wong YS (2013) Solar photocatalytic degradation of mono Azo methyl orange and diazo reactive green 19 in single and binary dye solutions: adsorbability vs photodegradation rate. Environ Sci Pollut Res 20:3405–3413. https://doi.org/10.1007/s11356-012-1286-1
8. Sun J, Hu Y, Hou B (2011) Electrochemical characteriztion of the bioanode during simultaneous azo dye decolorization and bioelectricity generation in an air-cathode single chambered microbial fuel cell. Electrochim Acta 56:6874–6879. https://doi.org/10.1016/j.electacta.2011.05.111

Workability and Density of Concrete Containing Coconut Fiber

Norlia Mohamad Ibrahim, Nur Liza Rahim, Roshazita Che Amat, Mustaqqim Abdul Rahim, Chin Kah Woo, Irnis Azura Zakarya, and Andreea Moncea

Abstract Use of natural fiber in concrete to enhance the strength of concrete have been used widely and become as part of an alternative building materials. For instance, the use of coconut fiber (CF) which are non-hazardous, environmental-friendly and can improves the engineering properties of concrete. The aim of this study is to identify the workability and density of CF modified concrete. CF were added into the mixture in 3 different amount that is 200 g, 400 g, and 600 g. The size of the cube samples is 100 × 100 × 100 mm and were cured for 14 days, and 28 days. To evaluate the effect of CF in improving the properties of concrete, the properties of ordinary concrete are used as a reference which consist 0% CF. The fresh and hardened densities for all samples also show that when more fiber was added into mixture, densities reduced. As summary, the study shows that by adding CF in concrete reduced the workability and density of concrete.

Keywords Concrete · Construction materials · Workability · Density · Coconut fiber

N. M. Ibrahim (✉) · N. L. Rahim · R. C. Amat · I. A. Zakarya
Sustainable Environment Research Group (SERG), Centre of Excellence Geopolymer and Green Technology (CEGeoGTech), Universiti Malaysia Perlis, 01000 Kangar, Perlis, Malaysia
e-mail: norlia@unimap.edu.my

N. L. Rahim
e-mail: nurliza@unimap.edu.my

R. C. Amat
e-mail: roshazita@unimap.edu.my

I. A. Zakarya
e-mail: irnis@unimap.edu.my

M. A. Rahim · C. K. Woo
Faculty of Civil Engineering Technology, Universiti Malaysia Perlis, 02600 Arau, Perlis, Malaysia
e-mail: mustaqqim@unimap.edu.my

A. Moncea
National Institute for Research and Development in Environmental Protection Bucharest (INCDPM), 294 Splaiul Independentei Street, 6th District, 060031 Bucharest, Romania

N. Mohamed Noor et al. (eds.), *Proceedings of the 3rd International Conference on Green Environmental Engineering and Technology*, Lecture Notes in Civil Engineering 214,
https://doi.org/10.1007/978-981-16-7920-9_41

1 Introduction

The use of concrete in construction industry has been widely accepted because it is a stable and durable material despite some of its weaknesses that is low in tensile strength and flexibility. Therefore. it is necessary to use other materials to enhance the tensile strength of concrete [1]. Considerable efforts have been made worldwide to add various types of fibers either natural fiber or manmade steel fiber into the concrete mixture to make them more reliable, durable, and economical. Natural fibers, such as banana fiber, sugarcane bagasse, bamboo coir and coconut fiber (CF) have specific physical and mechanical properties that can effectively use to develop reinforced concrete materials [2]. In most cases, these CF are dumped as agricultural wastages and are therefore readily available in large quantities and consequently inexpensive. The preprocessing of these natural fiber is less expensive and typically requires good sunlight for the drying process to take place.

Coconut fiber (CF) is an agricultural waste obtained during the processing of coconut oil. In Malaysia, CF can be found in coconut milk production industry and even as a by-product from coconut water selling stall. It can also be found in large quantities in the tropical regions of the world, especially in Africa, Asia, and most part of Americas. CF is not commonly used in the construction industry but usually dumped as agricultural waste. However, with the intention on providing more affordable house especially for rural areas, many researches have been made in preliminary stage to identify potential use of these agricultural wastes and residues as part or all of raw material in the building's materials especially for concrete production[2]. In countries where large amounts of agricultural waste emitted, these wastes can use as potential materials or alternative materials for the construction industry.

The method of CF usage in concrete varied from either in its natural appearance or properties or undergone a surface treatment such as through alkalization, washing and drying [3]. Many attempts also have been made to incorporate the use of CF either with combination with other waste materials [4, 5] or being added individually in the mixture. The use of fiber can affect the properties especially its workability and densities because the properties of the fiber itself.

2 Materials and Method

2.1 Material

In this study, cement, aggregates and water are the main material to produce the concrete whereas CF is used as additional material in the mixture. Ordinary Portland cement of grade 32.5 is used in this experiment. The CF used for the project was obtained locally from local coconut milk producing industry. The inner coir or fibers are mostly obtained by manually extracting fibers from the outer shell of the coconut.

These fibers have undergone prewashing process to remove any impurity or excessive dirt and debris attached to them, and then been sun-dried them for several days under the scorching sun. The size and length of coconut fiber varies from 5 to 25 mm depending on testing conducted and purpose of study. The dried CF is quite sensitive with the pulling actions in the process of separating them hence, the size is selected to be approximately 8 cm of the length of CF for the experimental work.

The casting of modified concrete with CF as the fiber reinforcement were based on the design mix in Jabatan Kerja Raya (JKR) Standard and mould with dimensions 100 × 100 × 100 mm were used for density testing. Concrete mix was designed by taking 1:1:2 as nominal mix and with water cement ratio of 0.5. Contradicts to most of the previous research, CF in this study was added into the concrete mixture rather than replacing the main components of other raw materials. Three different weight of CF which is 200 g, 400 g and 600 g were selected to be put into concrete mix. The main purpose of proportioning of concrete ingredients is to get strong, dense, and durable concrete with fibers.

2.2 *Method*

Workability of concrete can be measured in several ways depending on the types of concrete. Slump test is the most used method of measuring consistency of concrete at its fresh state. The workability of the fresh concrete was tested using slump test according to British Standard [6]. It can be conveniently used as a control test and can indicate the uniformity of concrete. By observing the direction and way of the concrete collapses the required amount of water to be used for optimum workability can be identified. The equipment for slump test is mainly composed of a metal mold in the form of a conical frustum, the internal size of the bottom diameter is 20 cm, the top diameter is 10 cm, and the height is 30 cm. Figure 1 shows the slump test conducted in this study.

Meanwhile, the density for all samples is measured at its fresh and hardened state. For fresh state, it was measured immediately after the mixing process [7] meanwhile for hardened state, density is measured at 7, 14 and 28 days in order to examine the trends following standard [8].

3 Result and Discussion

3.1 *Characteristics of CF Modified Concrete*

Based on Fig. 2, the slump value of fresh concrete for all sample shows a displacement from its original position, measured from the top of the container in the range of 25–37 mm. Slump value for control sample was 37 mm which describes that the

Fig. 1 Slump test for CF modified concrete

Fig. 2 Results of slump for CF modified concrete

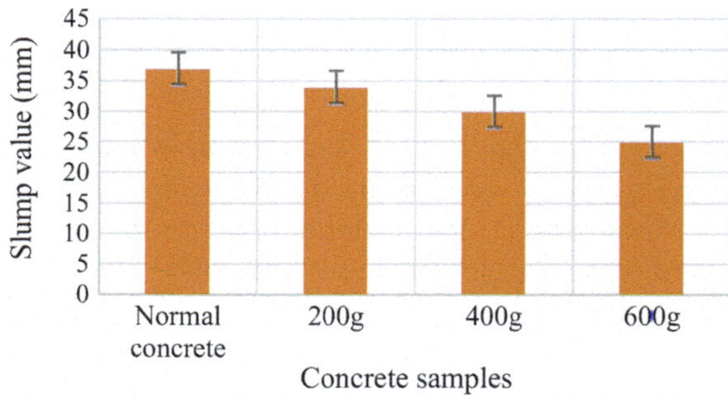

workability is good. Obviously, the lowest slump value is for sample that contained 600 gm of CF. For sample containing 200 gm, slump value is 34 mm and 30 mm for modified concrete containing 400 gm CF. Samples containing CF could absorb more water during casting process and increase moisture-retaining capacity. Samples containing CF could hold the concrete paste matrix together thus reduces the slump value proportionately to the amount of CF in the mixture.

The density for fresh concrete state and hardened concrete state for every mixture were determined. For denser concrete, it usually has higher compressive strength and less internal voids than ordinary concrete. For each type of sample, there are 3 cubes tested to get an average result which can be strengthen the accuracy of the results shown in Fig. 3. From the graph, it can be clearly seen that when the content of CF increase, the density of the fresh concrete will decrease. Density of fresh concrete range from the minimum value of 2351.11 kg/m^3 to 2513 kg/m^3. For 600 g of CF content in mixture, the maximum density at 7 days is 2392 kg/m^3 while the lowest density for CF at the same mixture is 2392 kg/m^3 which is at 28 days. Meanwhile, for sample containing 400 g and 200 g CF in mixture, density also lower than control

Fig. 3 Density of CF modified concrete

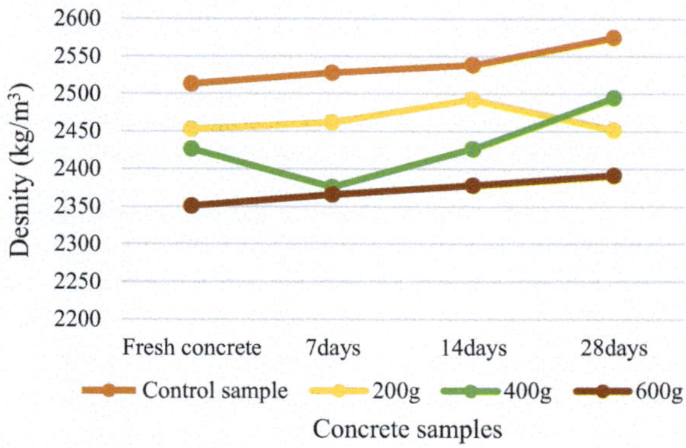

sample with 2495.33 kg/m^3 and 2452.67 kg/m^3, respectively. In this study, it can be concluded that by increasing the amount of CF causes a significant decrease in the density of concrete. The percentage of reduction in density for 600 g CF modified concrete is 7.13% if compared with the control sample at 28 days.

4 Conclusion

As conclusion, from the results obtained, it can be concluded that CF can be use as additional raw material in concrete production. The maximum amount of CF particularly in this study was 200 g to obtained comparable workability of concrete. Densities for all modified concrete also reduced when the CF was increased. For future works, it is recommended to vary the length of CF and age of CF in order to establish the full properties of concrete containing CF.

References

1. Naamandadin NA, Rosdi MS, Mustafa WA, Shahrol Aman MNS, Saidi SA (2020) Mechanical behaviour on concrete of coconut coir fiber as additive. IOP Conf Ser Mater Sci Eng 932:012098. https://doi.org/10.1088/1757-899X/932/1/012098
2. Quiñones-Bolaños E, Gómez-Oviedo M, Mouthon-Bello J, Sierra-Vitola L, Berardi U, Bustillo-Lecompte C (2021) Potential use of coconut fibre modified mortars to enhance thermal comfort in low-income housing. J Environ Manag 277:111503. https://doi.org/10.1016/j.jenvman.2020.111503
3. Andiç-Çakir Ö, Sarikanat M, Tüfekçi HB, Demirci C, Erdoğan ÜH (2014) Physical and mechanical properties of randomly oriented coir fiber-cementitious composites. Compos Part B Eng 61:49–54. https://doi.org/10.1016/j.compositesb.2014.01.029
4. Sathvik S, Edwin A, Basnett A, Sharma P, Carmicheal J (2019) Experiment of partial replacement of egg shell powder and coconut fibre in concrete. Int J Innov Technol Explor Eng 8:1034–1038. https://doi.org/10.35940/ijitee.F1213.0486S419

5. Sekar A, Kandasamy G (2018) Optimization of coconut fiber in coconut shell concrete and its mechanical and bond properties. Materials (Basel) 11(9):1726. https://doi.org/10.3390/ma11091726
6. BS 1881–116 (1983) Method for determination of slump. Br Stand 7–41
7. BS 1881–107 (1983) Method for determination of density of compacted fresh concrete. Br Stand 1–4
8. BS 1881–114 (1983) Methods for determination of density of hardened concrete. Br Stand. CS1:2010

Water and Wastewater

Assessment of Heavy Metal Pollution in Sediments and in *Phragmites Australis* from Argeş River

Ecaterina Marcu, György Deák, Irina-Elena Ciobotaru, Iasmina-Florina Burlacu, Carmen Tociu, and Abdul Haqi Ibrahim

Abstract There are several species of macrophyte which have the ability to absorb heavy metals from water and, therefore, are used to retain and remove them. In the present paper, the concentrations of heavy metals (Zn, Cu, Ni, Cd, Pb and Cr) were investigated in sediment samples from the Argeș River and their potential transfer from sediments to *Phragmites australis* was evaluated. The extent of sediment pollution with heavy metals and the potential risk to the aquatic environment were estimated based on the following indexes: bioaccumulation, geoaccumulation, ecological risk, translocation, contamination, etc. The metals concentrations in the analyzed sediments were, generally, below the limits of national legislation.

Keywords Heavy metals · Sediments · *Phragmites australis* · Argeș River

E. Marcu (✉) · G. Deák · I.-E. Ciobotaru · I.-F. Burlacu · C. Tociu
National Institute for Research and Development in Environmental Protection Bucharest (INCDPM), 294, Splaiul Independentei Street, 6th District, 060031 Bucharest, Romania
e-mail: ecaterina.marcu@incdpm.ro

I.-E. Ciobotaru
e-mail: irina.ciobotaru@incdpm.ro

I.-F. Burlacu
e-mail: iasmina.burlacu@incdpm.ro

C. Tociu
e-mail: carmen.tociu@incdpm.ro

A. H. Ibrahim
Centre of Excellence, Water Research and Environmental Sustainability Growth, Universiti Malaysia Perlis, 02600 Perlis, Malaysia
e-mail: abdulhaqi@unimap.edu.my

N. Mohamed Noor et al. (eds.), *Proceedings of the 3rd International Conference on Green Environmental Engineering and Technology*, Lecture Notes in Civil Engineering 214,
https://doi.org/10.1007/978-981-16-7920-9_42

1 Introduction

In recent decades, anthropic activities generated an increase in the content of heavy metals in the environment, and therefore, an increased pollution of many areas of the world. Heavy metals accumulate in soil, sediments and organisms, and and they are a serious risk to the health of the environment and people [1].

Pollution of water bodies has been intensively studied by measuring the heavy metals concentrations from the aquatic plants (leaves, stem, roots). Knowledge of the mechanisms of absorption, mobility and bioconcentration of metal ions by the macrophyte plants is important from a scientific and practical point of view, given that these plants are considered indicators of pollution [2, 3]. Due to their ability to bioaccumulate heavy metals, some macrophyte plants are used for wastewater treatment [4]. *Phragmites australis* is one of the most studied macrophytes in wetlands, with a capacity to hyperaccumulate chemical elements from both water and sediment, being often used in the treatment of urban and / or industrial wastewater polluted with metals. In general, organisms that have a large surface area in direct contact with sediments, such as rooted macrophytes, accumulate larger amounts of metals from sediments than from water. The bioaccumulation of metals occurs mainly in the roots, and to a lesser extent, by translocation in the upper parts of the plant [1, 3]. The transport of metals in the body of the plant is influenced by a series of factors such as: energy source for photosynthesis, relative air humidity, ambient temperature, etc. [3].

In this experimental study, the evaluation of sediment loading with heavy metal was done by calculating: the contamination factor, the contamination degree [5, 6] and the geo-accumulation index [7, 8]. The ecological risk factor and the potential ecological risk index were calculated in order to estimate the potential ecological risk caused by sediment pollution with the studied metals; the bioaccumulation factor was calculated to estimate both the bioaccumulation potential of metals in plants and their translocation factor [9].

2 Experimental Details

2.1 Examined Area

The examined area is located on the territory of 1 Decembrie commune, Ilfov County, on the lower course of the Arges River, at an altitude of 62 m above sea-level. The sampling area is located at 44.267371 N 26.069540 E [10]. Given the complex profile of socio-economic activities in the examined area, the sources of metal pollution in the area are very diverse.

For the sampling of sediment and biota, 5 points were selected, longitudinally distributed on the left bank of the river Argeş at approximately 500 m distance between them, on a river sector of approximately 2 km. Dragged sediment samples were taken with a plastic spade from about 1 m away from the shore. The samples of aquatic plants (*Phragmites australis*) were taken from the waterfront, from the same 5 points from which the other samples were collected. All samples were transported to the laboratory for processing on the same day.

2.2 Procedure for Analysis and Treatment of Samples

At their arrival in the laboratory, the samples were processed in order to determine the total metals concentrations, including steps such as drying, milling (SR EN 16,179), acid digestion (SR EN 16,174) and analysis by atomic absorption spectrometry (SR ISO 11047). The current standard methods in force were applied using a Binder ED53 oven, Retsch RM200 electric ceramic mortar and pestle, Berghoff MWS4 microwave digester, VELP Scientifica DK6 reflux digester, ContrAA700 HRCSAAS spectrometer. The analysed heavy metals were zinc (Zn), copper (Cu), nickel (Ni), cadmium (Cd), lead (Pb) and chromium (Cr).

3 Results and Discussions

Heavy metal levels determined in sediments registered values below the maximum limits from Order no. 161/2006 [11], with the exception of nickel at point P2 which slightly exceeded (36.4 mg/kg) the permitted limit (35 mg/kg), possible due to the impact of anthropogenic activities carried out in the proximity of this river. The loading of heavy metals (mg/kg) is higher at points P1 (189.38) and P2 (190.31) and lower at the other sampling points P3 (153.88), P4 (123.55) and P5 (171.843). Contamination factor and the heavy metal contamination degree of sediments were estimated both for each sampling point and metal analyzed, and on the basis of the average values of the concentrations of heavy metals in the sampling points. The degrees of heavy metal contamination of the sediments in the analyzed area were 2.90 (P1), 2.81 (P2), 2.39 (P3), 1.93 (P4) and 2.38 (P5), respectively < 8 at all sampling points. These results demonstrate a low contamination level with the above mentioned metals. According to the classification made by Hakanson [6] for the contamination factors, the sediment taken from point P2 from the Argeş River, registered a moderate contamination with Ni (1.040). For the other sampling points, the estimated contamination factor value for all analysed metals was <1 which indicates a low contamination factor of sediment with the studied metals.

Geoaccumulation index was calculated taking as background geochemical loading of the metal, the limit concentrations of the chemical quality standards for sediments from the national legislation (Order 161/2006). According to the classes of the geoaccumulation index, the calculated values of this index for each metal are < 0, which demonstrates that the sediments from all sampling points are uncontaminated with these studied heavy metals. Estimation of the contamination factor, the contamination degree and the geo-accumulation index showed that the sediments at all sampling points are practically uncontaminated with heavy metals. Moderate Ni contamination (1.040) at sampling point P2 may be due to the impact of anthropogenic activities on river quality. To assess the level of heavy metal sediment contamination, were calculated the potential ecological risk factor and index. The decreasing order of the potential ecological risk factors of heavy metals in sediments was: P1: Cd > Ni > Cu > Pb > Cr > Zn; P2: Ni > Cd > Cu > Pb > Cr > Zn; P3: Cd > Ni > Cu > Pb > Cr > Zn; P4: Cd > Ni > Cu > Pb > Cr > Zn; and P5: Ni > Cu > Cr > Pb > Zn > Cd. The potential ecological risk factors obtained for all analyzed metals (Zn, Cu, Ni, Cd, Pb, Cr) were less than 40, indicating a low ecological risk. All sediments analyzed had RI values below 150 (range = 8.97(P5)-17.52(P1)), suggesting a low ecological risk. Regarding the IR values obtained, the sampling sections can be classified in the following order: P1 > P6 > P4 > P2 > P5.

The translocation factor of heavy metals in sediments from *Phragmites australis* was calculated using the average metal concentrations in plants and sediments. Translocation factors greater than 1 were obtained for Zn and Pb from root to stem and leaves, and for Ni, Pb and Cr, only from root to stem. Therefore, *Phragmites australis* translocated heavy metals from root to stem rather than from root to leaf. The translocation of metals from root to stem is significant for Ni, Pb and Cr (3.294 – 11.145), while the translocation of metals from root to leaves is significant only for Zn (1.287) and Pb (15.213).

The process of binding heavy metals to plants takes place through the mechanisms of metal phytoextraction and bioaccumulation. The roots of *Phragmites australis* can help prevent the accumulation of metals in the environment.

The bioconcentration factor of metals to *Phragmites australis* was calculated using the average metal concentrations in the sediments and plants. Generally, the factor's values of the metals in *Phragmites australis* did not change much between the five sampling points on the Argeș River. The comparative analysis of the bioconcentration factors indicated that the plants showed a high bioconcentration capacity for Cd, the bioconcentration factors being higher than 1 in all parts of the plant. From the values obtained of the bioaccumulation factor in relation to the sediment, of the three parts of the aquatic plant, it is obvious that the root has the greatest capacity to accumulate metals in order Cd > Zn > Cu > Cr > Ni = Pb, the stem in order Cd > Zn > Cr > Pb > Cu > Ni, and leaves in order Cd > Zn > Pb > Cu > Cr > Ni.

4 Conclusions

Heavy metal pollution of sediments can adversely affect aquatic systems and the indices estimated in this study can be considered indicators of ecosystem health.

The results presented show that the values of Zn, Cu, Cd, Pb and Cr measured in sediments were below the maximum limits imposed by the quality standards for sediments in national legislation with the exception of Ni at point P2 which slightly exceeded the permitted limit (35 mg/kg), probably due to the impact of anthropogenic activities on the quality of Arges River. The estimated indices values for the assessment of heavy metal sediment contamination in the studied river sector indicate both a low sediment contamination, and a potential reduced ecological risk on the ecosystem.

The bioavailability of traces metals from sediments depends on their ionic forms (chemical and/or physical), the concentrations of metals present in the sediments as well as the exposure time of the biota to these concentrations.

Phragmites australis demonstrated a high bioconcentration capacity of Cd, the bioconcentration factors obtained being higher than 1 in all component parts of the plant. The translocation of metals from root to stem is significant for Pb, Ni and Cr (3.294 – 11.145), while the migration of metal from roots to leaves is significant only for Pb (15.213) and Zn (1.287).

Acknowledgements This paper was developed through Nucleu Programme—Contract no. 39N/2019 with the Romanian Ministry of Research and Innovation, within the project "Assessment of the state of aquatic ecosystems regarding the presence of inorganic micro-pollutants with a high degree of toxicity in the aquatic food chain".

References

1. Klink A, Dambiec M, Polechońska L (2019) Trace metal speciation in sediments and bioaccumulation in Phragmites australis as tools in environmental pollution monitoring. Int J Environ Sci Technol 16:7611–7622. https://doi.org/10.1007/s13762-019-02305-7
2. Singh N, Kaur M, Kaur JK (2017) Analysis on bioaccumulation of metals in aquatic environment of Beas River Basin: a case study from Kanjli wetland. Geohealth 1(3):93–105. https://doi.org/10.1002/2017GH000062.eCollection2017May
3. Štrbac S, Šajnović A, Grubin K, Vasić M, Dojčinović N, Simonović B, Jovančićević P (2014) Metals in sediment and Phragmites Australis (common reed) from Tisza River. Ecol Environ Res 12(1):105–122. http://www.aloki.hu, ISSN 1785 0037, Budapest, Hungary
4. Bai L, Liu XL, Hu J, Li J, Wang ZL, Han G, Li SL, Liu CQ (2018) Heavy Metal Accumulation in common aquatic plants in rivers and Lakes in the Taihu Basin. Int J Environ Res Public Health 15:2857. https://doi.org/10.3390/ijerph15122857
5. Ionescu P, Deák G, Diacu E, Marcu E, Ciobotaru IE (2020) Risk assessment of heavy metals in surface sediments from the aquatic environments located in crowded areas of Southern Romania. In: Proceedings of academicsera international conference, Rome, Italy
6. Hakanson L (1980) An ecological risk index for aquatic pollution control. A Sedimentol Approach Water Res 14:975–1001. https://doi.org/10.1016/0043-1354(80)90143-8
7. El-Sayed SA, Moussa EMM, El-Sabagh MEI (2015) Evaluation of heavy metal content in Qaroun Lake, El-Fayoum. Egypt Part I: Bottom Sediments J Radiat Res Appl Sci 8:276–285. https://doi.org/10.1016/j.jrras.2015.02.011
8. Nowrouzi M, Pourkhabbaz A (2014) Application of geoaccumulation index and enrichment factor for assessing metal contamination in the sediments of Hara Biosphere Reserve. Iran Chem Speciat Bioavailab 26(2):99–105. https://doi.org/10.3184/095422914X13951584546986
9. Takarina ND, Pin TP (2017) Bioconcentration factor (BCF) and translocation factor (TF) of heavy metals in mangrove trees of Blanakan fish farm. Makara J Sci 21(2):77–81. https://doi.org/10.7454/mss.v21i2.7308

10. https://earth.google.com/web/@44.26737100,26.06954000,54.64739676a,2295.172160 17d,35y,-0h,0t,0r
11. Ministry of Environment and Water Management, REGULATION no. 161 (2006) On the classification of surface water quality in order to establish the ecological status of water bodies, Official Gazette no. 511 bis of 13 June 2006

Assessment of Heavy Metals in Water and Sediment of Lower Klang River

Irma Noorazurah Mohamad, Nurul Nazurah Isa, and Siti Noor Afifah Ajid

Abstract Heavy metals toxicity in tidal river has proven to be a major threat and there are several health risks associated with it. The purposes of this study are to determine the concentration of heavy metals from anthropogenic sources in Lower Klang River. Water surface samples and surface sediment samples were analysed for cadmium (Cd), lead (Pb) and zinc (Zn). The concentration of heavy metals in surface sediment samples were found in decreasing order Pb > Zn > Cd. While the concentration of heavy metals in water samples are much lower than the heavy metals in surface sediment samples with decreasing sequence of Zn > Pb > Cd. This research shows the presence of the heavy metals contamination in water surface samples and surface sediment samples of Lower Klang River. However, monitoring program and further investigation on heavy metals accumulation due to anthropogenic activities should be carried out to protect Lower Klang River during future growth. The results of this study provide useful knowledge and information for sustainable estuary management in Lower Klang River.

Keywords Heavy metals · Sediment · Water · Tidal river

1 Introduction

Over the years, rivers had been significantly important as they act as water resources, provide food for the other animals, transportation, and defensive barrier. A certain river grows as they collect water from tributaries along the course and eventually flow into a larger of water body such as ocean. The major objective of river is as source of water supply. In Malaysia, it can be certainly founded that the main townships and early settlements are situated either at the riverbanks or estuaries. The result is that the river floodplains consistently occupied by settlements and industry. Hence,

I. N. Mohamad (✉) · N. N. Isa · S. N. A. Ajid
School of Civil Engineering, College of Engineering, Universiti Teknologi MARA, 40450 Shah Alam, Selangor, Malaysia
e-mail: irma1095@uitm.edu.my

N. Mohamed Noor et al. (eds.), *Proceedings of the 3rd International Conference on Green Environmental Engineering and Technology*, Lecture Notes in Civil Engineering 214,
https://doi.org/10.1007/978-981-16-7920-9_43

this makes flood control, and the control of bank erosion and meander mitigation is essentially importance.

Klang River Basin is considered as the most developed area in Malaysia which has become critical resources because of river pollution and inefficient of water management. It is the highest rate of urban growth which causing the lessen of accessibility of clean water due to the environment degradation [1].

Contamination of heavy metals in the aquatic environment has attracted global attention owing to its abundance, persistence and environmental toxicity [2]. Excessive number of heavy metals and other pollutant variables is a threat and affecting factor to the rivers water quality where it gives serious challenges in river basin management, and it could be declining the environment health of the river which may cause a serious pollution to the water ways. The level of heavy metals and pollutants was received by river water from either point sources which are industrial wastewater, municipal sewers, wet market, sand mining and landfill, whereas nonpoint sources from agricultural and urban runoff and other commercial activities and these over-supplied nutrients would have detrimental effects which will eventually lead to hypoxia and anoxia from eutrophication [3]. This must be taken seriously because Klang River is the main river that is being used by the dwellers throughout its span.

Most estuaries are important in terms of biological and economical for the surrounding's people which is why it is so polluted nowadays. It then leads to the disruption of food chain in the environment which is toxic to the aquatic animals and nearby people. These is necessary to determine the presence of heavy metal which could disrupting the environment and the people if the level of the contamination of the heavy metals is high in Lower Klang River. In addition, this study could help to assess the differences of heavy metals condition in water and sediment of Lower Klang River.

2 Materials and Method

2.1 Study Area

The data of the study were obtained by collecting samples of water and sediment in the Lower Klang River. The research study takes place at downstream part of Klang River which flows through Shah Alam and flows until the lower stretches of the river which is Port Klang and eventually flows into the Straits of Malacca to the west. Location of sampling stations were determined at selected points along the lower part of Klang River as shown in Fig. 1. These sampling stations were identified at the upstream point located in Pengkalan Batu, Bandar Klang, at the middle point which is near Department of Irrigation and Drainage (DID) station, Sungai Udang and downstream point in Jetty at Kampung Delek. The distance between Location 1

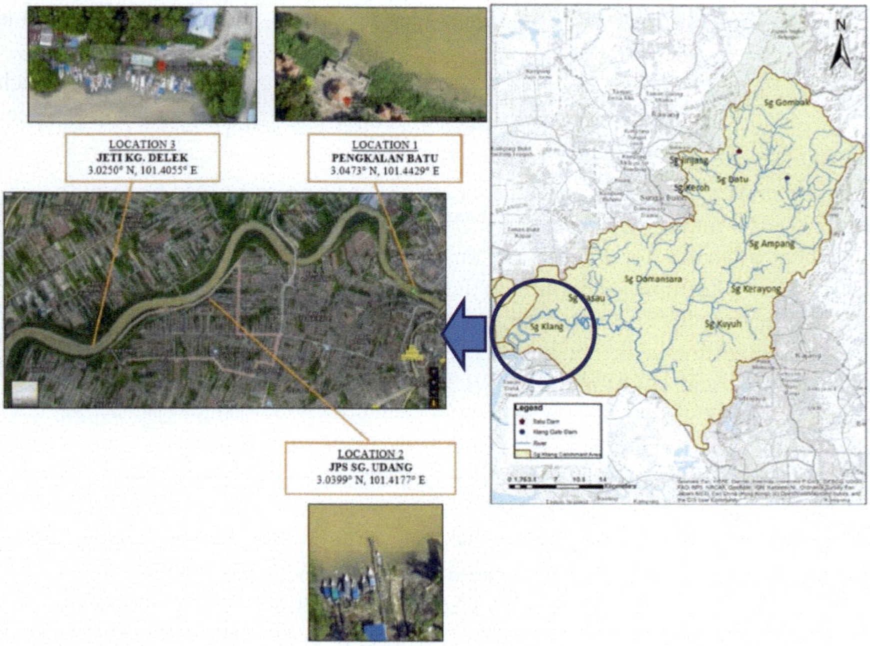

Fig. 1 Study area and location of sampling stations in the Lower Klang River

to Location 2 is about 5.4 km and between Location 2 to Location 3 is about 4.8 km (Google Map).

2.2 *Surface Water Sample*

Samples of surface water were collected at three (3) sampling locations along Lower Klang River namely Pengkalan Batu, DID Sg. Udang and Jeti Kg. Delek. Grab sampling was done to collect the water sample by using a pail. The water sample was then collected in plastic container. The bottle was kept in icebox filled with ice to keep the temperature to be 4 °C and below. The concentration of lead and cadmium were done using Dithizone Method which is suitable for water and wastewater by using Spectrophotometer DR2800 [4]. While zinc concentration can be done using Zincon Method by adding 20 ml ZincoVer® 5 Reagent Powder Pillow.

2.3 Surface Sediment Sample

Surface sediment (0–5 cm) samples were collected by using a clean plastic bottle. Then, samples were brought to the Environmental Laboratory and were dried by using an air-circulating oven at 1400 °C for minimum 24 h till further analysis were to be done. The grain size analysis was done by using series of sieves 600, 425, 300, 212, 150 and 63 μm apertures. The sediment samples then sieved by using a mechanical shaker for 10 min and analysed by using semi-log graph to get the classification of samples according to the grain size fractions [5].

For heavy metals concentration, the sediment samples were digested by using digesdahl digestion with Nitric Acid (HNO_3) or Hydrogen Peroxide (H_2O_2) or Hydrochloric Acid (HCl) mixture [6]. The samples for assessment were allowed to dry at 105 °C in an air circulating oven. Once the constant dry weight was attained, the dried sediment was ground by using a mortar and pestle into a fine powder. Then, samples were sieved through a 63 μm stainless steel sieve to remove any large organic matter and produce homogeneity. Next, the sieved samples were then made into sample of approximately 0.5 g in mass. Samples were then digested in 6 mL of Nitric Acid (HNO_3) solution (AnalaR grade, R&M 65%) and 10 mL of Perchloric Acid ($HClO_4$) solution (AnalaR grade, R&M 70%) in beakers. The samples were digested for four minutes at a temperature of 440 °C. Once the digestion was completed, it was diluted to 100 mL with distilled water. Lastly, the digested samples were filtered through Whatman No.1 filter paper into pre-cleaned 50 mL volumetric flasks.

3 Result and Discussion

3.1 Sediment Characteristic

The grain size analyses of sediments are illustrated in Fig. 2. The grain size distribution at all sampling locations showed that fine sand (63-425 μm) percentages appeared to be the main sediment type in Lower Klang River. It is proven that the sediment grain size are mostly fines grain and classified as silty sand. This indicate that the silty sand has high capability to adsorb trace metal contaminants as they had high specific area per unit quantity of particles. Hence, it is important to always monitor the concentration of heavy metals in the sediment as it is highly potential to adsorb the pollutants.

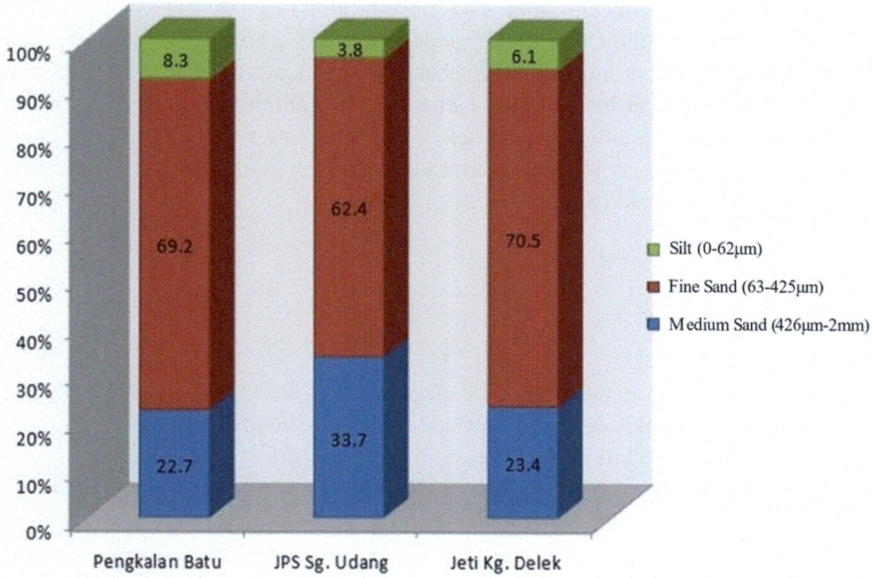

Fig. 2 Percentage of grain size composition/distribution silt (0–62 μm), fine sand (63–425 μm) and medium sand (426 μm–2 mm) sediment characteristics

3.2 *Heavy Metals Concentration*

The concentration of different heavy metals in sediment and water surface from Lower Klang River are summarised in Table 1. The data indicated that the concentration of Pb was high in sediment samples whereas Zn was high in water samples. In addition, Cd get least accumulation for both type of samples at all locations. The highest concentration of heavy metals is found at Location 2, DID Sg. Udang. This location is believed to contribute with high anthropogenic sources because this area very near to DID storage warehouse and jetty for the local people.

Results show that Cd in sediment samples was found in the range of 43.8–53.1 μg/L, Zn 180–250 μg/L and Pb 200–268.8 μg/L. On the other hand, the concentrations of the heavy metals in water samples were far lower than sediments

Table 1 Concentrations of heavy metals in sediment and water

Sampling location	Heavy metal concentrations in sediment (μg/L)			Heavy metal concentrations in water (μg/L)		
	Cd	Pb	Zn	Cd	Pb	Zn
Pengkalan Batu	43.8	200.0	180.0	5.1	37.0	150.0
DID Sg. Udang	53.1	268.8	220.0	3.7	17.5	185.0
Jeti Kg. Delek	52.5	225.0	250.0	4.5	12.0	75.0

samples except the Zn concentration. The concentration of Zn in water samples were found higher than Pb and Cd with the range of 75–185 μg/L, Pb 12–37 μg/L and Cd 4.5–5.1 μg/L. According to [7], Zn has the higher potential to accumulate in aquatic ecosystems than in sediment. Generally, concentration of heavy metals in the sediment samples are higher than heavy metals in water samples. This because of the heavy metals are poorly soluble in water and much easily adsorb to suspended particles then accumulated and settled as sediments [8].

4 Conclusion

Three selected heavy metals such as cadmium (Cd), lead (Pb) and zinc (Zn) in surface sediment samples and water surface samples were investigated from Lower Klang River. Heavy metals concentration in surface sediment samples were found in a decreasing sequence of Pb > Zn > Cd. However, the concentration of heavy metals in water samples are much lower than the heavy metals in surface sediment samples with decreasing sequence of Zn > Pb > Cd. Heavy metals concentrations were detected as the Lower Klang River is located at urbanised and industrialised area. Monitoring program is highly recommended to assess the spatial and temporal effects of anthropogenic sources into Lower Klang River ecosystem.

Acknowledgements Authors acknowledge the financial support from College of Engineering, Universiti Teknologi MARA, Shah Alam, Selangor. Technical support and guidance from lab assistants from School of Civil Engineering is also acknowledged.

References

1. Sharif SM, Kusin FM, Asha'ari ZH, Aris AZ (2015) Characterization of water quality conditions in the Klang River Basin, Malaysia using self organizing map and K-means algorithm. Procedia Environ Sci 30:73–78. https://doi.org/10.1016/j.proenv.2015.10.013
2. Ali MM, Ali ML, Islam MS, Rahman MZ (2018) Assessment of toxic metals in water and sediment of Pasur River in Bangladesh. Water Sci Technol 77(5):1418–1430. https://doi.org/10.2166/wst.2018.016
3. Jaishankar M, Tseten T, Anbalagan N, Mathew BB, Beeregowda KN (2014) Toxicity, mechanism and health effects of some heavy metals. Interdiscip Toxicol 7(2):60–72. https://doi.org/10.2478/intox-2014-0009
4. DR 2800 Spectrophotometer (2007) Procedures manual, 2nd edn. Hach Company, p 814
5. Liao J, Chen J, Ru X, Chen J, Wu H, Wei C (2017) Heavy metals in river surface sediments affected with multiple pollution sources, South China: distribution, enrichment and source apportionment. J Geochem Explor 176:9–19. https://doi.org/10.1016/j.gexplo.2016.08.013
6. USEPA Environmental Protection Agency United States (1996) Method 3050B - acid digestion of sediments, sludges, and soils. Revision 2. Washington, DC

7. Aydin-Onen S, Kucuksezgin F, Kocak F, Açik S (2015) Assessment of heavy metal contamination in Hediste diversicolor (O.F. Müller, 1776), Mugil cephalus (Linnaeus, 1758), and surface sediments of Bafa Lake (Eastern Aegean). Environ Sci Pollut Res 22:8702–8718. https://doi.org/10.1007/s11356-014-4047-5
8. Yang J, Zhang B, Peng X, Wang H, Li Z, Cai W, Fang H (2014) Sediment quality assessment for heavy metal contamination in the Dongzhai Harbor (Hainan Island, China) with pollution indice approach. Open Chem Eng J 8:32–37. https://doi.org/10.2174/1874123101408010032

Cloud-Based Approach for Water Quality Monitoring: A Review

Nazirul Mubin Zahari, Adel Saleh Salem Saeed, Mohd Hafiz Zawawi, Nursyadzatul Tasnim Roslin, Nurhanani Abd Aziz, and Farah Nurhikmah

Abstract Water quality monitoring is essential to check whether the water is suitable and safe to consume daily. Many researchers conducted current systems to measure the water quality by the smart system in which helps to reduce the time consuming of the monitoring system. One of the recent ways to monitor water quality is a cloud-based approach which is the space of data collected to check water quality. The cloud-based system can be created by using equipment that helps make the water quality information such as sensors, controllers, and communication technology which are essential equipment to allow users to access and read the information of water quality. This review paper investigated many recent works of researchers using the cloud-based approach to monitor water quality. It focused more on controllers, sensors, communication technology of previous researchers they used to evaluate it to improve the system of water quality monitoring.

Keywords Water quality · Cloud · Controller · Sensors · Communication technology · IoT

1 Introduction

Water is one of the significant components in the earth for life, which plays a vital role in the human body. In Malaysia, surface water accounts for 99% of the total water supply, with underground water accounting for the remaining 1% [1, 2]. In many river systems, changes in land use, such as deforestation, agriculture, and

N. M. Zahari (✉) · A. S. S. Saeed · M. H. Zawawi · N. T. Roslin
Department of Civil Engineering, College of Engineering, Universiti Tenaga Nasional, 43000 Kajang, Selangor Darul Ehsan, Malaysia

M. H. Zawawi
e-mail: mhafiz@uniten.edu.my

N. A. Aziz · F. Nurhikmah
UNITEN R&D Sdn. Bhd., Universiti Tenaga Nasional (UNITEN), 43000 Kajang, Selangor Darul Ehsan, Malaysia

N. Mohamed Noor et al. (eds.), *Proceedings of the 3rd International Conference on Green Environmental Engineering and Technology*, Lecture Notes in Civil Engineering 214,
https://doi.org/10.1007/978-981-16-7920-9_44

industrial and residential growth, have a significant influence on water quality. One of the most prevalent concerns in Malaysia is the water quality of rivers and lakes [3]. Most people, around 5–10 million per year, have been died due to water pollutions. Therefore this is a hazardous sign for the world to improve the water quality to enhance humans' lives and even animals and plants [4].

The importance of water quality is to determine whether the water is in good health or be the main reason for diseases and death. The water quality is usually affected by the sources of environmental pollutions such as effluents, sewage discharge, runoff, and others [5]. Moreover, lack of awareness and interest in water quality monitoring is also an essential contributor to water pollution, affecting life on the earth [6]. Water quality monitoring is defined as the data collection during a regular interval in specific locations to give information that can be used to clarify the current conditions, establish trends, and so on [7–9]. This information consists of the values of the parameters, which are the main part of monitoring the water quality, such as pH level, turbidity, temperature, dissolved oxygen (DO), water level, and others. In addition, the monitoring system analyses data collected in real-time and recommends appropriate corrective actions.

There are many ways to monitor water quality. Still, the recent methods have used the Internet of things (IoT) to monitor water quality which makes it a development solution to improve water quality monitoring [10–12]. The main objectives of these recent ways are to measure variations in parameters and give early warning identification of risks by measuring water quality parameters. This paper aims to review recent works carried out in the cloud-based approach in water quality monitoring in terms of parameters, controller, and communication technology they used.

2 The System Overviews

A cloud is a space of data that has been collected by using various equipment technologies to provide information about the water quality, which helps to monitor the water quality in easier ways [9, 13–15]. Based on Ahmad et al. [16], a cloud is a kind of record that create, and pass on through the cloud. A cloud approach is an administration movement model that allows organizations, end-users, and their applications to store, manage, and recover data on the cloud. It's a standard database strategy that's carried out by distributing database programming over a register/establishment cloud. According to Geetha and Gouthami [6], they mentioned that the most reason for creating a cloud is to send alert SMS messages programmed to monitor water quality if the parameters exceed the threshold limit.

The cloud can be created by connecting the sensors to the wireless communication device, and this wireless device is connected to the controller. The data storage of parameters monitored that can send a message to monitors through an application call it as a data management service for the monitoring system [17–19]. The summary of the cloud creating has been shown in Fig. 1.

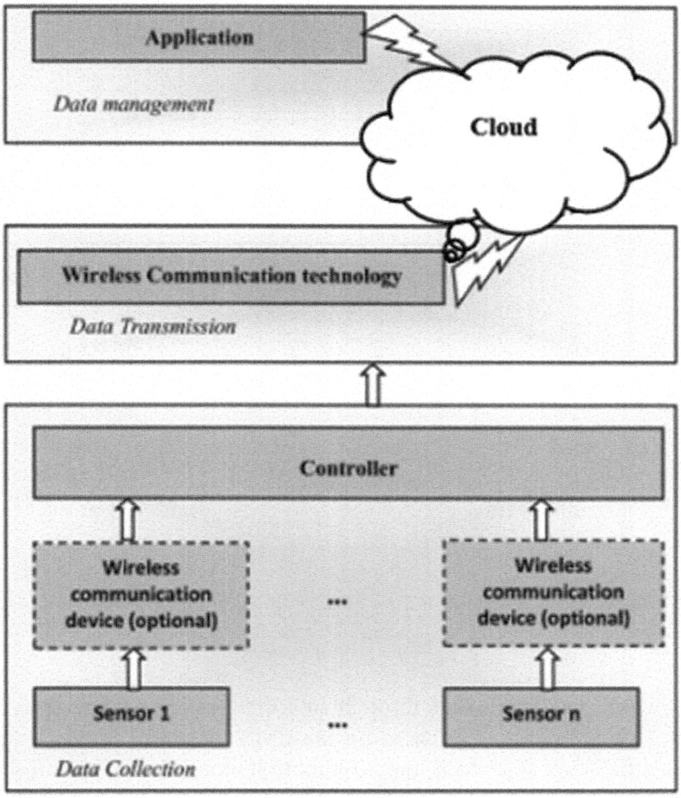

Fig. 1 The cloud approach process [6]

In addition, the cloud includes a variety of interfaces that make it simple to access and analyze data. Mobile users may easily access the cloud [20]. The following sections of the present review paper are to show the related works for water quality monitoring that has been done by the previous researchers who were using the cloud-based approach in water quality monitoring.

3 Related Studies of Water Quality Monitoring System

Based on the study from Pasika and Gandla [21], they used a DHT-11 sensor to detect the temperature and humidity of the surrounding air. Other sensors in this study including pH, turbidity, and water level in the tank. These four sensors are connected to the microcontroller that is Arduino Mega and links to the NodeMCU (low-cost wifi module), which passes all these data to the specific space named as a cloud (Fig. 2a). The data is then processed, analyzed, and displayed in ThingSpeak,

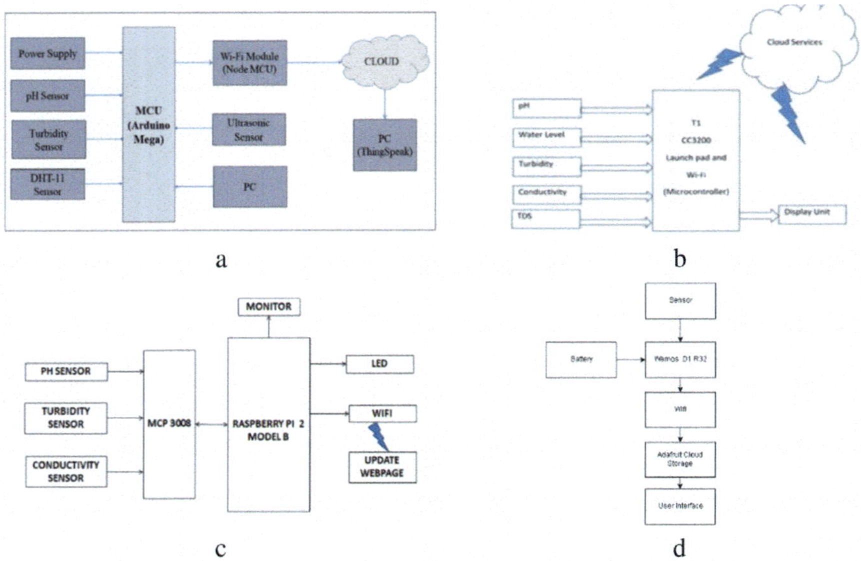

Fig. 2 The examples of the block diagram of the systems from the past studies; **a** [21], **b** [20], **c** [23] and **d** [27]

accessible to authorized end-users through smartphones and computers. According to Jan et al. [22], NodeMCU is based on the ESP8266 SoC, which has all of the necessary features such as an ADC unit, digital I/O, timers/counters, memory, 32-bit CPU, and wifi module to perform analogue signal conditioning, processing, logging, and monitoring. Therefore, it is advisable not to combine both Arduino Mega and NodeMCU. The researchers [11] have used an application called ThingSpeak to generate the results obtained from the cloud, and the users can check their results by using a PC.

In addition, the study of Ahmad et al. [16] has been concerned with the checking of stream water quality based on five sensors which are pH sensor, turbidity sensor, temperature sensor, salinity sensor, oxygen level sensor. The researcher used the Arduino unit connected with five sensors and also connected to the wifi module. The collected data on the cloud has been used to send alert messages to users by using a mobile app.

Meanwhile, according to Jha et al. [20], they have developed a cloud-based approach to monitoring water quality from groundwater samples at different places. The system consists of four sensors which are conductivity, pH, nitrate, and biochemical oxygen. Microcontrollers such as T1CC3200 are connected to the cloud named Ubidots, or mobile devices use a decision tree to predict sensed data by Zigbee/wifi technology, as shown in Fig. 2b. Data is saved in a CSV file, submitted to the Machine Learning (ML) module. A decision Tree-based classifier is utilized in their proposed system to categorize instances as drinkable or not drinkable.

Another study of a cloud-based approach based on Satish Kumar et al. [23], the system is consists of three sensors which are pH, turbidity, and conductivity, to monitor the water quality by taking samples of water. These three sensors are connected to the MCP 3008. The Raspberry Pi 2 consists of LED indication, which is used in any of the parameters that exceed the specified threshold. According to Laluma et al. [24], Raspberry Pi's superiority is low cost and compact size. The automatic operation entails monitoring and controlling sensors, actuators, relays, and motors. The Raspberry Pi gathers or reads data from many sensors in parallel and stores it in local memory until it is sent to cloud storage [25]. The block diagram of the proposed system is shown in Fig. 2c.

Bhagat et al. [26] developed IoT-based water monitoring to monitor water quality. They measure four parameters which are pH, temperature, turbidity, and carbon dioxide (CO_2) in the water. The MQ135 sensor detects CO_2 on the water's surface CO_2. The Arduino Mini Pro (microcontroller) is sent to the cloud server Thinger.io using the ESP8266 wifi module. The end-user can view real-time water parameters on their PC or smartphone. According to Jan et al. [22], the use of Thinger.io, an open-source IoT platform that provides users with a free lifetime freemium subscription. Customers with limited restrictions can use this account to begin learning and developing their projects.

According to Pantjawati et al. [27], researchers proposed and used a system of cloud-based on three sensors which are pH, turbidity, and total dissolved solids (TDS) sensors, to monitor the river water quality in Citarum River in Katapang area, Bandung (Indonesia). The researchers used a microcontroller that consists of a Wemos D1 R32 module for analogue data that comes from the sensors to be processed into digital data. Also, the data was connected to the wifi to send the data into space which is a cloud. This result can be viewed in the web browser, as shown in Fig. 2d.

4 Discussion

Several controllers have been used in the literature for cloud-based approach systems, as shown in Table 1, to monitor the water quality. All of these controllers have specific features. If the controller has few parts, it can be affected the quality of the system [16, 21, 27, 28]. Also, most of these controllers used are works with an external GPRS/wifi module for connectivity to the data storage or application [16, 21, 23, 28]. In addition, wireless networking signals are exposed to a broad range of interference as well as complicated dispersion effects that are beyond the control of the network administrator. Therefore, the GPRS /wifi module plays a role in the side of dependability.

Moreover, the number of sensors connected to the controller is essential to present the data on water quality. If there are fewer sensors connected in the system of water quality monitoring, that will provide minimal data on water quality which is not enough information can provide for users to decide the quality of water is ok or not [23, 27, 28]. Also, some of the previous researchers used sensors in the system but

Table 1 Comparison of the system

References	Water quality parameters	Controllers	Wifi module	Servers
Pasika and Gandla [21]	pH, turbidity, temperature, water level, humidity	Arduino Mega	NodeMCU/ESP8266	ThingSpeak
Ahmad et al. [16]	pH, turbidity, temperature, salinity, oxygen level	ATmega 2560	Ethernet LAN	Mobile app
Jha et al. [20]	Conductivity, pH, nitrate, biochemical oxygen	T1CC3200 wi-fi launchpad board	Zigbee/Wi-Fi	Ubidots
Satish Kumar et al. [23]	pH, turbidity, conductivity	MCP 3008 and Raspberry PI 2 (Model B)	Wi-Fi Router (built-in wi-fi)	URL (webpage)
Bhagat et al. [26]	pH, turbidity, temperature, CO_2	Arduino Pro Mini card	ESP8266 Wi-Fi transceiver	Thinger.io
Pantjawati et al. [27]	pH, turbidity, TDS	Wemos D1 R32	external Wi-Fi Module	Adafruit cloud

doesn't include the sensor of turbidity, which is one of the crucial sensors to monitor the water quality [20, 28]. Furthermore, most of the researchers did not include more servers for users to access the information on water quality like the server of GSM [21, 23, 27, 28]. Providing more servers helps users to easier access the data and make the system continue to work like if one server facing some problems, may other servers possible for users to see all the data of water quality, it helps to improve the quality of the system.

According to Jan et al. [22] and Vergina et al. [29], for optimal outcomes, developers should use machine learning (ML) approaches. ML algorithms construct a mathematical model that can make decisions or predictions based on data samples. For example, in this study [30], they include a section on sensing the external temperature near the water storage and controlling the heater or cooler based on the temperature in which the system forecasts weather conditions based on previously labelled data and adjusts the heater and chiller to match external weather conditions.

Moreover, Jan et al. [22] also mentioned that Ubidot, ThingSpeak, Blynk, Pachube, Thinger.io, Google Firebase, and Adafruit are all popular IoT systems. The past researchers, such as [20, 21, 26, 27, 31], implemented these IoT in their studies. These cloud platforms are beneficial to developers, but the IoT developers should, if feasible, create their cloud servers.

5 Conclusion

In summary, this review paper is a detailed review of the parameters, controller, and communication technology that previous researchers used in the existing cloud base approach in water quality monitoring systems. Many different controllers have been shown in the present paper used by previous researchers. It can be concluded that if the controller has a limited feature, it will not help to increase more about the quality of the system that supports monitoring the water quality. Furthermore, the parameters are the main factor to measure the quality of the water. In addition, the recent methods of water quality monitoring done by previous researchers didn't include more sensors to measure the water quality. Researchers need to focus more on the number of sensors that must be included in the water quality monitoring system because more sensors have more data that helps to check more on water quality. Machine learning approaches help to improve the efficiency of water quality monitoring. Moreover, the recent water quality monitoring system needs more to develop the communication technology to provide faster data and more accessible access to it.

Acknowledgements The authors would like to acknowledge Universiti Tenaga Nasional for providing the facilities and financial assistance under BOLD Research Grant 2021.

References

1. Adu-Manu KS, Tapparello C, Heinzelman W, Katsriku FA, Abdulai J-D (2017) Water quality monitoring using wireless sensor networks: current trends and future research directions. ACM Trans Sens Netw 13:1–41. https://doi.org/10.1145/3005719
2. Ong NTJ, Yee SK (2020) Review on water quality monitoring technologies. Indones J Electr Eng Comput Sci 18:1416–1423. https://doi.org/10.11591/ijeecs.v18.i3.pp1416-1423
3. Afroz R, Masud MM, Akhtar R, Duasa JB (2014) Water pollution: challenges and future direction for water resource management policies in Malaysia. Environ Urban ASIA 5:63–81. https://doi.org/10.1177/0975425314521544
4. Ramadhan AJ (2020) Smart water-quality monitoring system based on enabled real-time internet of things. J Eng Sci Technol 15:3514–3527
5. Central Ground Water Board Mandate (2009) Central Ground Water Board. 2009
6. Geetha S, Gouthami S (2016) Internet of things enabled real time water quality monitoring system. Smart Water 2:1–19. https://doi.org/10.1186/s40713-017-0005-y
7. Cloete NA, Malekian R, Nair L (2016) Design of smart sensors for real-time water quality monitoring. IEEE Access 4:3975–3990. https://doi.org/10.1109/ACCESS.2016.2592958
8. Banna MH, Imran S, Francisque A, Najjaran H, Sadiq R, Rodriguez M, Hoorfar M (2014) Online drinking water quality monitoring: review on available and emerging technologies. Crit Rev Environ Sci Technol 44:1370–1421. https://doi.org/10.1080/10643389.2013.781936
9. Dong J, Wang G, Yan H, Xu J, Zhang X (2015) A survey of smart water quality monitoring system. Environ Sci Pollut Res 22:4893–4906. https://doi.org/10.1007/s11356-014-4026-x
10. Prasad AN, Mamun KA, Islam FR, Haqva H (2016) Smart water quality monitoring system. In: 2015 2nd Asia-Pacific world congress on computer science and engineering (APWC CSE 2015). https://doi.org/10.1109/APWCCSE.2015.7476234

11. Schindhya R, Simi VP, Sathya E, Janani B (2017) Real time water monitoring system using IoT. IJARCCE 6:378–380. https://doi.org/10.17148/ijarcce.2017.6386
12. Fleisch E (2010) What is the internet of things? An economic perspective. Econ Manag Financ Mark 5:125–157
13. Thombre S, Ul Islam R, Andersson K, Hossain MS (2016) Performance analysis of an IP based protocol stack for WSNs. In: Proceedings - IEEE INFOCOM 2016-September, pp 360–365. https://doi.org/10.1109/INFCOMW.2016.7562102
14. Abedin MZ, Chowdhury AS, Hossain MS, Andersson K, Karim R (2017) An interoperable IP based WSN for smart irrigation systems. In: 2017 14th IEEE annual consumer communications & networking conference (CCNC 2017), 2017-January. https://doi.org/10.1109/CCNC.2017.8013434
15. Rundel PW, Rundel PW, Graham EA, Allen MF, Fisher JC, Harmon TC (2009) Ecol Res 589–607
16. Ahmad A, Shekhar S, Roy A (2020) Cloud based water reservoir quality monitoring system. Int J Recent Technol Eng 8:4108–4113. https://doi.org/10.35940/ijrte.f8948.038620
17. Islam RU, Andersson K, Hossain MS (2017) Heterogeneous wireless sensor networks using CoAP and SMS to predict natural disasters. In: 2017 IEEE conference on computer communications workshops (INFOCOM WKSHPS 2017), pp 30–35. https://doi.org/10.1109/INFCOMW.2017.8116348
18. Alam Siddiquee KNE, Khan FF, Andersson K, Hossain MS (2017) Optimal dynamic routing protocols for agro-sensor communication in MANETs. In: 2017 14th IEEE annual consumer communications & networking conference (CCNC 2017). https://doi.org/10.1109/CCNC.2017.8013436
19. Alam ME, Shamim Kaiser M, Hossain MS, Andersson K (2019) An IoT-belief rule base smart system to assess autism. In: 4th international conference on electrical engineering and information & communication technology (iCEEiCT 2018), pp 672–676. https://doi.org/10.1109/CEEICT.2018.8628131
20. Jha BK, Sivasankari GG, Venugopal KR (2020) Cloud-based smart water quality monitoring system using IoT sensors and machine learning. Int J Adv Trends Comput Sci Eng 9:3403–3409. https://doi.org/10.30534/ijatcse/2020/141932020
21. Pasika S, Gandla ST (2020) Smart water quality monitoring system with cost-effective using IoT. Heliyon 6:e04096. https://doi.org/10.1016/j.heliyon.2020.e04096
22. Jan F, Min-allah N, Dü D (2021) IoT based smart water quality monitoring: recent techniques, trends and challenges for domestic applications. 1–37
23. Satish Kumar K, Sarojini M, Ranga VP (2016) Iot based real time monitoring of water quality. Int J Prof Eng Stud 7:174–179
24. Laluma RH, Giantara R, Sugiarto B, Gunawan, Siregar CA, Risnanto S (2019) Automation system of water treatment plant using raspberry Pi.3 model B+ based on internet of things (IoT). In: TSSA 2019 - 13th international conference on telecommunication systems, services, and applications proceedings, pp 72–76. https://doi.org/10.1109/TSSA48701.2019.8985516
25. Das Gupta S, Zambare MS, Kulkarni NM, Shaligram AD (2020) IoT based real-time water quality monitoring and classification. Lect Notes Electr Eng 630:661–670. https://doi.org/10.1007/978-981-15-2305-2_53
26. Bhagat PS, Gulhane DVS, Rohankar PTS (2019) Implementation of internet of things for water quality monitoring. Int J Trend Sci Res Dev 3:306–311. https://doi.org/10.31142/ijtsrd23655
27. Pantjawati AB, Purnomo RD, Mulyanti B, Fenjano L, Pawinanto RE, Nandiyanto ABD (2020) Water quality monitoring in Citarum River (Indonesia) using IoT (internet of thing). J Eng Sci Technol 15:3661–3672
28. Perumal T, Sulaiman MN, Leong CY (2016) Internet of things (IoT) enabled water monitoring system. In: 2015 IEEE 4th Global Conference on Consumer Electronics (GCCE 2015), pp 86–87. https://doi.org/10.1109/GCCE.2015.7398710
29. Vergina SA, Kayalvizhi S, Bhavadharini RM, Kalpana Devi S (2020) A real time water quality monitoring using machine learning algorithm. Eur J Mol Clin Med 7:2035–2041

30. Kumar Koditala N, Shekar Pandey P (2018) Water quality monitoring system using IoT and machine learning. In: Proceedings of 2018 3rd IEEE international conference on research in intelligent and computing in engineering (RICE 2018). https://doi.org/10.1109/RICE.2018.8509050
31. Manoharan S, Sathiyaraj G, Thiruvenkadakrishnan K, Vetriselvan GV, Kishor P (2019) Water quality analyzer using IoT. Int J Innov Technol Explor Eng 8:34–37

Constructed Wetland-Microbial Fuel: Biotechnology for Simultaneous Wastewater Treatment and Electricity Generation

Tean Peng Teoh, Soon An Ong, Yee Shian Wong, Li Ngee Ho, Norazian Mohamed Noor, and Monica Matei

Abstract Shortage of energy and pollution of water are two concern global issues. The treatment of polluted water usually requires a lot of energy. The demand for wastewater treatment and energy shortage has attracted the interest of research on converting pollutants to energy. Constructed wetland (CW) is environmentally friendly technology developed for the purification of polluted water and microbial fuel cell (MFC) is biotechnology that could turn organic pollutants into electricity. Hence, this paper aimed to identify the advantage of integrating MFC into CW through the study of the closed and open circuit on the wastewater treatment performance as well as the effect of spacing between cathode and anode on the generation of electricity in up-flow constructed wetland coupled microbial fuel cell (UFCW-MFC) system. The wastewater treatment performance was evaluated through the analysis of COD. The total COD removal efficiency was 99%. The closed circuit improved 10.9% compared to the open circuit at the anaerobic compartment (S1) in terms of COD removal. The effect of electrode spacing was investigated by

T. P. Teoh (✉) · S. A. Ong · Y. S. Wong
Centre of Excellence Water Research and Environmental Sustainability Growth (WAREG), Universiti Malaysia Perlis, 02600 Arau, Perlis, Malaysia

Y. S. Wong
e-mail: yswong@unimap.edu.my

Faculty of Civil Engineering Technology, Universiti Malaysia Perlis (UniMAP), 02600 Arau, Perlis, Malaysia

L. N. Ho
Faculty of Chemical Engineering Technology, Universiti Malaysia Perlis (UniMAP), 02600 Arau, Perlis, Malaysia
e-mail: lnho@unimap.edu.my

N. M. Noor
Sustainable Environment Research Group (SERG), Centre of Excellence Geopolymer and Green Technology (CeGeoGTech), Universiti Malaysia Perlis, 02600 Arau, Perlis, Malaysia
e-mail: norazian@unimap.edu.my

M. Matei
National Institute for Research and Development in Environmental Protection Bucharest (INCDPM), 294, Splaiul Independentei Street, 6th District, 060031 Bucharest, Romania

N. Mohamed Noor et al. (eds.), *Proceedings of the 3rd International Conference on Green Environmental Engineering and Technology*, Lecture Notes in Civil Engineering 214,
https://doi.org/10.1007/978-981-16-7920-9_45

comparing the maximum power density produced by three different electrode spacing between cathode and anode (15, 30, 45 cm). The highest maximum power density was achieved at the smallest spacing of electrode (15 cm), which recorded 73.48 ± 0.31 mW/m^3. In conclusion, the MFC feature (closed circuit) improved the organic matter degradation and smaller electrode spacing benefits the energy production in UFCW-MFC.

Keywords CW · MFC · Wastewater treatment · Electricity generation

1 Introduction

Constructed wetland (CW) is a method that utilizes the natural process to purify wastewater. Treatment mechanisms of CW included microbial activities, chemical, and physical [1]. CW has been developed to treat a wide range of wastewater, which is not limited to municipal wastewaters but also included industrial wastewaters, stormwater runoff and landfill leachate because of its advantages such as low construction and maintenance cost [2].

Microbial fuel cell (MFC) is promising biotechnology that has shown capability in organic wastewater treatment and produces electricity simultaneously. MFC utilizes microorganisms to degrade organic pollutants and convert them to electricity. MFC consists of a cathode chamber and an anode chamber. The organic pollutants in the wastewater are used by the microbes for the metabolism activities and produce electrons at the anode. These electrons are then flowing from anode chamber to cathode chamber through an external circuit to complete the loop [3]. The electrons then react with the protons and electron acceptor at the cathode chamber to complete the circuit [4].

The integrating system of constructed wetland and microbial fuel cell (CW-MFC) is a technology that embeds the electrodes of MFC in CW. The utilization of microbial activities for the breakdown of organic pollutants has driven the combination of these two systems. Moreover, CWs could provide the redox gradient required for electricity generation in MFCs based on the depth of the bioreactor and flow direction of the wastewater [5]. The advantages of integrating MFC into CW have not been clarified. Therefore, this paper aims to identify the advantage through the study of the closed and open circuit on the treatment performance of wastewater as well as the effect of spacing between cathode and anode on the production of electricity in up-flow constructed wetland coupled microbial fuel cell (UFCW-MFC) system.

2 Material and Method

The UFCW–MFC reactor was developed using an acrylic column (height of 75 cm and diameter of 18 cm). The reactor was established outdoor (temperature of 28 ± 4 °C). The bioreactor was packed with glass beads (size of 10 mm) up to 3 cm to establish an even flow of wastewater. The supporting medium of the wetland bed used in this study was gravel. The bioreactor was developed with five sampling points at the height of 7, 21, 36, 51, 66 cm from the bottom [6]. The macrophyte employed in this study was water hyacinth (Eichhornia crassipes). Figure 1 depicts the structure of the bioreactor.

The cathode and anode used were activated carbon. The total volume of activated carbon for each cathode and anode was 2405.59 cm^3. The bioreactor was embedded with three layers of the anode at the lower region (8 cm, 23 cm, and 38 cm from the bottom of the wetland bed), while the cathode was developed at the upper region of the bioreactor (53 cm from the bottom of the bioreactor). Electrode spacing between the cathode and three anodes were 15 cm, 30 cm and 45 cm, respectively. The cathode and anode were connected by stainless steel, carbon rod, and insulated copper wires with an external resistance of 1000 Ω. The mixed culture sludge obtained from a glove manufacturing factory wastewater plant was used for the inoculation of gravels and activated carbon.

The synthetic wastewater consists of 1024.5 mg/L CH_3COONa, 535.5 mg/L C_6H_5COONa, 4.0 mg/L $CaCl_2{\cdot}2H_2O$, 3.4 mg/L $MgCl_2{\cdot}6H_2O$, 36.7 mg/L K_2HPO_4 and 7.0 mg/L NaCl. A peristaltic pump was used to feed the wastewater to the bioreactor with a hydraulic retention time (HRT) of 1 day. The wastewater from influent

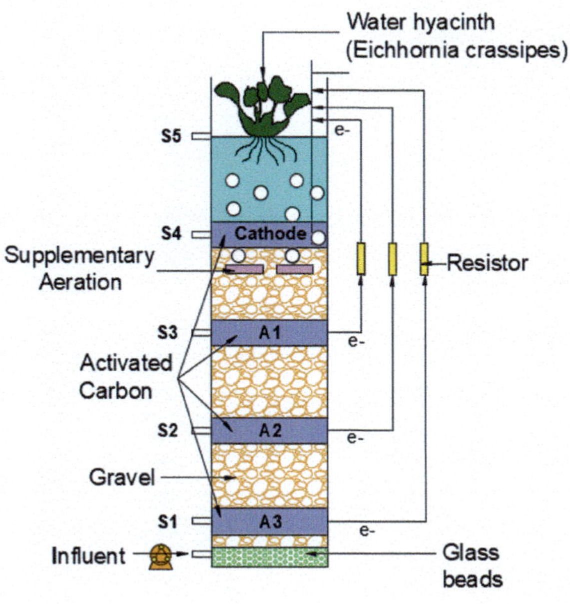

Fig. 1 Structure of UFCW–MFC

and all sampling points were collected for the analysis of COD. The voltage output in the system was recorded using a data logger. Calculations were done on current (I = V/R) to obtain power (P = IV). The power density was determined by dividing the power with the volume (v) of the anode chamber (Pd = P/v).

3 Results

3.1 Removal of Chemical Oxygen Demand (COD)

The influent COD concentration of the synthetic wastewater was 1395 ± 22 mg/L. Figure 2 demonstrates the COD profile of UFCW-MFC for both closed circuit and open circuit. As shown in Fig. 2, there was a drastic drop from influent to S1 followed by S2 to S3, which contribute 98.2% and 97.6% for closed and open circuits, respectively. The results showed that COD removal was mainly occurred in the anaerobic region (S1–S3). The existing microorganisms at the anaerobic region utilized the organic matter as energy sources for the microbial activity, which resulted in the decrement of COD [7]. Further removal of unutilized organic substrates was observed at sampling points S4 and S5, which achieved a COD removal efficiency of 99%. Supplementary aeration and macrophyte could enhance the biodegradation of organic matter by promoting the growth of the aerobes. Therefore, oxygen released from the supplementary aeration and E. crassipes in the aerobic region (S4–S5) may contribute to further COD reduction in the upper bed of reactors.

The effect of circuit connection on the COD removal efficiency was shown in Fig. 2. The effect of circuit connection on total COD removal was insignificant in this study since most of the organic pollutants were treated at S3. The significant effect of circuit connection on COD removal was observed at S1, which achieved

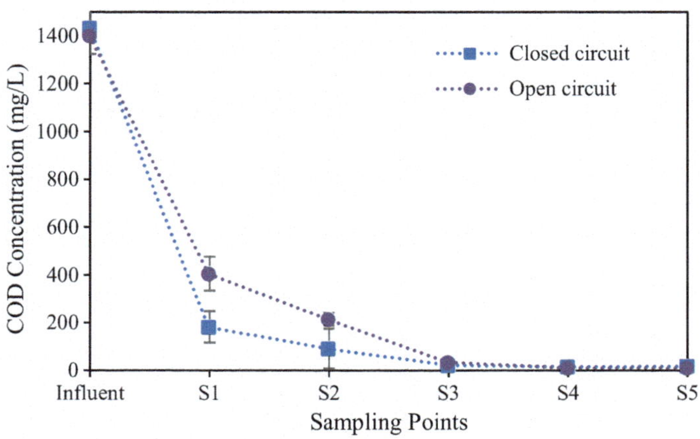

Fig. 2 COD removal profile along UFCW–MFC

removal efficiency of 81.9% and 71% in closed circuit and open circuit system, respectively. The closed circuit system showed 10.9% higher compared to the open circuit system in the anaerobic compartment. This indicated that in a closed circuit system, the transfer of electrons could enhance the degradation rate of the organic pollutants as well as promoting the growth of electrogenic bacteria. Oon et al. [8] reported a similar finding, where closed circuit improved the COD reduction in the anaerobic compartment by 8% compared to open circuit. Srivastava et al. [9] also reported that closed circuit enhanced 12–20% compared to open circuit. This study revealed that circuit connection could play an essential role in improving the COD removal in UFCW-MFC.

3.2 Electricity Generation Performance

The effect of spacing between cathode and anode on power production was investigated by developing three anodes in the reactor. The electrode spacing between the cathode and three anodes were 15 (C-A1), 30 (C-A2), and 45 cm (C-A3), respectively. Figure 3 demonstrates the power density generated by the three different electrode spacings. The highest maximum power density was achieved at C-A1 (73.48 ± 0.31 mW/m^3), followed by C-A2 (58.3 ± 6.26 mW/m^3) and C-A3 (51.71 ± 8.91 mW/m^3). The internal resistance showed a similar trend of power densities. The internal resistance of the hybrid system at A1, A2, and A3 were 620, 620, and 720 Ω, respectively. The results suggested that electrode spacing could affect the power density generated as lower internal resistance was observed with smaller electrode spacing. Oon et al. [10] reported a similar finding, where the smallest electrode spacing produced the highest maximum power density.

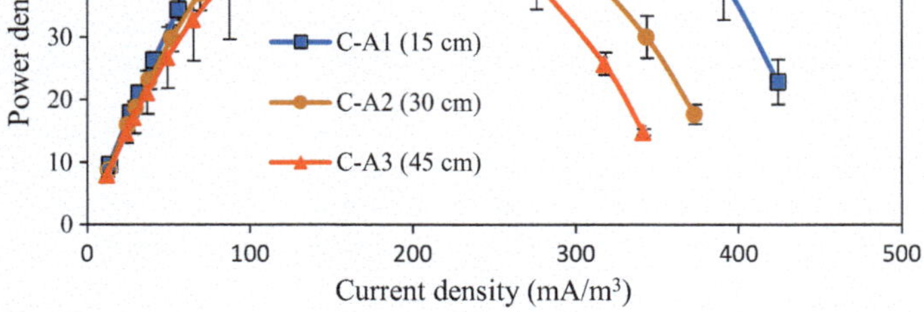

Fig. 3 Power density-current density at different electrode spacings

4 Conclusion

This paper demonstrated the benefit of integrating MFC in CW through the study of the closed and open circuit on the treatment performance of wastewater as well as the effect of spacing between cathode and anode on the production of electricity. MFC feature (closed circuit) improved the degradation of organic pollutants as the transfer of electrons stimulated the breakdown of organic pollutants as well as promoting the growth of electrogenic bacteria. In terms of electricity generation, smaller spacing between cathode and anode could lower the internal resistance, which increases the electricity output of UFCW-MFC.

References

1. Vymazal J (2007) Removal of nutrients in various types of constructed wetlands. Sci Total Environ 380:48–65. https://doi.org/10.1016/j.scitotenv.2006.09.014
2. Vymazal J (2011) Constructed wetlands for wastewater treatment: five decades of experience. Environ Sci Technol 45:61–69. https://doi.org/10.1021/es101403q
3. Fang Z, Song HL, Cang N, Li XN (2013) Performance of microbial fuel cell coupled constructed wetland system for decolorization of azo dye and bioelectricity generation. Bioresour Technol 144:165–171. https://doi.org/10.1016/j.biortech.2013.06.073
4. Liu R, Zhao Y, Doherty L, Hu Y, Hao X (2015) A review of incorporation of constructed wetland with other treatment processes. 279:220–230
5. Yadav AK, Dash P, Mohanty A, Abbassi R, Mishra BK (2012) Performance assessment of innovative constructed wetland-microbial fuel cell for electricity production and dye removal. Ecol Eng 47:126–131. https://doi.org/10.1016/j.ecoleng.2012.06.029
6. Teoh TP, Ong SA, Ho LN, Wong YS, Oon YL, Oon YS, Tan SM, Thung WE (2020) Up-flow constructed wetland-microbial fuel cell: influence of floating plant, aeration and circuit connection on wastewater treatment performance and bioelectricity generation. J Water Process Eng 36:101371. https://doi.org/10.1016/j.jwpe.2020.101371
7. Oon YL, Ong SA, Ho LN, Wong YS, Oon YS, Lehl HK, Thung WE (2015) Hybrid system up-flow constructed wetland integrated with microbial fuel cell for simultaneous wastewater treatment and electricity generation. Bioresour Technol 186:270–275. https://doi.org/10.1016/j.biortech.2015.03.014
8. Oon YL, Ong SA, Ho LN, Wong YS, Dahalan FA, Oon YS, Lehl HK, Thung WE (2016) Synergistic effect of up-flow constructed wetland and microbial fuel cell for simultaneous wastewater treatment and energy recovery. Bioresour Technol 203:190–197. https://doi.org/10.1016/j.biortech.2015.12.011
9. Srivastava P, Yadav AK, Mishra BK (2015) The effects of microbial fuel cell integration into constructed wetland on the performance of constructed wetland. Bioresour Technol 195:223–230. https://doi.org/10.1016/j.biortech.2015.05.072
10. Oon YL, Ong SA, Ho LN, Wong YS, Dahalan FA, Oon YS, Lehl HK, Thung WE, Nordin N (2017) Role of macrophyte and effect of supplementary aeration in up-flow constructed wetland-microbial fuel cell for simultaneous wastewater treatment and energy recovery. Bioresour Technol 224:265–275. https://doi.org/10.1016/j.biortech.2016.10.079

Crop Water Requirement of Paddy Cultivation in Kedah, Malaysia

Nur Anis Syarafinaz Abd Manan, Wan Amiza Amneera Wan Ahmad, Nik Meriam Nik Sulaiman, and Noor Zalina Mahmood

Abstract In the coming decades, climate change is predicted to have an enormous impact on the water requirements of paddy crops worldwide. Water scarcity will become more severe in the agricultural sector. The purpose of this study is to estimate the water requirements of paddy farming in Kedah for two seasons, namely the main and off-season, using historical data. Cropwat 8.0 model was used to calculate crop water requirements for rice agriculture to estimate crop water requirements for paddy cultivation. Climate data for 2015–2019, crop data and soil data from the Malaysian Meteorological Department (MMD), website, and interviewed farmers were used as input data for Cropwat 8.0 software to assess crop water requirement. From the result, crop water requirement for the main season and off-season is 454.1 and 560.3 mm, respectively. The main season water requirement is sufficient and met the condition with the rainfall rates. In contrast, water requirement for the off-season is supported by the irrigation requirement to meet the need of paddy due to lower rainfall rates during the season.

Keywords Agricultural · Crop water requirement · Cropwat 8.0 · Paddy cultivation

N. A. S. Abd Manan · W. A. A. Wan Ahmad (✉)
Faculty of Civil Engineering Technology, Universiti Malaysia Perlis (UniMAP), 02600 Arau, Perlis, Malaysia
e-mail: amneera@unimap.edu.my

W. A. A. Wan Ahmad
Centre of Excellence Water Research and Environmental Sustainability Growth (WAREG) Universiti Malaysia Perlis (UniMAP), 02600 Arau, Perlis, Malaysia

W. A. A. Wan Ahmad · N. M. Nik Sulaiman
Department of Chemical Engineering, Faculty of Engineering, University of Malaya, 50603 Kuala Lumpur, Malaysia
e-mail: meriam@um.edu.my

N. Z. Mahmood
Institute of Biological Sciences, Faculty of Science, University of Malaya, 50603 Kuala Lumpur, Malaysia
e-mail: alin@um.edu.my

N. Mohamed Noor et al. (eds.), *Proceedings of the 3rd International Conference on Green Environmental Engineering and Technology*, Lecture Notes in Civil Engineering 214,
https://doi.org/10.1007/978-981-16-7920-9_46

1 Introduction

Rice is a staple food in Malaysian households. Malaysian paddy farmers met up to 70% of the country's needs, according to official statistics. Consequently, rice and paddy are essential components of food security, socio-cultural development, and government strategic interventions in many developing countries [1]. Sustaining rice production is critical to maintaining a positive balance in the economy's social and economic sectors [2]. As a result, maintaining sustainable rice production in the country is extremely important to satisfy demand as the country's population grows and domestic consumption increases.

Weather conditions in the rice field region, such as temperature and rainfall, significantly impact rice crop growth. According to the Muda Agriculture Development Authority (MADA), the region's yield forecast results revealed a decrease in both the main and off seasons until 2030. The maximum and minimum temperatures and rainfall throughout the most season are anticipated to rise by around 0.05 °C and 0.12 mm per year, respectively, over the next few decades. It was predicted that the yield would fall by 12.2% during the next 18 years during harvest season and that this would occur within the main season. Numerous times, minimum and maximum temperatures were anticipated to climb by 0.19 and 0.08 °C annually throughout the off-season [3]. As a result, changes in the atmosphere have an impact on agricultural activity [4].

On the other hand, precipitation is expected to decrease by approximately 0.18 mm per year between 2013 and 2030 [5]. According to the projections, in the off-season, the yield will fall by 45.5% over the next 18 years, which is more than in the main season compared to the standard [6]. This will be possible to determine the water requirements for the season's growth. Natural disasters, such as drought, impact the timing and output of rice paddy planting [7].

Crop water requirements are the amount of water required by plants for life, growth, and the volume of water lost through evaporation and transpiration over the entire cultivation period. To meet the requirement, either naturally occurring precipitation or artificial irrigation must be used. A study is required to address the optimum use of limited water resources while improving the economic return on water usage. This necessitates assessing crop water requirements, cropping systems, paddy output, and meteorological and rainfall data [8]. This study aims to determine the crop water requirements in Kedah, Malaysia, based on five years of climate data by using Cropwat 8.0, one of the methods developed for agriculture to estimate water consumption in agricultural production.

2 Materials and Method

Cropwat 8.0 software was used in determining crop water requirements. Meteorological data were obtained from the meteorological department for up to five years in this study (2015–2019). Climate variables such as maximum and minimum temperatures, sunshine hours, rainfall, relative humidity, and wind speed were input data for the software to determine reference evapotranspiration (ET_o). Crop water requirement is then calculated using Eq. 1 for the Crop Water Requirement (CWR) value.

$$CWR = ET_o * K_c \tag{1}$$

Where ET_o = Evapotranspiration, K_c = Crop Coefficient.

3 Result and Discussion

The amount of water lost due to crop evapotranspiration (ETc) is referred to as the crop water requirement. That is to say, the crop water requirement is assigned an equal value. The amount of evapotranspiration produced by plants shows the importance of rice's regular or decadal crop water requirements. According to the result, the effective rainfall, which is output data from the cropwat 8.0 software, is 479.9 mm and 350.1 mm in the main and off-season, respectively. The total crop water need is 454.1 mm and 560.3 mm in main and off-season, respectively shown in Table 1. This data shows that crop water requirements are higher in the off-season (dry season) than in the main season (rainy season). The FAO [9] also reported that crops grown in the dry season require more water than crops grown in the rainy season.

Water requirement can affect the rice yield, as shown in Figs. 1 and 2. The productivity in the main season is higher than the production in the off-season due to many factors, especially evaporation and climatic factors highly influenced by effective rainfall [10]. Figure 1 shows the correlation between CWR and paddy production for main season. Main season for rice is usually from August to January. The paddy production and water need for 2015 were higher because in late 2015, Malaysia suffered from El- Nino. In late 2016 which is the main season for the paddy, the rains begin to fall and reduce the effects of the drought. Paddy production shows a reduction from 2017 until 2019 because, in these three consecutive years, flood happens at the end of the year, which was also the main season for paddy. Hence, it shows that

Table 1 Comparison of crop water requirement and paddy production (2015–2019)

Season	Kedah crop water requirement (mm)	Kedah paddy production (kg/ha)
Main season	454.1	25511
Off season	560.3	24875
Total	1014.4	50386

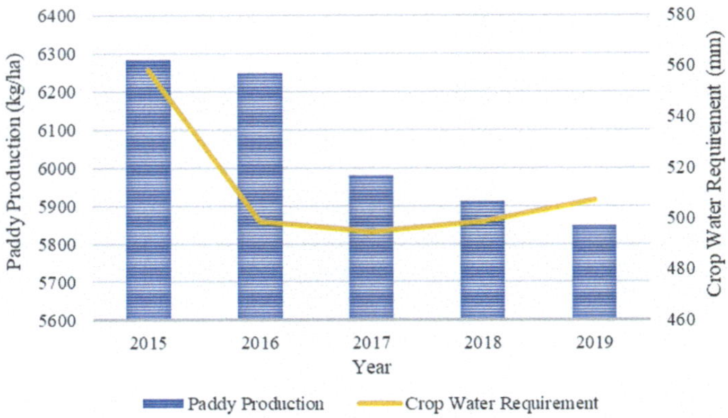

Fig. 1 Relationship paddy production during main season and CWR

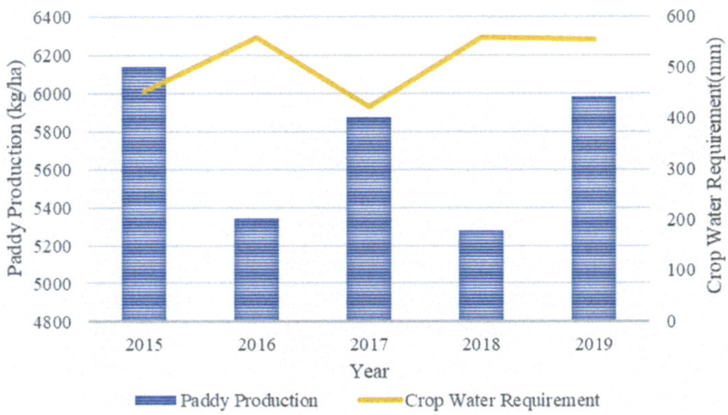

Fig. 2 Relationship paddy production during off-season and CW

there was a correlation between paddy production and CWR. Flood disasters result in a lack of certain nutrients needed by paddy plants to produce yields and a lack of zinc availability in submerged agricultural areas causing paddy yields to decline [11].

4 Conclusion

The conclusion found a greater quantity of effective rainfall available to grow the paddy crops in Kedah, suggesting less need for irrigation water requirement. Higher effective rainfall suggests an adequate rainfall rate for Kedah paddy cultivation. The total water requirement for the main and off-season was 454.1 mm and 560.3 mm, respectively. This was due to some factors that affect the total amount, like climatic and environmental factors. The factors include rainfall pattern, soil characteristic, and types of paddy as there are few paddy variants currently grown with maturities ranging from 105 to 120 days. All those factors will affect crop yield production.

References

1. Omar SC, Shaharudin A, Tumin SA (2019) The status of the paddy and rice industry in Malaysia. Khazanah Research Institute, Kuala Lumpur, Malaysia
2. Papademetriou MK (2000) Rice production in the Asia-Pacific region: issues and perspectives. In: FAO report. http://www.fao.org/3/x6905e/x6905e04.htm. Accessed 30 Jun 2021
3. MADA (2017) Rice industry development program – official website muda agricultural development authority. http://www.mada.gov.my/?page_id=13761&lang=en. Accessed 10 Dec 2020
4. Md Yusof Z, Misiran M, Baharin NF, Yaacob MF, Abdul Aziz NA, Sanan NH (2019) Projection of paddy production in Kedah Malaysia: a case study. Asian J Adv Agric Res 10(3):1–6. https://doi.org/10.9734/AJAAR/2019/v10i330130
5. Farrell WE, Clark JA (1976) On postglacial sea level. Geophys J R Astron Soc 46:647–667. https://doi.org/10.1111/j.1365-246X.1976.tb01252.x
6. Vaghefi V, Shamsudin MN, Radam A, Rahim KA (2013) Impact of climate change on rice yield in the main rice growing areas of Peninsular Malaysia. Res J Environ Sci 7(2):59–67. https://doi.org/10.3923/rjes.2013.59.67
7. Maidin KH, Ithnin Mohamad CK, Othman SK (2015) Impacts of natural disasters on the paddy production and its implications to the economy. FFTC Agricultural Policy Articles 2012–2015
8. Brouwer C, Heibloem M (1986) Irrigation water management: irrigation water needs. Train Man 3
9. Surendran U, Sushanth CM, Mammen G, Joseph EJ (2015) Modelling the crop water requirement using FAO-CROPWAT and assessment of water resources for sustainable water resource management: a case study in Palakkad district of humid tropical Kerala, India. Aquat Procedia 4:1211–1219. https://doi.org/10.1016/j.aqpro.2015.02.154
10. Kang Y, Khan S, Ma X (2009) Climate change impacts on crop yield, crop water productivity and food security - a review. Prog Nat Sci 19(12):1665–1674. https://doi.org/10.1016/j.pnsc.2009.08.001
11. Nor NM, Masdek NNM, Maidin MKH (2015) Youth inclination towards agricultural entrepreneurship. Econ Technol Manag Rev 10:47–55

Identification of Antibiotics as Emerging Contaminants and Antimicrobial Resistance in Aquatic Environment of the Arges-Vedea, Buzau-Ialomita, Dobrogea-Litoral River Basins in Romania

Mihaela Ilie, György Deák, Florica Marinescu, Gina Ghita, Carmen Tociu, Monica Silvia Matei, Ioana Savin, Madalina Boboc, Marius Constantin Raischi, and Miruna Arsene

Abstract Water pollution by emerging contaminants such as antibiotics is a major and current problem in the field of environmental protection. The presence of these micropollutants in surface waters, even in low concentrations, disturbs the ecological balance and leads to the selection and increase of the antibiotic resistant bacteria incidence, so that the evaluation of these substances' presence in environment has become a necessity. The identification of six widely used antibiotics: azithromycin, chloramphenicol, clarithromycin, doxycycline, erythromycin and oxytetracycline, respectively was performed by UHPLC-MS/MS in surface water, wastewater and biota (fish) matrix from the Arges-Vedea, Buzau-Ialomita and Dobrogea-Litoral river basins in Romania. Clarithromycin was most commonly detected (75.00%), while erythromycin was detected in the smallest number of samples, with a frequency of 15.63%. Detection frequencies were 62.50% for oxytetracycline and 25.00% for doxycycline, while azithromycin was detected with a frequency of 59.38% and chloramphenicol with a frequency of 46.88%. The investigation of antibiotic resistance reveled significant high resistance rates in *E. coli* strains, particularly: ampicillin (75%), penicillin (75%), clarithromycin (75%), cefazolin (50%), cefoxitin (50%), amoxicillin/clavulanic acid (50%) and erythromycin (42%). The bacterial strains isolated from the WWTP effluent and downstream of the effluent discharge of the treatment plant exhibited the highest resistance rates to β-lactam antibiotics and other classes of antibiotics, which could suggest that WWTP creates favourable conditions for the transfer of resistance genes between different bacteria which are further disseminated in the receiving aquatic environment.

Keywords Active pharmaceuticals ingredients (APIs) · Antibiotics · Antimicrobial resistance (AMR) · Aquatic ecosystems

M. Ilie · G. Deák · F. Marinescu · G. Ghita (✉) · C. Tociu · M. S. Matei · I. Savin · M. Boboc · M. C. Raischi · M. Arsene
National Institute for Research and Development in Environmental Protection Bucharest (INCDPM), 294, Splaiul Independentei Street, 6th District, 060031 Bucharest, Romania

N. Mohamed Noor et al. (eds.), *Proceedings of the 3rd International Conference on Green Environmental Engineering and Technology*, Lecture Notes in Civil Engineering 214,
https://doi.org/10.1007/978-981-16-7920-9_47

1 Introduction

Emerging pollutants such as active pharmaceutical ingredients (APIs) are a major global problem due to the potential risk to human health and environment [1]. The presence of these emerging micropollutants in surface waters, even in low concentrations, endangers the life cycle of aquatic organisms and disturbs the ecological balance [2]. Potential human health hazards caused by antibiotics through exposure to drinking water as a source of surface water supply [3–5] require special attention and sustained scientific research because conventional water treatment and wastewater treatment are not effective in removing them.

Over 3,000 APIs as human or veterinary drugs are traded in the European Union [6]. Although the environmental impact of most of these substances is largely unknown, recent studies show that pharmaceuticals can accumulate in aquatic organisms and especially fish [7]. Four antibiotic types: quinolones, sulfonamides, tetracyclines, and macrolides, which are thought to be genetically toxic for fish have been reported to bioaccumulate in fish tissues, as well as in aquatic environments [8]. The adverse effects of these antibiotics are known to cause damage to developmental, cardiovascular, and metabolic systems, as well as in altering anti-oxidant and immune responses, in fish [8].

Water environment that receives urban and medical effluents is particularly threatened by antibiotic pollutants due to poor treatment and the presence of pharmaceuticals in the final step of classic wastewater treatment plant (WWTP), which finally discharges into aqueous ecosystems. It may be related with the emergence and spread of multidrug-resistant pathogenic bacteria, and it can lead to unpredictable ecological impacts and responses, and may have an impact on human health [9–14]. At present, effluent water of WWTP is discharging directly into the aqueous environment without using advanced techniques for removing antibiotic residues and antibiotic resistance genes, and this behaviour is considered as the main source of antibiotic residues spreading in the environment, which finally leads to accumulation and increase of antibiotics resistant bacteria. Moreover, studies report that the antibiotics can resist biodegradation due to their antimicrobial nature and for this reason, antibiotics represent emerging pollutants with continual input into the environment and permanent presence [15–18].

For a better understanding of the direct impact on the environment and human health as a result of exposure to pharmaceutical substances and residues in the environment as well as how they enter and persist in aquatic ecosystems, further research is needed.

There are also serious concerns about the toxicity of antibiotics in fish, the main problem related to their presence in fish tissues is the possibility of inducing bacterial resistance genes [19] so further research and strategies are needed to prevent fish exposure to pharmaceutical residues.

In this paper, the identification of six widely used antibiotics: clarithromycin, erythromycin, oxytetracycline, chloramphenicol, doxycycline and azithromycin, respectively was performed by UHPLC-MS/MS, in surface water, wastewater and biota (fish) matrix from the Arges-Vedea, Buzau-Ialomita and Dobrogea-Litoral river basins in Romania. The prevalence of antibiotic resistance has also been studied for strains of *Escherichia coli* isolated from wastewater and surface water samples.

2 Materials and Method

2.1 Research Area

The identification of the emerging pharmaceutical micro-pollutants from the antibiotic group was made from different water matrices, such as: wastewater sampled from the wastewater treatment plants of Slobozia, Fundulea, Lehliu, Calarasi and Ploiesti, drinking water samples from the treatment plants of Fundulea, Lehliu and Calarasi as well as surface water samples collected from Dambovita R.—Budesti area, Glina Wastewater Treatment Plant area, upstream and downstream of Glina WWTP, Ialomita R.—upstream and downstream of Slobozia WWTP, Arges R.—Hotarele and Clatesti area, Sabar R., Ciorogarla R.—upstream and downstream of Domnesti locality, Prahova R.—upstream and downstream of Campina Wastewater Treatment Plant, Dambu R.—upstream and downstream of Ploiesti Wastewater Treatment Plant, Danube River—Orsova area, Tulcea (upstream, downstream), Bastroe, Sulina, Sf. Gheorghe, Fetesti (km 43), Bala (km 9.5), Epurasu Branch, Izvoarele, Sulina. Determinations have also been made for identifying the antibiotics from biota samples (fish) from the area of Arges R.—Clatesti, Dambovita R.—Trout farm from Stoenesti (Fig. 1).

At the level of the 3 river basins, water and biota samples were collected during May 2019–May 2020, along the main rivers and their important tributaries, from 27 representative monitoring sections, 20 sections for surface water respectively and 7 sections at the wastewater treatment plants of cities such as Bucharest, Slobozia, Fundulea, Lehliu, Calarasi, Ploiesti and Campina.

2.2 Method

At the level of the 3 river basins, samples were collected from 27 representative monitoring sections along the main rivers and their important tributaries, 20 sections for surface water respectively and 7 sections at the treatment plants of cities such as Bucharest, Slobozia, Fundulea, Lehliu, Calarasi, Ploiesti and Campina. Water samples were collected as 1 L in glass bottles by means of water sampler. After collection, all samples were stored in coolboxes and delivered on the sample collection day to the laboratory where water samples were stored at 4 °C until analysis. All

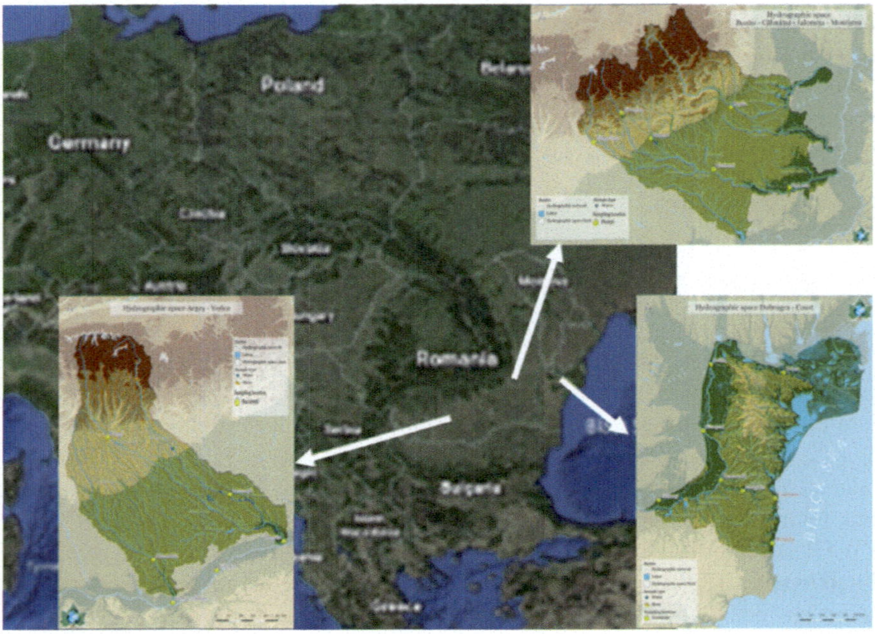

Fig. 1 Sampling locations—river basins of Argeş-Vedea, Buzău-Ialomiţa, Dobrogea-Litoral and the Danube River

water samples were filtered through 0.2 μm polyethersulfone (PES) filters (Sartorius Stedim Biotech GmbH, Germany) before direct injection into UHPLC-MS/MS system.

Antibiotics chemical analysis. Method developed in this study for identification of six antibiotics: clarithromycin, chloramphenicol, erythromycin, oxytetracycline, doxycycline and azithromycin respectively, by Ultra-High-Performance Liquid Chromatography coupled with Triple Quadrupole Mass Spectrometry (UHPLC-MS/MS) meets the requirements at the EU level regarding the detection limit, i.e. it should be at least the same as the predicted no-effect concentration (PNEC) values in the corresponding matrix. The equipment used for identification of antibiotics was SPE-online-UHPLC-MS/MS Thermo Fisher Scientific™ EQuan MAX Plus™UltiMate 3000 system coupled with TSQ Quantiva triplequadrupole mass spectrometer. Chromatographic separation of the analyses was performed with Hypersil GOLD C18 column (50 mm × 2.1 mm, 1.9 μm) from Thermo Scientific™. The mobile phase was a mixture of water with the addition of 0.1% formic acid (A) and acetonitrile with the addition of 0.1% formic acid (B), eluting at a flow rate 0.3 mL/min. The method used is the most sensitive and selective compared to other frequently used analytical methods, allowing the detection and quantification of emerging pollutants in the category of pharmaceuticals at ng/L level.

Antibiotic susceptibility testing. The prevalence of antibiotic resistance was studied for strains of *Escherichia coli* (n = 22) isolated from wastewater and surface water samples. Representative colonies were randomly selected from each sample and were purified on TrypticaseSoy Agar. The bacterial isolates were further identified by biochemical tests—API 20E (bioMérieux).

The analysis of the antibiotic sensitivity spectrum was performed by the standardized disc-diffusion method (Kirby-Bauer), according to the criteria provided in CLSI 2020 (Clinical Laboratory Standard Institute). The antibiotics used for the bacterial strains studied included a number of 22 antibiotics, namely: ampicillin (AMP), cefazolin (KZ), aztreonam (ATM), ceftriaxone (CRO), imipenem (IMP), ceftazidime (CAZ), cefoxitin (FOX), amoxicillin + clavulanic acid (AMC), amikacin (AK), ertapenem (ERT), trimethoprim (W), tetracycline (TE), piperacycline (PRL), ciprofloxacin (CIP), chloramphenicol (C), penicillin (P), doxycycline (DO), gentamicin (CN), clarithromycin (CLR), sulfamethoxazole (RL), azithromycin (AZM) and erythromycin (E) (Oxoid).

3 Results and Discussion

3.1 Identification of Pharmaceutical Compounds—Antibiotics in the Area of the Studied River Basins

In the analyzed surface water and biota samples, the screening for antibiotics identification displayed their presence in variable concentrations from ng/L to μg/L. (Figs. 2 and 3).

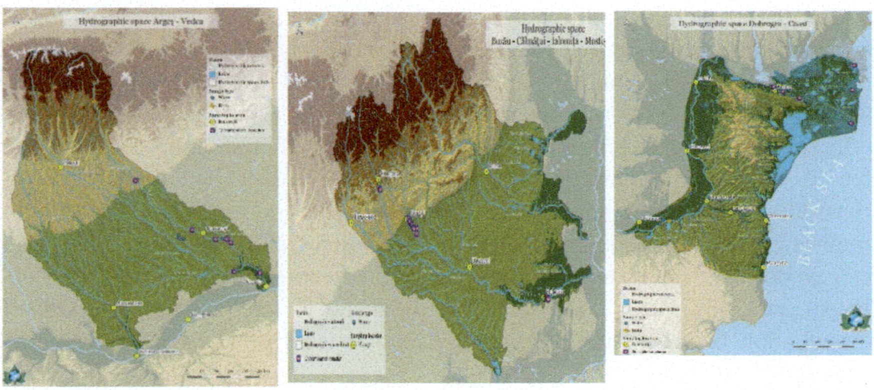

Fig. 2 Distribution of chloramphenicol in aquatic environments from the areas under study

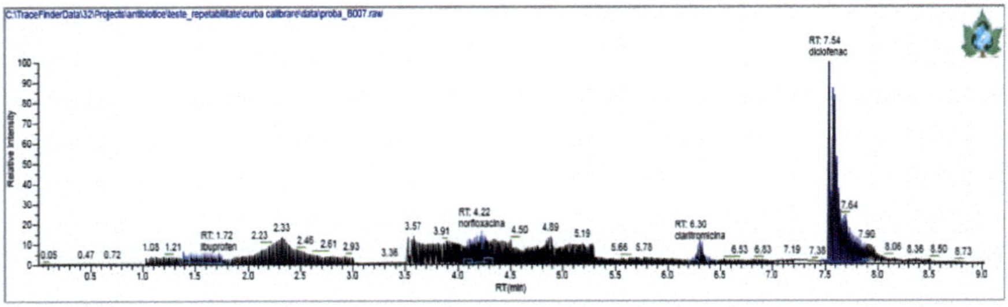

Fig. 3 Identification of antibiotics (UHPLC-MS/MS)–biota (fish) sample—Danube River, km 238

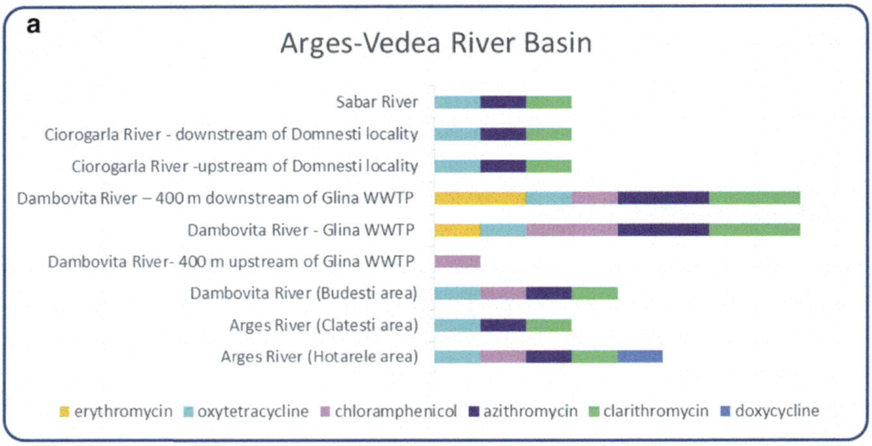

Fig. 4a Identification of pharmaceutical compounds in samples collected from the Arges-Vedea River Basin

Identification of antibiotics in the area of the studied river basins: Arges-Vedea, Buzau-Ialomita and Dobrogea-Litoral is shown in Figs. 4a, 4b and 4c.

Among the pharmaceutical micropollutants identified in the antibiotic category, clarithromycin (75.00%) was detected most frequently in the total samples, while erythromycin was detected in the lowest number of samples, with a frequency of 15.63%. Detection frequencies were 62.50% for oxytetracycline and 25.00% for doxycycline, while azithromycin was detected with a frequency of 59.38% and chloramphenicol with a frequency of 46.88%. Azithromycin and clarithromycin were detected downstream by the wastewater treatment plants of Bucharest, Slobozia and Ploiesti, in Dambovita R., Ialomita R. and Dambu R., respectively. The highest concentration of antibiotics identified was in the area of Danube River, Chilia Branch, of 8312 ng/L for doxycycline. The analysis for identifying antibiotics in biota samples revealed a concentration of 2.21 ng/g of clarithromycin in the biota sample (fish)—Danube River, km 238.

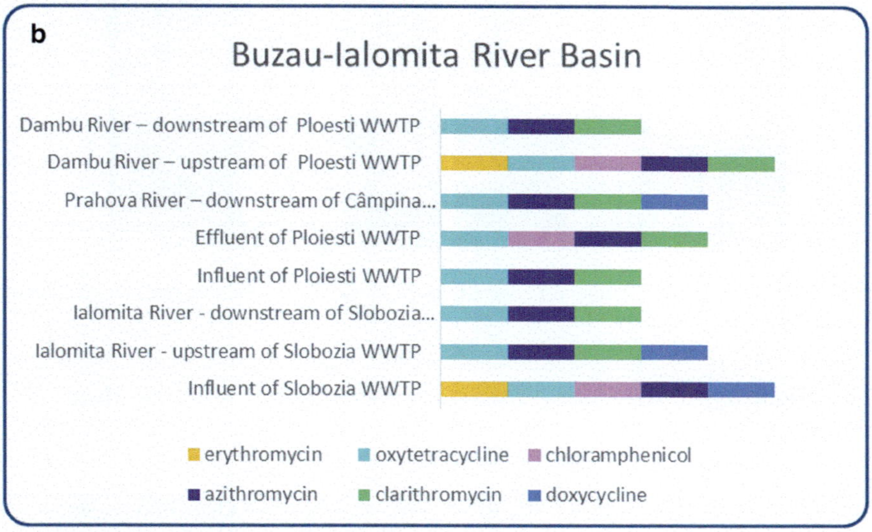

Fig. 4b Identification of pharmaceutical compounds in samples collected from the Buzau-Ialomita River Basin

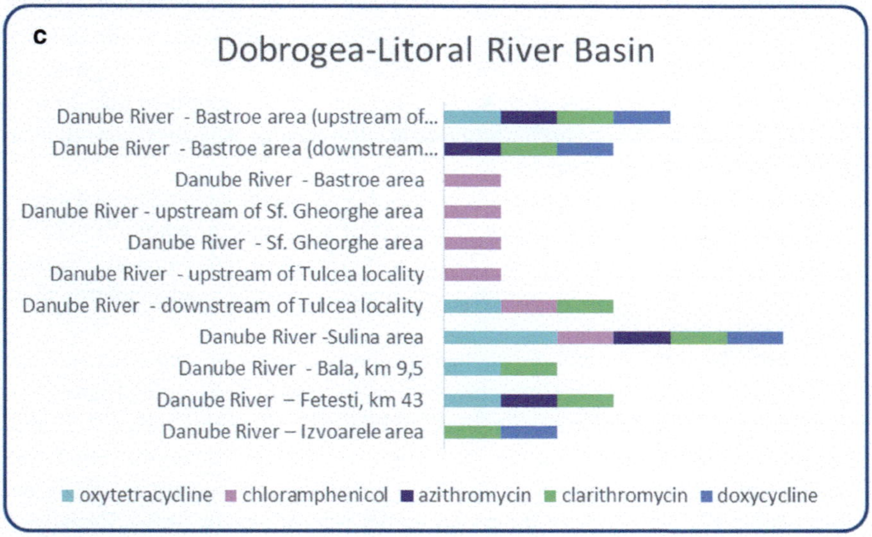

Fig. 4c Identification of pharmaceutical compounds in samples collected from the Dobrogea-Litoral River Basin

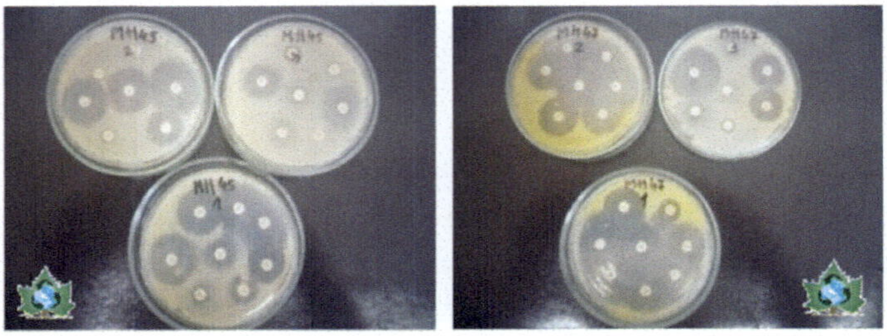

Fig. 5 Diffusimetric antibiogram on *E. coli* strains (C45 and C47)—isolated from the effluent of Ploiesti Wastewater Treatment Plant

3.2 Phenotypic Antibiotic Resistance Profiles of Escherichia Coli *Strains*

After 24 h of incubation at 37 °C, the *E. coli* strains were classified as sensitive (S), intermediate (I) or resistant (R) based on CLSI 2020 break points. Intermediate strains were included in the resistant class (Fig. 5).

Resistance levels of *E. coli* strains were interpreted taking into account the resistance ranges proposed for bacterial strains isolated from the clinic by the European Center for Infectious Disease Control (ECDC/European Resistance Surveillance Network EARS-Net). Thus, the following ranges were used: < 1% = insignificant, 1 to < 5% = very low, 5 to < 10% = low; 10 to < 25% = intermediate; 25 to < 50% = high; ≥ 50% = very high. (http://ecdc.europa.eu/en/activities/surveillance/EARS-Net/database/Pages/maps).

Therefore, in the case of β-lactam antibiotics, *E. coli* strains isolated from wastewater and surface water showed a very high level of resistance to ampicillin (75%), penicillin (75%) and clarithromycin (75%), a high level of resistance to cefazolin (50%), cefoxitin (50%) and amoxicillin + clavulanic acid (50%) and an intermediate level to piperacycline (17%).

Resistance to third-generation cephalosporins (ceftriaxone and ceftazidime), imipenem and ertapenem was not present in the studied *E. coli* strains. The resistance level to aztreonam was high (25%) (Fig. 6). Resistance to other classes of antibiotics (non-β-lactam) was high in essential metabolite inhibitors (trimethoprim/sulfamethoxazole—25/33%), tetracyclines (tetracycline—25%) and erythromycin (42%) and low in fluoroquinolones (ciprofloxacin—8%). Resistance to amikacin and chloramphenicol was not observed in the studied *E. coli* strains. The level of doxycycline resistance was low (8%).

Analyzing the levels of antibiotic resistance of *E. coli* strains correlated with isolation sources (Fig. 7), it is observed that the highest rates of resistance to β-lactam antibiotics and other classes of antibiotics were mainly recorded in strains isolated from water sources subjected to selective pressure of antibiotics, namely: effluent of the treatment plant (50–100%), as well as downstream from the discharge of effluent from the treatment plant—Dambu River (25–100%).

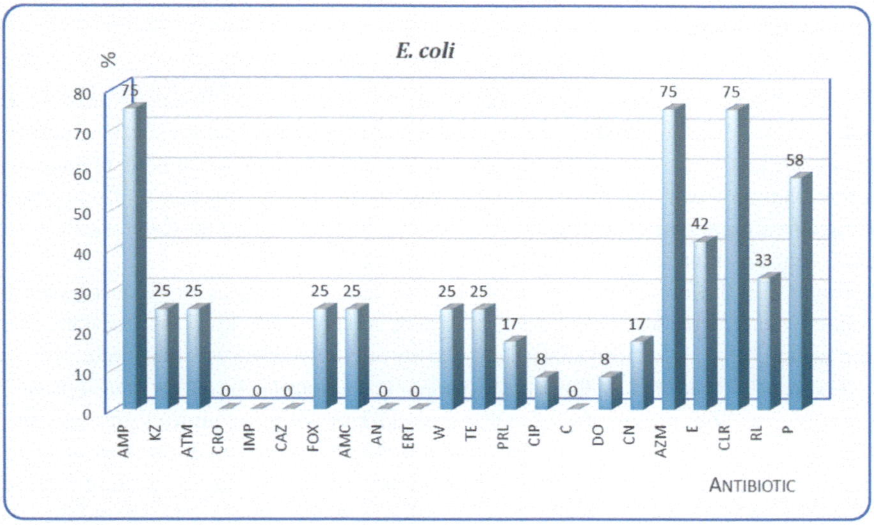

Fig. 6 Distribution of antibiotic resistance markers (%) in *E. coli* strains isolated from wastewater and surface water samples

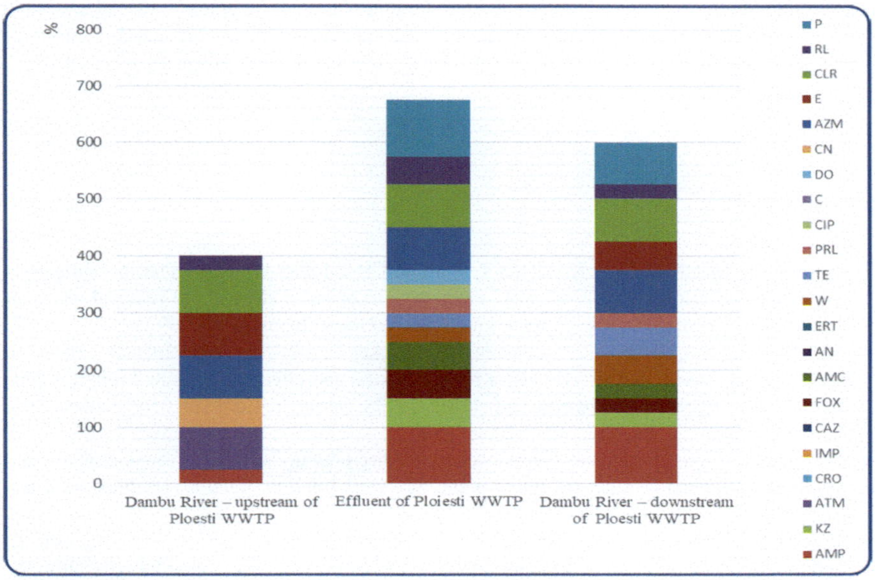

Fig. 7 Distribution of antibiotic resistance markers in *E. coli* strains correlated with isolation sources

4 Conclusion

The investigations carried out at the level of Arges-Vedea, Buzău-Ialomita and Dobrogea River basins and of the Danube River highlighted the presence of antibiotics both in the wastewater at the entrance and exit of the treatment plants and in the Danube River and the main tributaries in the area of discharges from urban sewage treatment plants of Bucharest, Slobozia, Fundulea, Lehliu, Calarasi, Ploiesti and Campina, in variable concentrations from ng/L to μg/L.

Regarding the prevalence of antibiotic resistance for strains of E. coli bacteria isolated from wastewater and surface water, a very high level of resistance to ampicillin (75%), penicillin (75%) and clarithromycin (75%) was observed.

The results obtained in this study represent a scientific basis for the approach according to the provisions of Directive 39/EU/2013—regarding the priority substances in the field of water policy and the Decisions for Selection of substances for the Watch List under the Water Framework Directive, based on their hazard properties as well as the availability of reliable safety thresholds, including the contribution to antimicrobial resistance for antibiotics and the availability of relevant analytical methods for monitoring in the appropriate environmental matrix.

Acknowledgements This paper was developed through Nucleu Programme – Contract number 39N/2019 carried out with the Romanian Ministry of Research and Innovation support, within the project "Research on the ecotoxicity of pharmaceutical micropollutants on aquatic ecosystems".

References

1. Patel M, Kumar R, Kishor K, Mlsna T, Pittman CU, Mohan D (2019) Pharmaceuticals of emerging concern in aquatic systems: chemistry, occurrence, effects, and removal methods. Chem Rev 119(6):3510–3673. https://doi.org/10.1021/acs.chemrev.8b00299
2. Grzesiuk M, Spijkerman E, Lachmann SC, Wacker A (2018) Environmental concentrations of pharmaceuticals directly affect phytoplankton and effects propagate through trophic interactions. Ecotoxicol Environ Saf 156:271–278. https://doi.org/10.1016/j.ecoenv.2018.03.019
3. Persistance of pharmaceutical compounds and other organic wastewater contaminants in a conventional drinking-water-treatment plant. Sci Total Environ 329:99–113. https://doi.org/10.1016/j.scitotenv.2004.03.015. PMID: 15262161
4. Loraine GA, Pettigrove ME (2006) Seasonal variations in concentrations of pharmaceuticals and personal care products in drinking water and reclaimed wastewater in Southern California. Environ Sci Technol 40(3):687–695. https://doi.org/10.1021/es051380x. PMID: 16509304
5. Servos MR, Smith M, McInnis R, Burnison BK, Lee BH, Backus S (2007) The presence of selected pharmaceuticals and the antimicrobial triclosan in drinking water in Ontario, Canada. Water Qual Res J Can 42(2):130–137. https://doi.org/10.2166/wqrj.2007.016
6. COM (2019) 128 final: communication from the commission to the European parliament, the council and the European economic and social committee, European union strategic approach to pharmaceuticals in the environment
7. Milieu Ltd, et al (2019) Options for a strategic approach to pharmaceuticals in the environment: final report. Retrieved from https://policycommons.net/artifacts/280678/opt

ions-for-a-strategic-approach-to-pharmaceuticals-in-the-environment/ on 09 Jul 2021. CID: 20.500.12592/z0ck90

8. Yang C, Song G, Lim W (2020) A review of the toxicity in fish exposed to antibiotics. Comp Biochem Physiol C Toxicol Pharmacol. 237:108840. https://doi.org/10.1016/j.cbpc.2020.108840. Epub 2020 Jul 5. PMID: 32640291
9. Marinescu F, Ilie M, Ghita G, Savin I, Tociu C, Anghel AM, Marcu E, Marcus I (2019) Antibiotic resistance profile and chemical quality assessment of groundwater sources from periurban area of Bucharest, Romania. Rev Chim 70(10):3549–3554
10. Ilie M, Deák G, Marinescu F, Ghita G, Tociu C, Matei M, Covaliu CL, Raischi M, Yusof SY (2020) IOP Conf Ser: Earth Environ Sci 616:012016
11. Gulkowska A, Leung HW, So MK, Taniyasu N, Yamashita N, Yeung L, Richardson B, Lei AP, Giesy JP, Lam P (2008) Removal of antibiotics from waste water by sewage treatment facilities in Hong Kong and Shenzhen, China. Water Res 42(1–2):395–403. https://doi.org/10.1016/j.watres.2007.07.031. Epub 2007 Jul 27. PMID: 17706267
12. Cordoba MA, Roccia IL, De Lucca MM, Pezzani B, Basualdo JA (2001) Resistance to antibiotics in injured coliforms isolated from drinking water. Microbiol Immunol 45(5):383–386. https://doi.org/10.1111/j.1348-0421.2001.tb02634.x
13. Sapkota A, Curriero F, Gibson K, Schwab K (2007) Antibiotic-resistant enteroccoci and fecal indicators in surface water and groundwater impacted by a concentrated swine feeding operation. Environ Health Perspect 115(7):1040–1045. https://doi.org/10.1289/ehp.9770
14. Garcia S, Wade B, Bauer C, Craig C, Nakaoka K, Lorowitz W (2007) The effect of wastewater treatment on antibiotic resistance in Escherichia coli and Enterococcus sp. Water Environ Res 79(12):2387–2395. https://doi.org/10.2175/106143007x183826. PMID: 18044355
15. Ma J, Zhai G (2014) Antibiotic contamination: a global environment issue. J Bioremed Biodegrad 5:e157. https://doi.org/10.4172/2155-6199.1000e157
16. Haggard B, Bartsch L (2009) Net changes in antibiotic concentrations downstream from an effluent discharge. J Environ Qual 38:343–352. https://doi.org/10.2134/jeq2007.0540
17. Serwecińska L (2020) Antimicrobials and antibiotic-resistant bacteria: a risk to the environment and to public health. Water 12(12):3313. https://doi.org/10.3390/w12123313
18. Abbassi BE, Saleem MA, Zytner RG, Gharabaghi B, Rudra R (2016) Antibiotics in wastewater: their degradation and effect on wastewater treatment efficiency. J Food Agric Environ 14(3&4):95–99
19. Elizalde-Velazquez A, Gomez-Olivan LM, Galar-Martinez M, Islas-Flores H, Garcia OD, SanJuan-Reyes N (2016) Amoxicillin in the aquatic environment, its fate and environmental risk, Chap 10. In: Larramendy ML, Soloneski S (eds) Environmental health risk - hazardous factors to living species. IntechOpen, London

Influence of Glucose Supplementation on the Organic Removal and Biomass Growth in Anaerobically Digested Vinasse (AnVE) by Using Aerobic Sequencing Batch Reactor (SBR)

Wei-Chin Kee, Yee-Shian Wong, Soon-An Ong, Nabilah Aminah Lutpi, Sung-Ting Sam, Audrey Chai, and György Deák

Abstract This paper analyses the effect of glucose supplementation in the treatment of anaerobically digested vinasse (AnVE) through reduction performance and biomass growth by using aerobic SBR system. The result revealed that the COD and ammonium reduction with the addition of glucose supplementation (65.3 ± 6.0 and 30.0 ± 5.5%) was greater than the reactor without glucose supplementation (23.6 ± 10.8 and 18.7 ± 6.0%). This can be due to the additional glucose as the carbon source for microbe metabolism nutrients. The biomass growth in the reactor with glucose supplementation increased rapidly over time, while there was no obvious biomass growth shown in the reactor without the addition of glucose. It is because glucose supplementation increased biomass accumulation in biomass production. The increment of biomass in the reactor with glucose supplementation can be further explained by the F/M ratio and biomass yield. The F/M ratio reduced when the biomass growth increased. Thus, the biomass yield started to decrease after the operation in week 2,

W.-C. Kee (✉) · Y.-S. Wong · S.-A. Ong · N. A. Lutpi · A. Chai
Faculty of Civil Engineering Technology, Universiti Malaysia Perlis, 02600 Arau, Perlis, Malaysia

Y.-S. Wong
e-mail: yswong@unimap.edu.my

S.-A. Ong
e-mail: saong@unimap.edu.my

N. A. Lutpi
e-mail: nabilah@unimap.edu.my

Centre of Excellence Water Research and Environmental Sustainability Growth (WAREG), Universiti Malaysia Perlis, 02600 Arau, Perlis, Malaysia

S.-T. Sam
Sustainable Environment Research Group (SERG), Centre of Excellence Geopolymer and Green Technology (CeGeoGTech), Universiti Malaysia Perlis, 02600 Arau, Perlis, Malaysia
e-mail: stsam@unimap.edu.my

G. Deák
National Institute for Research and Development in Environmental Protection Bucharest (INCDPM), 294, Splaiul Independentei Street, 6th District, 060031 Bucharest, Romania

N. Mohamed Noor et al. (eds.), *Proceedings of the 3rd International Conference on Green Environmental Engineering and Technology*, Lecture Notes in Civil Engineering 214,
https://doi.org/10.1007/978-981-16-7920-9_48

until it reached maximum biomass growth. Therefore, it concluded that the addition of glucose supplementation in the treatment of AnVE showed a positive effect on the removal performance and biomass growth by using aerobic SBR.

Keywords SBR · Glucose supplementation · Vinasse · Organic removal · Biomass growth

1 Introduction

Sugarcane vinasse has the characteristic of acidic, intensive brown-colored, and high content of organic compounds [1]. The characteristic of sugarcane vinasse can cause soil and water pollution if it is disposed without proper treatment. Anaerobic digestion of vinasse is known as the primary treatment option because a high percentage of organic matter can be removed and converted into biogas as an alternative fuel source [2]. However, the further treatment process is necessary for the anaerobically treated vinasse (AnVE) since the high content of the organic compounds have still remained in the residue. Aerobic digestion can be one of the effective post treatment process to remove the remaining organic substances and colour in AnVE. Maximum COD reduction (65.2 $\pm$ 2.5%) was achieved with the anaerobically digested molasses wastewater containing 30 g/L glucose under aerobic SBR system at HRT 7 days [3].

Nutrient supplementation is crucial in promoting the removal performance and biomass growth in the biological wastewater treatment. It can enhance the process with the addition of suitable microbial nutrients to the wastewater treatment to improve the degradation potential and stimulate bacteria growth [4]. Carbon source availability is one of the factors to promote the biomass growth in the sequencing batch reactor (SBR) system. Paper, textile, manufacturing, and molasses industries require external nutrients as the supplementation to the SBR due to the nutrient-deficient wastewater generated [5]. The types of carbon sources commonly used in aerobic SBR are sodium succinate, sodium acetate, and glucose [6]. These carbon sources can enhance the process performance by promoting the functional microbial community. The high removal performances of COD, nitrogen and phosphorus was achieved in the SBR with glucose supplementation [7]. Besides that, F/M ratio is a process variable that can be easily adjusted in operating bioreactors. F/M ratio could be linked to the biomass growth and the sludge loading rate.

In this paper, the effect of glucose supplementation as the additional carbon source was determined in the treatment of anaerobically digested vinasse (AnVE) by using aerobic SBR system. The reduction efficiency and biomass growth of the SBR system were compared through the operation of reactors with and without glucose supplementation. Food to microorganism (F/M) ratio and biomass yield were also determined to investigate the relationship between both parameters and biomass growth in the SBR system with glucose supplementation. To the best of authors' knowledge, the finding of the glucose supplementation in the aerobic digestion system was limited, as compared to the anaerobic digestion system. This paper can be one of the

Table 1 The characteristic of AnVE wastewater

Parameters	Value	Parameters	Value
pH	8.11 ± 0.3	MLSS (mg/L)	690 ± 105
COD (mg/L)	3202 ± 424	MLVSS (mg/L)	100 ± 25
NH_4^+ (mg/L)	18 ± 3		

references to understand the role of glucose supplementation in contributing to the biological wastewater treatment system.

2 Methodology

2.1 Seed Sludge and AnVE

The seed sludge was used as inoculum in this study. It was taken from an anaerobic pond located in Malpom Industries Berhad, Penang. The AnVE from the final pond in Fermpro Sdn. Bhd, Perlis was used as the wastewater to be treated in the operation. The wastewater samples were kept at a temperature of 4 ± 1 °C to minimize the biodegradation in the wastewater. Glucose with a concentration of 5.0 g/L was added as an external carbon source to the AnVE for the system with the glucose supplementation. The AnVE wastewater was characterised in Table 1.

2.2 Reactor Set-Up and Operation

The aerobic SBR with the 22 cm height and 20 cm internal diameter was operated for the treatment of AnVE as shown in Fig. 1. The working volume is 4 L, which consists of 1 L of seed sludge and 3 L of AnVE. Aeration was supplied (100 mL/min) at the

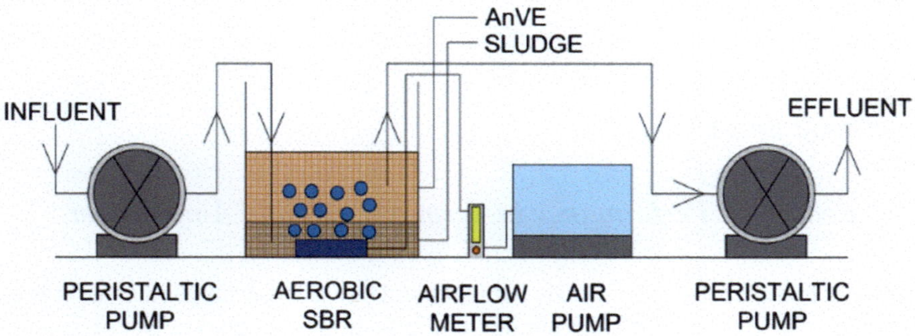

Fig. 1 Reactor configuration of aerobic SBR

bottom of the reactor through the aeration bar by using Atinan HP-4000 aquarium air pump. The dissolved oxygen (DO) concentration in the SBR was maintained at 2.40 ± 0.6 mg/L. The feed and decant processes were conducted by using peristaltic pumps.

The operation of the SBR was conducted for 28 operational days. The hydraulic retention time (HRT) was fixed at 4 days throughout the operation. The operation consists of four phases in one cycle. First, 1 L of AnVE was filled into the reactor (0.5 h). Then, the aeration was supplied in the reactor to maintain aerobic condition for 21 h. The mixed wastewater was settled for 2 h to separate the sludge and supernatant before the draw phase was carried out (0.5 h). One complete cycle for the operation is 24 h.

2.3 Analytical Method

COD was analyzed by the Dichromate Reactor Digestion Method of Wastewater Analysis. Ammonium ion (NH_4^+) was analyzed by using the Standard Method 4500-NH_3 D. Moreover, MLSS and MLVSS were carried out with the Standard Method 2540-D method and Standard Method 2540-B, E, respectively. The results were obtained and tested in repetition, which expressed as the mean ± standard deviation. Food to microorganism (F/M) ratio was also determined through Eq. (1).

$$F/M\ ratio = \frac{S_0}{\tau \cdot X} \tag{1}$$

Where S_0 is influent COD concentration, τ is hydraulic retention time (HRT), and X is mixed liquor volatile suspended solid (mg/L). Biomass yield can be defined as the ratio of the amount of biomass produced to the amount of substrate consumed [8], and was shown in Eq. (2).

$$Biomass\ yield,\ Y = \frac{MLVSS\ (g\ biomass\ produced)}{COD\ (g\ substrate\ utilized)} \tag{2}$$

3 Result and Discussion

3.1 Reduction Performance in Term of COD and Ammonium

The reduction performance of SBR with and without glucose supplementation was shown in Fig. 2. The COD and ammonium reduction with glucose supplementation achieved greater performance when compared to the system without glucose supplementation. In the perspective of COD reduction, the system with glucose

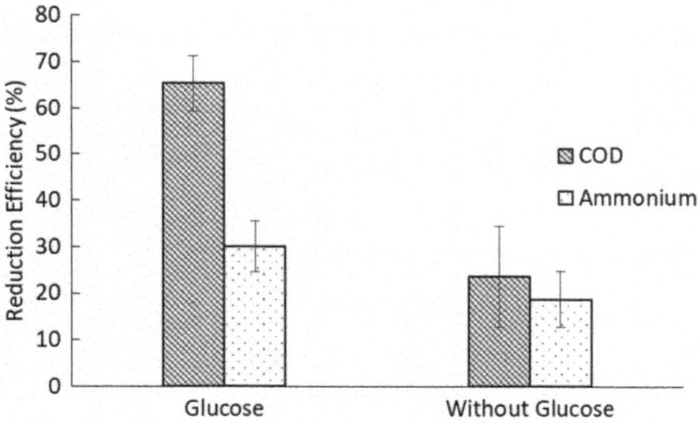

Fig. 2 COD and ammonium reduction in the aerobic SBR with and without glucose

supplementation (65.3 ± 6.0%) was 2.7 times greater than without glucose supplementation (23.6 ± 10.8%). The growth of heterotrophic bacteria was promoted by the carbon sources from the glucose supplementation, which lead to the high COD reduction efficiency in the system [9]. While for ammonium reduction, the system with glucose supplementation achieved 30.0 ± 5.5% reduction, which was higher when compared to the system without glucose supplementation (18.7 ± 6.0%). It can be explained by the additional glucose as the carbon source for microbe metabolism nutrients. A similar result was obtained when the glucose was added and contributed to higher TN removal efficiency [10]. The co-metabolic enzymes required for the transformation of refractory compounds are induced. The energy quickly regenerates with the rapid metabolism of glucose, which, in turn, can be used to transform refractory compounds [11]. Evans et al. [12] reported that the addition of glucose in the wastewater can improve the NH_3-N removal efficiency, which the removal efficiency was higher than the system without glucose [12].

3.2 *Biomass Growth*

Table 2 shows the MLVSS, F/M ratio and biomass yield during the operation of aerobic SBR with and without glucose supplementation. The biomass growth in terms of MLVSS showed a gradual increase from 4540 mg/L to 8780 mg/L in 28 operational days. This result revealed that the addition of glucose into the activated sludge system can significantly improve the biomass growth. The result can be further explained from the system without glucose addition. The MLVSS concentration was 2485 ± 108 mg/L in the system without glucose supplementation, and minor decrement of the biomass growth observed in the reactor. A significant enhancement in algal biomass yields was reported by Xiong et al. [13] when the glucose was added in the culturing medium. A similar result was reported by Gabriele et al. [14] that the biomass in algal production was accumulated when the glucose supplementation was provided. The glucose supplementation promoted algal growth and the accumulation

Table 2 MLVSS, F/M ratio and biomass yield in the operation of aerobic SBR with and without glucose supplementation

Week	Glucose			Without glucose		
	MLVSS (mg/L)	F/M ratio	Biomass yield	MLVSS (mg/L)	F/M ratio	Biomass yield
1	4540	0.51	0.08	2640	0.33	−0.10
2	5320	0.36	0.14	2480	0.26	−0.28
3	7220	0.28	0.35	2420	0.28	−0.13
4	8780	0.23	0.30	2400	0.27	−0.05

of chlorophyll can be prevented. However, excessive biomass growth may lead to the formation of more sludge in the condition with glucose supplementation; Thus, The MLVSS concentration can be maintained at desired concentration by applying daily sludge discharge in the reactor [15].

The F/M ratio was inversely proportional to the biomass growth in the aerobic SBR with glucose supplementation. The result showed that the F/M ratio decreased from 0.51 to 0.23 in the four weeks operation. This was due to the fact that the intensive growth of the biomass with the fixed organic loading throughout the operation of aerobic SBR with the addition of glucose. A low F/M ratio can improve the sludge flocculation and the removal of organic substances [16]. Thus, the biomass settleability was also increased. However, the failure of cell growth and sludge deflocculation may also occur if the F/M ratio becomes too low. F/M ratio can also be linked with biomass yield. Biomass yield achieved the lowest in week 1 (0.08). It can be explained by the microorganism in the process of adapting to the environment. The biomass yield increased to 0.35 in week 3, which can be related to the rapid biomass growth after the early stage of the acclimatization phase. However, the biomass yield showed a downward trend in week 4, which was 0.30. It can be due to the reactor reached maximum growth rate and low F/M ratio; Thus, the biomass yield was reduced. The F/M ratio in the aerobic SBR without glucose supplementation was in between 0.26 to 0.33, since the biomass concentration was slightly reduced. The reduction of biomass growth can also reveal from the biomass yield, which was negative throughout the operation.

4 Conclusion

This research focused on the effect of glucose supplementation through the determination of reduction performance and biomass growth in the treatment of anaerobically digested vinasse (AnVE) by using aerobic SBR system. The COD and ammonium reduction with the addition of glucose supplementation were 65.3 ± 6.0 and 30.0 ± 5.5%, respectively. The results were greater than the condition without glucose supplementation, which only achieved 23.6 ± 10.8% of COD reduction and 18.7

$\pm$ 6.0% of ammonium reduction. The glucose supplementation was added as the carbon source for microbe metabolism nutrient. Thus, it can improve the removal performance. A similar result was observed in the biomass growth. The biomass growth in the reactor with glucose supplementation increased over time, while the biomass growth in the reactor without the addition of glucose was negligible. The F/M ratio and biomass yield were determined in the reactor with glucose supplementation. When the biomass growth increased, the F/M ratio reduced and the biomass yield started to decrease after the operation in week 2 until it reached maximum biomass growth. Therefore, it can conclude that the addition of glucose supplementation showed a positive effect on the removal performance and biomass growth in the treatment of AnVE by using aerobic SBR.

References

1. Rulli MM, Villegas LB, Colin VL (2020) Treatment of sugarcane vinasse using an autochthonous fungus from the northwest of Argentina and its potential application in fertigation practices. J Environ Chem Eng 8:104371. https://doi.org/10.1016/j.jece.2020.104371
2. Janke L, Leite A, Nikolausz M, Schmidt T, Liebetrau J, Nelles M, Stinner W (2015) Biogas production from sugarcane waste: assessment on kinetic challenges for process designing. Int J Mol Sci 16:20685–20703. https://doi.org/10.3390/ijms160920685
3. Sirianuntapiboon S, Prasertsong K (2008) Treatment of molasses wastewater by acetogenic bacteria BP103 in sequencing batch reactor (SBR) system. Bioresour Technol 99:1806–1815. https://doi.org/10.1016/j.biortech.2007.03.040
4. O-Thong S, Prasertsan P, Intrasungkha N, Dhamwichukorn S, Birkeland NK (2007) Improvement of biohydrogen production and treatment efficiency on palm oil mill effluent with nutrient supplementation at thermophilic condition using an anaerobic sequencing batch reactor. Enzyme Microb Technol 41:583–590. https://doi.org/10.1016/j.enzmictec.2007.05.002
5. Johnson K, Kleerebezem R, vanLoosdrecht MCM (2010) Influence of the C/N ratio on the performance of polyhydroxybutyrate (PHB) producing sequencing batch reactors at short SRTs. Water Res 44:2141–2152. https://doi.org/10.1016/j.watres.2009.12.031
6. Peng H, Zhang Q, Tan B, Li M, Zhang W, Feng J (2021) A metagenomic view of how different carbon sources enhance the aniline and simultaneous nitrogen removal capacities in the aniline degradation system. Bioresour Technol 335:125277. https://doi.org/10.1016/j.biortech.2021.125277
7. Li S, Fei X, Cao L, Chi Y (2019) Insights into the effects of carbon source on sequencing batch reactors: performance, quorum sensing and microbial community. Sci Total Environ 691:799–809. https://doi.org/10.1016/j.scitotenv.2019.07.191
8. Metcalf & Eddy, Inc (2003) Wastewater engineering: treatment and reuse, 4th edn. McGraw Hill, New York
9. Shen QH, Jiang JW, Chen LP, Cheng LH, Xu XH, Chen HL (2015) Effect of carbon source on biomass growth and nutrients removal of Scenedesmus obliquus for wastewater advanced treatment and lipid production. Bioresour Technol 190:257–263. https://doi.org/10.1016/j.biortech.2015.04.053
10. Wang D, Ji M, Wang C (2014) The stimulating effects of the addition of glucose on denitrification and removal of recalcitrant organic compounds. Brazilian J Chem Eng 31:9–18. https://doi.org/10.1590/s0104-66322014000100002

11. Wang SG, Liu XW, Zhang HY, Gong WX, Sun XF, Gao BY (2007) Aerobic granulation for 2,4-dichlorophenol biodegradation in a sequencing batch reactor. Chemosphere 69:769–775. https://doi.org/10.1016/j.chemosphere.2007.05.026
12. Evans L, Hennige SJ, Willoughby N, Adeloye AJ, Skroblin M, Gutierrez T (2017) Effect of organic carbon enrichment on the treatment efficiency of primary settled wastewater by Chlorella vulgaris. Algal Res 24:368–377. https://doi.org/10.1016/j.algal.2017.04.011
13. Xiong W, Gao C, Yan D, Wu C, Wu Q (2010) Double CO2 fixation in photosynthesis-fermentation model enhances algal lipid synthesis for biodiesel production. Bioresour Technol 101:2287–2293. https://doi.org/10.1016/j.biortech.2009.11.041
14. Conceição GR, Xavier LMBD, Matos JBTL, deAlmeida PF, deMoura-Costa LF, Chinalia FA (2019) Glucose and nitrogen amendments can mitigate wastewater-borne bacteria competition effect against algal growth in wastewater-based systems. J Phycol 55:1050–1058. https://doi.org/10.1111/jpy.12902
15. Li AJ, Li XY (2009) Selective sludge discharge as the determining factor in SBR aerobic granulation: numerical modelling and experimental verification. Water Res 43:3387–3396. https://doi.org/10.1016/j.watres.2009.05.004
16. Liu Y, Liu H, Cui L, Zhang K (2012) The ratio of food-to-microorganism (F/M) on membrane fouling of anaerobic membrane bioreactors treating low-strength wastewater. Desalination 297:97–103. https://doi.org/10.1016/j.desal.2012.04.026

Natural Zeolite Modified with Oxidizing Agent

Salwa Mohd Zaini Makhtar, Nor Amirah Abu Seman, Mahyun Ab Wahab, Ain Nihla Kamarudzaman, and György Deák

Abstract Zeolite is an abundant mineral in Indonesia, but is not maximally utilized. Natural zeolite possesses several properties, such as dehydration, adsorption, ion exchange, catalysis, and separation. In general, natural zeolite contains both organic and inorganic impurities that cover their pores. Thus, natural zeolite should be initially activated to improve its absorption. This study aimed to determine the effect of NaOH solution concentration, temperature, volume, and zeolite diameter on the adsorption capacity of zeolite. The highest adsorption capacity of 7.03% was obtained at the optimal concentration of 1.0 N NaOH solution (100 °C) with a volume of 50 mL at a diameter of 2.00 mm during chemical activation.

Keywords Zeolite · Oxidizing agent · Chemical activation

S. M. Z. Makhtar (✉) · N. A. A. Seman · M. A. Wahab · A. N. Kamarudzaman
Centre of Excellence Water Research and Environmental Sustainability Growth (WAREG), Universiti Malaysia Perlis, 02600 Arau, Perlis, Malaysia
e-mail: salwa@unimap.edu.my

N. A. A. Seman
e-mail: noramirah@unimap.edu.my

M. A. Wahab
e-mail: mahyun@unimap.edu.my

A. N. Kamarudzaman
e-mail: ainnihla@unimap.edu.my

Faculty of Civil Engineering Technology, Universiti Malaysia Perlis, Perlis, Malaysia

G. Deák
National Institute for Research and Development in Environmental Protection Bucharest (INCDPM), 294, Splaiul Independentei Street, 6th District, 060031 Bucharest, Romania

N. Mohamed Noor et al. (eds.), *Proceedings of the 3rd International Conference on Green Environmental Engineering and Technology*, Lecture Notes in Civil Engineering 214,
https://doi.org/10.1007/978-981-16-7920-9_49

1 Introduction

The name zeolite is derived from the word "zein," which means to boil, and "lithos", which denotes rock, and the mineral is called as such because of its boiling characteristics or expansion when heated. This describes the mineral behavior, which quickly releases water when heated, thereby appearing to boil. Zeolite is a hollow crystal formed by a network of 3D tetrahedral silica alumina. It has a relatively regular structure with a cavity filled with alkali or alkaline earth metal as its load balancer. Cavities constitute a complete system in which channels are occupied by water molecules [1].

Natural zeolite possesses several properties, such as dehydration, adsorption, ion exchange, catalysis, and separator. Dehydration mainly removes water molecules from the zeolite framework, thus enhancing zeolite activity with the heating process. Dehydration causes the pore structure of the zeolite to open and results in a large internal surface area, thereby adsorbing a large number of substances other than water and being able to separate the molecules of a substance based on molecular size and polarity. Natural zeolites present a frame structure, containing an empty space occupied by cations and water molecules that freely allow ion exchange or chemisorption [2]. Given its crystalline intracavity, zeolite can be used as a catalyst. The catalytic reaction is influenced by the size of the oral cavity and flow system, because this reaction depends on the diffusion of reactants and reaction products.

Natural zeolite must be activated before being used as adsorbent to open more pores and increase surface area. Zeolite activation process can be conducted using physical and chemical methods. Physical activation involves heating the zeolite with the intention of evaporating the water trapped in the pores of the zeolite crystal. Thus, the surface area of the pores increases. Heating is performed in a conventional oven at a temperature of 300 °C to 400 °C (for the laboratory scale). Chemical activation is conducted using a solution of H_2SO_4 acid or alkaline NaOH for the purpose of cleaning the pore surface, removing impurities and rearranging the location of atoms. A chemical reagent is added to the zeolite, which is placed in a round bottom boiling flask within a specified period. The zeolite is washed with deionized water until neutral and then dried [3].

In general, natural zeolite contains organic and inorganic impurities covering the pore. Thus, zeolite must be initially activated to improve absorption [4]. This study was carried out by observing the ability of zeolite in adsorbing synthetic metal concentration in the water and assess the effect of activation on the performance of zeolite as metal adsorbent. This study also aimed to determine the effects of temperature, volume, and NaOH solution concentration on the adsorption of synthetic metal concentration in the water, as well as to observe the characteristics of zeolite in absorbing metal concentration at various temperatures.

2 Materials and Method

Rock zeolite, which underwent processing stages prior to screening, and NaOH were used in this study. An oven, an evaporating dish, a round bottom boiling flask, control heating mantle, and a simple reflux apparatus were also employed such as in Fig. 1.

Zeolite is a sorption material, and its size and weight are fixed. Thus, the chemical activation period lasted for 3 h. The independent variable in this study was the chemical activation temperature (without heating, 25 °C, 50 °C, 75 °C, and 100 °C). The concentrations of NaOH solution used were 0.25, 0.5, 0.75, 1.0, and 1.25 N, and the volumes of NaOH solution were varied between 50 and 250 mL at 50 mL increments (50, 100, 150, 200, and 250 mL). Experiments were conducted using a descriptive method by constructing a chart to determine the optimum value of the independent variable used.

Zeolite was crushed, flaked, and passed through a sieve to obtain the desired size (i.e., 2.00 and 3.35 mm). Chemical activation was then performed using NaOH solution for 3 h. Zeolite was washed with deionized water until neutral, pH 7 and then dried in an oven at 105 °C for 24 h.

Experiments were conducted at room temperature using the constant conditions of all pertinent factors, such as dosage, agitation speed, and contact time, to study metal (Zn) concentration removal by adsorbent [5]. Adsorption tests were conducted in the reaction mixture consisting of 200 mL of Zn solution with a concentration of 20 mg/L.

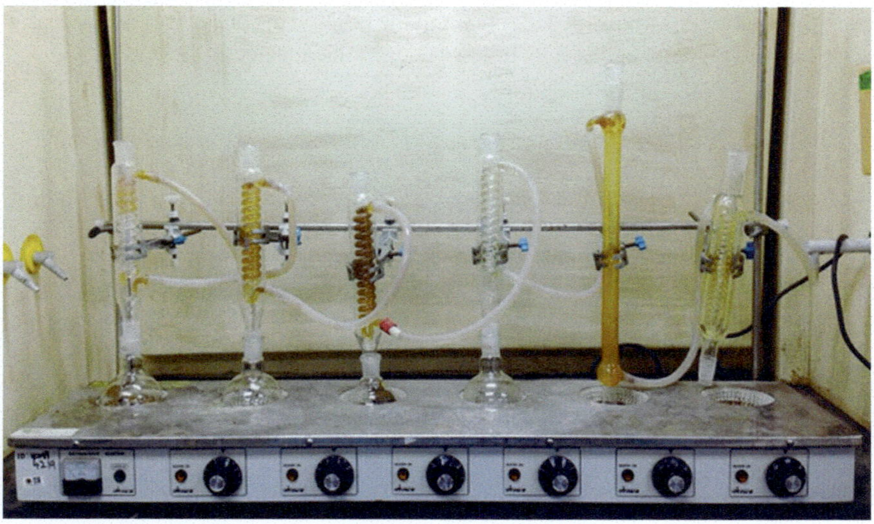

Fig. 1 Simple reflux apparatus with a control heating mantle

3 Result and Discussion

3.1 Effect of NaOH Solution Concentration on the Adsorption Capacity of Zeolite

The increased concentration of NaOH solution increased the adsorption capacity of zeolite. The optimum concentration for use in a chemical activation step using alkaline solution, or NaOH in this case.

Figure 2 shows that the optimal concentration of NaOH solution was 1.0 N (Volume of NaOH solution to each fixed concentration = 250 mL at 50 °C), which obtained the highest metal concentration adsorption capacity (6.01%). Chemical activation using alkaline solution could dissolve the alkaline-soluble impurities (e.g., silica and alumina) that are located outside of the crystal structure and cover the pores of the zeolite surface.

As shown in Fig. 2, the metal concentration adsorption capacity of zeolite decreased when the concentration of NaOH solution was 1.25 N. When the NaOH solution exceeded the optimal concentration, the elements outside the surface and the sides of the crystal were dissolved. The dissolution of elements reduced the surface area of zeolite crystal and, consequently, the adsorption capacity of the zeolite itself.

Results showed that 1.0 N was the optimal NaOH solution concentration to dissolve the impurities on the zeolite crystal. However, 1.25 N NaOH solution concentration considerably increased the viscosity of the solution, resulting in reduced diffusivity of the solution into the pores. Reduced diffusivity in NaOH solution causes numerous impurities to remain in the pores. This phenomenon decreases the crystal surface area and results in low adsorption capacity of the zeolite [6].

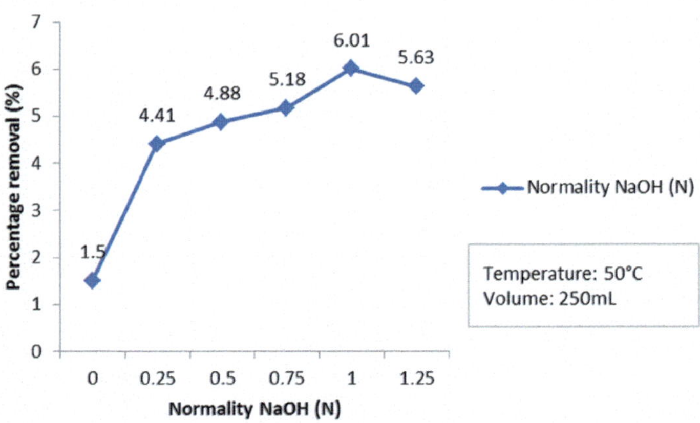

Fig. 2 Effect of NaOH solution concentration on the adsorption capacity of zeolite

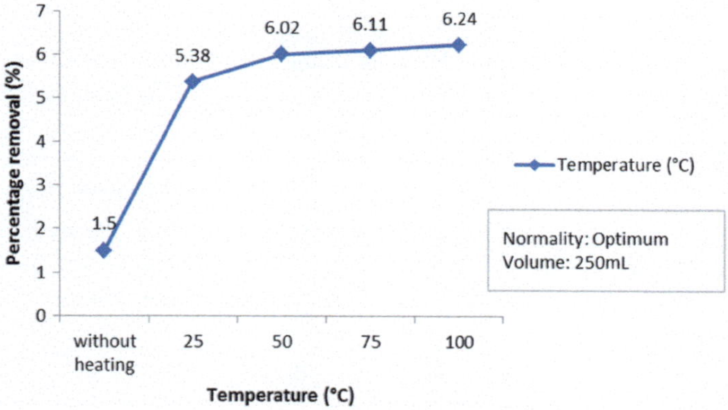

Fig. 3 Effect of NaOH heating temperature on the adsorption capacity of zeolite

3.2 *Effect of Heating Temperature on the Adsorption Capacity of Zeolite*

Figure 3 shows that metal concentration adsorption capacity of zeolite increased with increasing heating temperature during chemical activation. Among the four temperatures (25 °C, 50 °C, 75 °C, 100 °C) used, 100 °C (Volume of NaOH solution to each fixed concentration = 250 mL at 1.0 N NaOH) resulted in the highest water vapor adsorption capacity.

At 100 °C, more zeolite crystals are evaporated, thereby resulting in a more open zeolite pore structure, as well as wide internal structure that is capable of adsorbing a large variety of substances, such as water vapor. Thus, zeolite can function as a gas or a liquid absorbent.

As shown in Fig. 3, the adsorption capacity of zeolite without heat treatment is extremely low (1.5%). This low capacity was due to cavities in the zeolite crystals, which some already contained water. This phenomenon enables zeolite to approach saturation, decreasing the ability of zeolite to absorb moisture from the environment, compared with zeolite with most water removed in their cavities [7].

3.3 *Effect of NaOH Solution Volume on the Adsorption Capacity of Zeolite*

The highest metal concentration adsorption capacity of zeolite was obtained after using 1.0 N NaOH with a volume of 50 mL during chemical activation, as shown in Fig. 4 (NaOH concentration equal to all volume = 1.0 N at 100 °C).

Small volume increased the diffusivity of a solution. A 50 mL volume resulted in highest adsorption capacity of zeolite. This volume (the lowest in the NaOH volume series) produced the largest diffusivity, resulting in a more efficient and

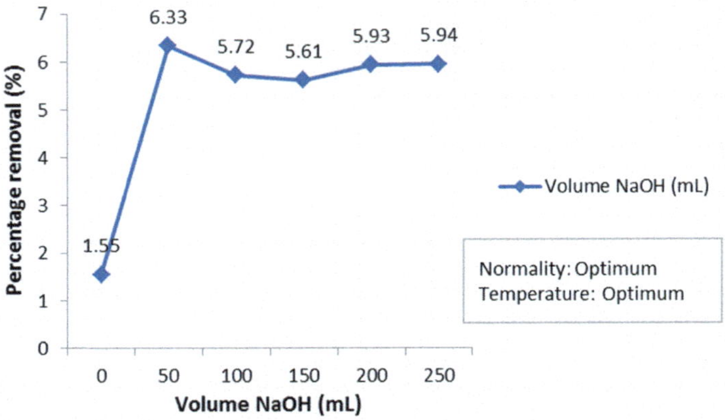

Fig. 4 Effect of NaOH solution volume on the adsorption capacity of zeolite

better dissolution of impurities on the zeolite crystal surface, as well as in the pores of the zeolite itself. Thus, the zeolite surface increased, resulting in increased metal concentration adsorption capacity of zeolite.

Figure 4 shows the absorption capacities of zeolite [lowest (5.61%) and highest (6.33%)] and the independent variable of NaOH normality [lowest (4.41%) and highest (6.01%)]. This result indicates the small absorption capacity of the former (0.72%) compared with the latter (1.6%). The effect of independent variable volume of NaOH is smaller than that of NaOH concentration [8].

3.4 Effect of Diameter Zeolite on the Adsorption Capacity of Zeolite

Chemically activated zeolite with 2.00 mm diameter exhibited the highest metal concentration absorption of 7.03% (Fig. 5). This result is attributed to the large surface area caused by the small diameter of the zeolite. A large surface area leads to more free space and zeolite pores. The empty spaces in the zeolite function as a place to accommodate the adsorbed metal concentration, and the zeolite pores function as an entry point of adsorbed metal concentration. Thus, empty spaces and higher pore number enable the zeolite to adsorb more metal concentration [9].

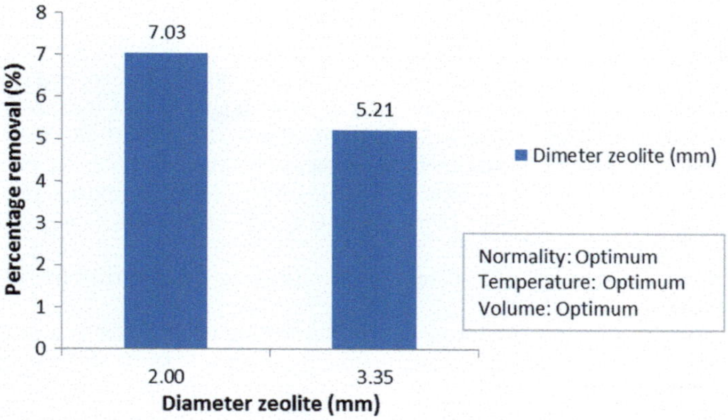

Fig. 5 Effect of diameter zeolite on the adsorption capacity of zeolite

4 Conclusion

The highest adsorption capacity of 6.01% was obtained at the optimal concentration of NaOH solution with normality of 1.0 N, temperature of 50 °C, and volume of 250 mL. This result indicates that metal concentration adsorption of zeolite increased with increasing NaOH solution concentration.

The highest adsorption capacity of 6.24% was obtained at the optimal temperature of 100 °C, NaOH solution concentration of 1.0 N, and a volume of 250 mL. This result indicates that metal concentration adsorption of zeolite increased with increasing temperature during chemical activation.

The highest adsorption capacity of 6.33% was obtained at the optimal NaOH solution volume of 50 mL, NaOH solution concentration of 1.0 N, and temperature of 100 °C. This result indicates that metal concentration adsorption of zeolite decreased with increasing NaOH solution volume.

Finally, the highest adsorption capacity value of 7.03% was achieved with a zeolite diameter of 2.00 mm after chemical activation. This result indicates that metal concentration adsorption of zeolite increased with small zeolite size.

References

1. Fei Z, Zhao D, Geldbach TJ, Scopelliti R, Dyson PJ, Antonijevic S, Bodenhausen G (2005) A synthetic zwitterionic water channel: characterization in the solid state by X-ray crystallography and NMR spectroscopy. Angew Chem Int Ed 44(35):5720–5725. https://doi.org/10.1002/anie.200500207
2. Bhatia S (2020) Zeolite catalysts: principles and applications. CRC Press
3. Basaldella EI, Vázquez PG, Iucolano F, Caputo D (2007) Chromium removal from water using LTA zeolites: effect of pH. J Colloid Interf Sci 313(2):574–578. https://doi.org/10.1016/j.jcis.2007.04.066

4. Li M, Zhu X, Zhu F, Ren G, Cao G, Song L (2011) Application of modified zeolite for ammonium removal from drinking water. Desalination 271(1–3):295–300
5. Igberase E, Osifo P, Ofomaja A (2017) The adsorption of Pb, Zn, Cu, Ni, and Cd by modified ligand in a single component aqueous solution: equilibrium, kinetic, thermodynamic, and desorption studies. Int J Anal Chem, 6150209. https://doi.org/10.1155/2017/6150209
6. Tanaka S, Fujita K, Miyake Y, Miyamoto M, Hasegawa Y, Makino T, Van der Perre S, Remi JCS, Van Assche T, Baron GV, Denayer JFM (2015) Adsorption and diffusion phenomena in crystal size engineered ZIF-8 MOF. J Phys Chem C 119(51):28430–28439. https://doi.org/10.1021/acs.jpcc.5b09520
7. Lima C, Bieseki L, Melguizo PV, Pergher SBC (2019). Environmentally friendly zeolites. Springer International Publishing
8. Somna K, Jaturapitakkul C, Kajitvichyanukul P, Chindaprasirt P (2011) NaOH-activated ground fly ash geopolymer cured at ambient temperature. Fuel 90(6):2118–2124. https://doi.org/10.1016/j.fuel.2011.01.018
9. Anonymous (2002) Activated alumina and molecular sieves. http://www.cabestisrl.com.ar/Axens%20AluminaMolecular%20Sieves.pdf. Accessed: January 4, 2020

Organic Constituent, Color and Ammoniacal Nitrogen Removal from Natural Rubber Wastewater Using Kenaf Fiber

Nur Faizan Mohamad Rais, Zawawi Daud, Mohd Arif Rosli, Halizah Awang, and Amir Detho

Abstract The high quality and grade of latex or rubber skim for rubber industries products need to chemical added. Ammonia used in the preservation and processing ware contributed the high concentration of chemical oxygen demand (COD), color and ammoniacal nitrogen in natural rubber wastewater. Adsorption using agriculture material is one of new method in wastewater treatment for the removal pollutants and also environmentally friendly. In this study kenaf fiber has been used as adsorbent in wastewater treatment system and focus on COD, color and ammoniacal nitrogen removal from natural rubber wastewater. Alkali modification and acid modification on kenaf fiber were performed using 0.1 M sodium hydroxide (NaOH) and 0.6 M citric acid (CA). The efficiency of COD, color and ammoniacal nitrogen adsorption was investigated. Percentage removal of COD, color and ammoniacal nitrogen between kenaf fiber treated using NaOH and kenaf fiber treated with citric acid was compared. The maximum removal of COD, color and ammoniacal nitrogen were obtained at dosage 4 g of citric acid treated kenaf fiber with maximum adsorption is 80%, 75% and 86%, respectively. The removal of COD, color and ammoniacal nitrogen by kenaf fiber treated using citric acid more effective than the kenaf fiber which treated using NaOH in COD, color and ammoniacal nitrogen removal from natural rubber wastewater.

Keywords Kenaf fiber · Ammoniacal Nitrogen · Adsorption · Natural Rubber wastewater · COD and Color

---

N. F. M. Rais · Z. Daud (✉) · M. A. Rosli · H. Awang · A. Detho
Faculty of Civil Engineering and Built Environment, Universiti Tun Hussein Onn Malaysia, 86400 Parit Raja, Batu Pahat, Johor, Malaysia
e-mail: zawawi@uthm.edu.my

M. A. Rosli
e-mail: mohdarif@uthm.edu.my

H. Awang
e-mail: halizah@uthm.edu.my

N. Mohamed Noor et al. (eds.), *Proceedings of the 3rd International Conference on Green Environmental Engineering and Technology*, Lecture Notes in Civil Engineering 214,
https://doi.org/10.1007/978-981-16-7920-9_50

1 Introduction

Natural rubber, an elastic hydrocarbon polymer, originally derived from a milky colloidal suspension or latex of *Hevea brasiliensis.* The processing of raw natural rubber can be divided into two types of processes; the production of latex concentration and the production of Standard Malaysian Rubber (SMR) [1]. The Natural rubber processing sector is an industry which produces raw materials for manufactory of rubber industrial products, automotive products, latex product and many kinds of adhesives. The major used of natural rubber are tire and footwear industries [2]. Large amount of effluent was produced from raw natural rubber since it required large amount of water for its operation. In general, rubber wastewater has high concentration of ammonia, COD, BOD, nitrate, phosphorus and total suspended solid [3]. High concentration of nitrogen component in rubber wastewater, mainly the ammoniacal nitrogen as a result of the use of ammonia in the preservation to control the quality and grade of latex. High concentration of nitrogen and ammonia in natural rubber wastewater if discharge to water bodies, it could contribute to undesirable eutrophication and lead to death of some aquatic organisms living in the water [4].

Kenaf (*Hibiscus cannabinus L*), is one of the members of hibiscus family. Kenaf is a type of fast-growing plants which is capable of growing to over three meters within three months. Kenaf has become the choice in industry because this material is easily stored and stable in most condition [5].

In this study, kenaf fiber was treated with Sodium hydroxide (NaOH) and citric acid (CA) to produce adsorbent for the removal COD, color and ammoniacal nitrogen from natural rubber wastewater.

2 Materials and Method

2.1 Material

Kenaf core fibre was obtained from Lembaga Kenaf and Tembakau Negara, Malaysia. Kenaf with size of 150–300 μm was selected and used. Two types of chemical modification of kenaf were prepared. First, kenaf was modified with 0.1 M of NaOH and secondly, kenaf was modified with 0.6 M of citric acid. The NaOH and citric acid treatment of the kenaf fibres was performed according to previous method reported by Sajab et al. [6].

2.2 Method

The chemical compositions of kenaf fiber was analyzed by using TAPPI standards method T 222 om-88, chlorination method and Kurschner-Hoffner methods for determine lignin, hemicellulose and cellulose content from sample. In order to analyze the morphological features and surface structure, fiber will be tested using Field Emission Scanning Electron (FESEM).

The adsorption studies were carried out to determine the adsorption of COD, color and ammoniacal nitrogen from natural rubber wastewater by treated kenaf. The COD, color and ammoniacal nitrogen concentrations were determined using closed reflux and Colorimetric Method (5220-D). The adsorption studies was carried out with treated kenaf fiber dosage ranging 0.5 g to 8.0 g by remained the original pH of wastewater, speed and contact time at pH 8, 225 rpm and 120 min, respectively. Color measurement was reported as true color assayed at 455 nm using a HACH/DR6000 spectrophotometer and reported as platinum-cobalt (Pt–Co) method. The ammoniacal nitrogen measurement was reported as Nessler's method, sing an ultraviolet visible spectrophotometer (Model HACH DR6000). The present of COD, ammoniacal nitrogen and color removal was calculated using Eq. (1) where C_o and C_e are the initial and equilibrium concentration of COD and ammoniacal nitrogen (mg/L).

$$\text{Percentage removal}(\%) = \frac{(\text{Co} - \text{Ce})}{\text{Co}} \text{X100} \tag{1}$$

3 Results and Discussion

3.1 Characteristics of Adsorbent

The chemical kenaf fiber composition analysis such as cellulose, lignin and hemicelluloses were determined accordance with respective TAPPI standards method T 222 om-88 for lignin content, Kurcnher-Hoffner method for cellulose and chlorination method for determination of hemicellulose in sample [7]. From Table 1, determined that kenaf fiber contains high in cellulose content (52.4%), followed by 28.7% of hemicellulose contents. kenaf fiber gives low lignin content with 18.9%. Lower lignin

Table 1 Chemical composition of the kenaf fiber

Properties	Percentage (%)
Cellulose	52.4
Hemicellulose	28.7
Lignin	18.9

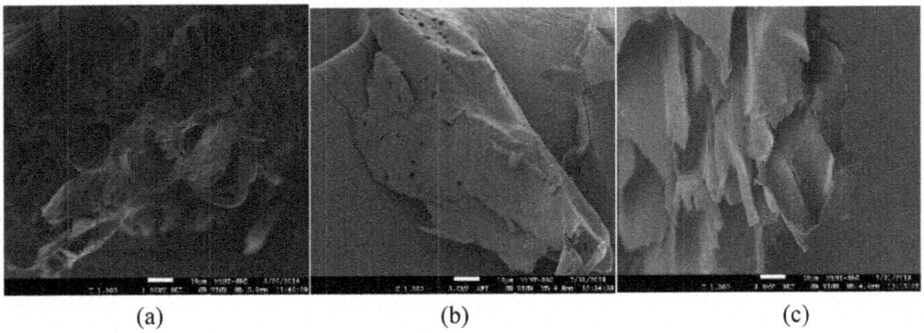

Fig. 1 FESEM micrographs of **a** untreated kenaf fibre **b** Alkali-treated kenaf **c** Acid- treated kenaf

content is normally found in nonwooden fiber, function as adhesive to bind the cellulose in fiber and makes the fiber strength increase and difficult break [8]. At the beginning of this study, one parameter was studied, with all the other parameters kept constant.

The morphology of the untreated kenaf and treated kenaf was observed and analyzed with Field Emission Scanning Electron (FESEM) [9]. Figure 1a–c shows the FESEM image micrograph of the untreated kenaf, alkali-treated kenaf by NaOH and Acid-treated kenaf by citric acid. The micrographs reveal the change occurred after treatments. The surface of untreated kenaf fiber looks rough and some has irregular pore and without smooth surface and lots of impurities (Fig. 1a). The alkali-treated kenaf fiber (Fig. 1b) had a smoother surface likely due to remove of certain amounts of wax, hemicelluloses, lignin and other impurities. The study by Siti [9] also yielded a similar result when fiber treatment with NaOH removed undesirable material and separated fiber bundle into individual fibers. However (Fig. 1c) acid-treated kenaf fiber show the surface has a cleaner and finer surface than alkali-treated kenaf fiber.

3.2 Comparison of NaOH Treated and Citric Acid Treated Kenaf Fiber

The COD, color and ammoniacal nitrogen adsorption experiments effect of dosage was conducted. Figure 2 the graph showed the percentage of COD removal by alkali-treated and acid-treated kenaf was increased when the dosage of absorbent was continuing increased until it was achieved the optimum value where the maximum percentage of COD for alkali-treated and acid-treated kenaf is 80% with 4 g dosage and 75% with 5 g dosage, respectively. However, when the optimum value was achieved, the percentage of removal will decrease when the dosage is increased. The study by Padmavathy et al. [10] also yielded a similar result whereas adsorbent dosage increases keeping all the other parameters at constant value removal efficiency first increases, reaches maximum and then decreases.

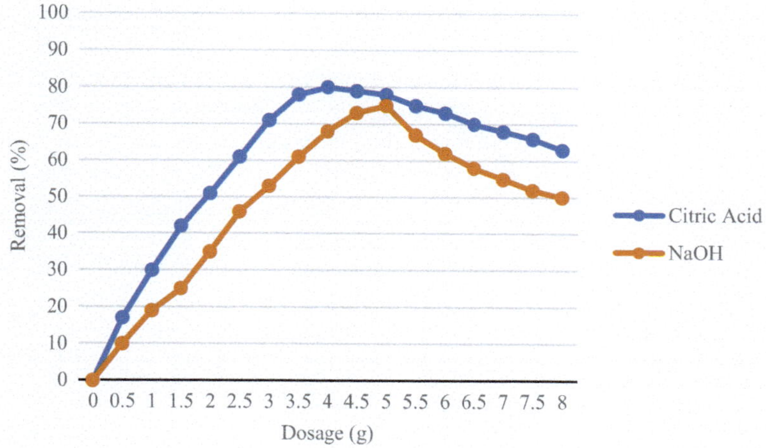

Fig. 2 Effects of adsorbent dosage in alkali-treated and acid-treated kenaf on COD nitrogen removal

Figure 3 shows the situation of percentage of ammoniacal nitrogen removal by acid-treated kenaf were similar with alkali-treated kenaf where the removal efficiency first increases, reaches maximum and then decreases. However, the higher percentage removal of ammoniacal nitrogen by acid-treated was higher than alkali-treated were obtained at dosage 4 g with percentage removal is 86%, 79%, respectively.

The effect of different optimum dosage varying from (0 g to 8 g) towards the percentage reduction of color as shown in Fig. 4. The reduction percentage efficiencies of color increases significantly with of adsorbent dosage from 0.5 g to 5.0 g. The best optimum result was considered at 4 g for acid-treated kenaf and 5 g for

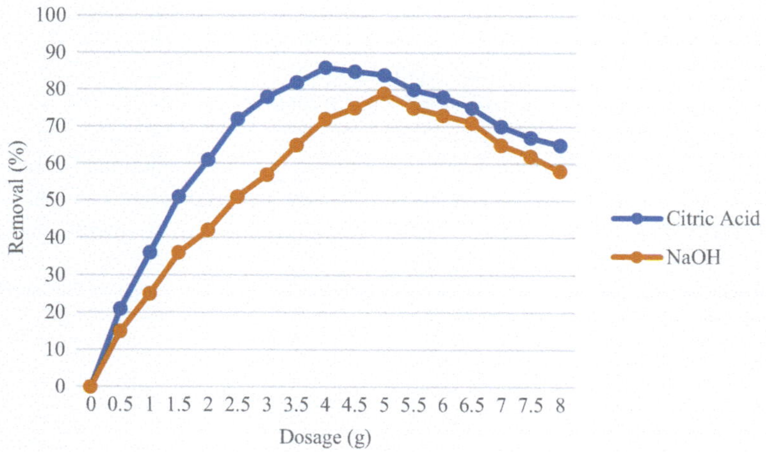

Fig. 3 Effects of adsorbent dosage in alkali-treated and acid-treated kenaf on Ammoniacal nitrogen removal

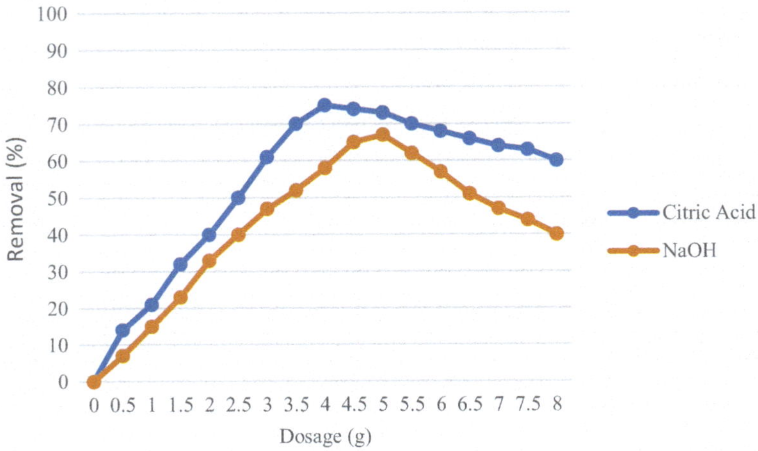

Fig. 4 Effects of adsorbent dosage in alkali-treated and acid-treated kenaf on color removal

alkali-treated kenaf with removal percentage of color were 75% and 67%, respectively. Subsequently, further increments in the dosage does not increase in removal percentage of factors because utmost all color was adsorbed in rubber effluent when further increment in dose was 4.5 g. More rapidly increasing in adsorption with increasing the amount of adsorbent due to higher surface area and accessibility more adsorption site [11].

4 Conclusions

Kenaf fiber have been successfully modified using citric acid. The citric acid treated kenaf fiber showed higher adsorption percentage removal towards COD, color and ammoniacal nitrogen than that of kenaf fiber treated using NaOH. The maximum adsorption of COD, color and ammoniacal nitrogen is 80%, 75% and 86%, at 4 g adsorbent dosage with pH 8 and 120 min of contact time were obtained from experiments. Kenaf fiber can be one of alternative adsorbent in wastewater treatment system since it is effective, environmentally friendly and can reduce environmental pollution.

Acknowledgements The authors acknowledge the financial support provided by the ministry of higher education through Mybrain scholarship.

References

1. Sulaiman NMN, Ibrahim S, Abdullah SL (2010) Membrane bioreactor for the treatment of natural rubber wastewater. Int J Environ Eng 2(1–3):92–109. https://doi.org/10.1504/IJEE.2010.029823
2. Jai SPHP, Girish K (2010) Rubber processing industry treatment using a bacterial consortium. Int J Curr Microbiol Appl Sci 3(10): 775–782. https://doi.org/10.2166/wst.2016.019
3. Noor HR, Aznah NA, Shreeshivadasan C, Mohd MD, Zaini U (2014) Characteristic and performance of aerobic granular sludge treating rubber wastewater at different hydraulic retention time. Bioresour Technol 161, 155–161. https://doi.org/10.1016/j.biortech.2014.03.047
4. Mohammadi M, Che Man H, Hassan MA, Yee PL (2010) Treatment of wastewater from rubber industry in Malaysia. Review Afr J Biotechnol 9(38):6233–6243
5. Wong YC, Lim SH, Atiqah NA (2013) Remediation of industry wastewater effluent by using kenaf as wax absorbent. Present Environ Sustain Dev 7(1):290–295
6. Sajad MS, Chin HC, Sarani Z, Saad MJ, Mohd KA, Kah Leong C, Poi SK, Wee SC (2011) Citric acid modified kenaf core fiber for removal of methylene blue from aqueous solution. Bioresour Technol 102: 7237–7243. https://doi.org/10.1016/j.biortech.2011.05.011
7. Zawawi D, Mohd ZMH, Angzzas SMK, Halizah A, Ashuvila MA (2014) Exploring of Argo waste (pineapple leaf, corn stalk, and Napier grass) by chemical composition and morphology study. BioResources 9(1):872–880
8. Tran AV (2005) Chemical analysis and pulping study of pineapple crown leaves. Ind Crops Prod 24:66–74
9. Yuli H, Ismojo EY, Azmi AN, Tina E, Mochamad C (2019) The effect of bleaching treatment on the mechanical strength of PP-Kenaf composite. AIP Conf Proceed 2175:020051. https://doi.org/10.1063/1.5134615
10. Padmavathy KS, Madhu G, Haseena PV (2016) A study on effects of pH, adsorbent dosage, time, initial concentration and adsorption isotherm study for removal of hexavalent chromium (Cr (VI)) from wastewater by magnetic nanoparticles. Sci Direct Procedia Technol 24:585–594. https://doi.org/10.1016/j.protcy.2016.05.127
11. Kamil R, Sarra B, Selma H (2013) Adsorption of methylene blue from aqueous solution by kaolin and zeolite. Appl Clay Sci 83–84:99–105. https://doi.org/10.1016/j.clay.2013.08.015

Potential of Pretreated Spent Coffee Ground as Adsorbent for Oil Adsorption

Nur Farhana Najwa Nasaruddin, Hairul Nazirah Abdul Halim, Siti Khalijah Mahmad Rozi, Zulfakar Mokhtar, Lian See Tan, and Nurfatehah Wahyuny Che Jusoh

Abstract Spent Coffee Ground (SCG) has the potential of becoming a low-cost adsorbent for oil removal. However, the effectiveness of SCG as an oil adsorbent is limited by its hydrophilicity. In this work, chemical pretreatment of SCG with acid and alkali was evaluated as a means of enhancing its performance for adsorbing palm cooking oil. The surface morphology of raw and pretreated SCG samples was characterised using a Scanning Electron Microscope (SEM). The functional groups were analysed using Fourier Transform Infrared (FTIR) spectroscopy. A comparison between adsorption performances was performed in terms of contact time in batch adsorption experiments to determine the adsorption capacity of SCG-based adsorbents with oil. The SEM results showed that more developed pores were present on the acid- and alkali-treated SCG samples compared to on the raw SCG sample, while the FTIR results indicated the elimination of hydrophilic cellulose and hemicellulose on these samples. According to these results, both acid- and alkali-treated SCG have better tendencies of adsorbing oil compared to raw SCG. The

N. F. N. Nasaruddin · H. N. A. Halim (✉) · S. K. M. Rozi · Z. Mokhtar
Faculty of Chemical Engineering Technology, Universiti Malaysia Perlis, Kompleks Pusat Pengajian Jejawi 3, 02600 Arau, Perlis, Malaysia
e-mail: hairulnazirah@unimap.edu.my

S. K. M. Rozi
e-mail: khalijahrozi@unimap.edu.my

Z. Mokhtar
e-mail: zulfakar@unimap.edu.my

H. N. A. Halim
Centre of Excellence for Biomass Utilization, Universiti Malaysia Perlis, Kompleks Pusat Pengajian Jejawi 3, 02600 Arau, Perlis, Malaysia

L. S. Tan · N. W. C. Jusoh
Department of Chemical and Environmental Engineering (CHEE), Malaysia-Japan International Institute of Technology (MJIIT), Universiti Teknologi Malaysia (UTM), Jalan Sultan Yahya Petra (Jalan Semarak), 54100 Kuala Lumpur, Malaysia
e-mail: tan.liansee@utm.my

N. W. C. Jusoh
e-mail: nurfatehah@utm.my

N. Mohamed Noor et al. (eds.), *Proceedings of the 3rd International Conference on Green Environmental Engineering and Technology*, Lecture Notes in Civil Engineering 214,
https://doi.org/10.1007/978-981-16-7920-9_51

maximum oil adsorption capacity (2.549 g/g) occurred after 150 min of contact time between alkali-treated SCG adsorbents and oil.

Keywords Biomass adsorbent · Acid and alkali pretreatment · Oil adsorption capacity

1 Introduction

The beverage industry in Malaysia is dominated by coffee drinks based on an analysis on beverage consumptions in 2016. With approximately 60.83% of respondents claiming to drink coffee daily [1], huge amounts of SCG, which are solid residues that are collected after the ground coffee beans have been brewed, would be disposed of as either domestic or industrial effluent [2]. As the waste of a high commodity product, SCG are readily available at a low cost, which allows them to be exploited for other purposes.

With the threat of climate change, researchers are working hard at finding ways to mitigate oil pollution problems that arise from industrial activities. Several technologies have been implemented, such as adsorption using commercial adsorbents for the removal of oil from industrial effluent. Alternative adsorbents based on biomass have also been studied, such as rice husks, kapok fibre, and potato peels [3] in the effort to reduce the expensive cost of commercial adsorbents. However, the utilisation of SCG as oil adsorbent has rarely been reported. The moisture content of this potential adsorbent may range between 50% w/w and 85% w/w, which can be influenced by the brewing method [2]. In addition, SCG present a higher water holding capacity than other biomass adsorbents, such as rice husk, sugar cane bagasse and corn crops because SCG comprise of higher fibre content. Raw SCG have higher water holding capacity (5.73 ± 0.10) compared to oil holding capacity (5.20 ± 0.30) [4].

The modifications of adsorbents using acid and alkali pretreatments are some of the techniques in improving surface polarity and hydrophobicity. A recent study by [5] reported that after coking coal was pretreated with 1 M H_2SO_4, increases in surface area, pore volume of micropore, as well as hydrophobic properties of the adsorbent were observed. According to [6] and [7], an alkali pretreatment can increase an adsorbent's capacity for adsorbing non-polar materials by increasing its alkali groups and changing its surface polarity. For example, [7] reported that the surface properties and adsorption capacity of activated carbon was enhanced via NaOH pretreatment. Therefore, this study has focused on utilising acid and alkali pretreatment to enhance the adsorption capacity of SCG for oil. A comparative study on the raw and pretreated SCG samples was conducted to analyse any improvements on adsorption performance.

2 Materials and Method

2.1 *Material and Chemical*

In this work, spent coffee grounds were used as adsorbents, which were collected from McDonald's Kangar, Perlis. Concentrated sulphuric acid (H_2SO_4, 96%) and sodium hydroxide pellets (NaOH) were purchased from Fisher Scientific, Malaysia, and were used in the SCG pretreatment process. Palm cooking oil was used as the adsorbate in this study.

2.2 *Method*

2.2.1 Acid and Alkali Treated Spent Coffee Ground

First, the SCG were rinsed with distilled water and placed in an oven (Binder, Malaysia) for 24 h at 70 °C. The dried sample was sieved into uniform particle of size (500 μm - 2 mm). Then, 20 g of the dried SCG was immersed in 100 mL of 1 M H_2SO_4 and continuously stirred on a hotplate for 1 h, at room temperature. Afterwards, the sample was left unstirred for 24 h. Acid residue was repeatedly rinsed out of the sample using distilled water until neutral pH was achieved, and then, the clean sample was dried at 70 °C overnight. The SCG were also treated with 1 M NaOH under similar conditions [8, 9].

2.2.2 Characterization of Adsorbents

A scanning electron microscope (SEM) (Jeol, Japan) was used for the raw, and the acid- and alkali-treated SCG samples to analyse their surface textures and the development of porosity. The functional groups were analysed using Fourier Transform Infrared (FTIR) spectroscopy (Perkin Elmer, Massachusetts, USA).

2.2.3 Comparison on the Performance of Adsorbents

The performances of raw, and acid- and alkali-treated SCG samples were compared in terms of different contact times in a batch adsorption process. The experiments were conducted for 30–240 min, with an increment of 30 min for every time interval [10]. Approximately 1 g of each sample was weighed and wrapped in a 5 × 7 cm nylon teabag, which was immersed in 20 g of oil in a 100 mL glass beaker without agitation. The experiments were conducted at different contact times at a constant room temperature. After a specific time, the teabag was taken out from the glass beaker and placed on a filter funnel for 15 min to let the oil drip off. Then, the sample was weighed, and the adsorption capacity (g/g) was calculated using the following Eq. (1) [11]. Each experiment was conducted three times under identical conditions.

$$q = \frac{M_F(g) - M_I(g)}{M_I\ (g)} \tag{1}$$

where,

M_I is the mass of adsorbent before adsorption and

M_F is the mass of adsorbent after adsorption

3 Results and Discussion

3.1 *Characterization of Adsorbents*

3.1.1 Textural Characterization of Raw, Acid and Alkali Treated SCG

Figure 1A–C shows the textural characterisation of the raw SCG, acid-treated SCG, and alkali-treated SCG, respectively. The SEM image of raw SCG showed a fairly dense morphology composed of thin sheets of materials that resembled sawdust, as similarly reported by [4]. However, large pores were clearly seen on the surfaces of the acid- and alkali-treated SCG samples. These results showed that H_2SO_4 and NaOH treatments were effective for creating well-developed pores on the surface of SCG [8, 12].

3.1.2 FTIR Analysis

Figure 2A-C shows the FTIR spectra of raw SCG, and acid- and alkali-treated SCG. The spectrum of raw SCG in Fig. 2A shows a fairly broad peak between 3,600 and 3,200 cm^{-1}, which can be relate to the hydroxyl group of O–H stretching vibration. Meanwhile, the bands at 2968 and 2897 cm^{-1} corresponded to the asymmetric stretching of C–H bonds of the methyl group ($-CH_3$) in the caffeine molecule [4, 13]. The small bands located at 1788 and 1751 cm^{-1} can be assigned to the $C = O$ stretching of hemicellulose. These results showed that the raw SCG sample contained several functional groups, such as hydroxyl, alkanes, and carbonyls.

Figure 2B shows the spectrum of acid-treated SCG sample. Similar bands between 3,600 and 3,200 cm^{-1} appears to be less intense, very broad, and has shifted slightly compared to the band for the raw SCG sample (see Fig. 2A). This difference showed reduced hydrogen bonds in cellulosic hydroxyl groups, which decreased the hydrophilic behaviour of pretreated SCG. Both peaks at 2924 and 2858 cm^{-1} can be attributed to the C–H stretching vibration. A similar observation was reported

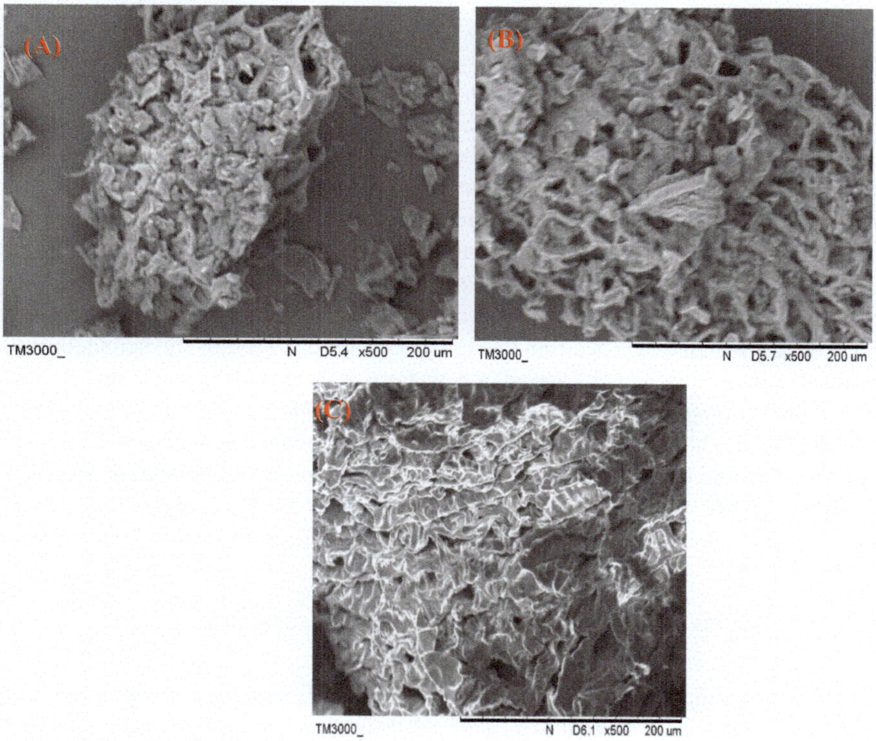

Fig. 1 SEM images at 500 × magnification (**A**) of raw SCG (**B**) acid-treated SCG (**C**) alkali-treated SCG

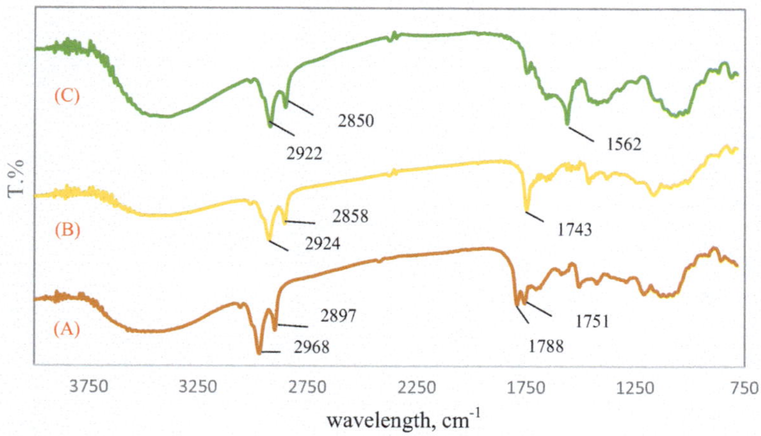

Fig. 2 FTIR spectra of **A** raw SCG **B** acid treated SCG **C** alkali treated SCG

in another study [14]. The new peak at 1743 cm^{-1} can be associated to the $C = O$ stretching vibrations. This band is related to the formation and increase of carbonyl groups via the degradation of β-linkages by acid treatment [14].

From the spectrum of alkali-treated SCG, as shown in Fig. 2C, a similarly broad, and the most intense band can be seen between 3,600 and 3,200 cm^{-1}, compared to the bands of the other two samples. This shift could be due to the hydrolysis effect of the NaOH treatment. The consequent breakdown of ester or ether bonding led to the removal of lignin or hemicelluloses from SCG and created a free O–H group [15]. Double peak at 2922 and 2850 cm^{-1} can also be seen, which represented the C–H stretching vibration. However, it can be observed that carbonyl band was not present in the alkali-treated SCG spectrum due to the dissolution of lignin and hemicellulose in the alkaline solution [14]. Instead, a new, fairly strong peak at 1562 cm^{-1} can be seen, which corresponded to $C = C$ stretching. This phenomenon showed that alkali-treated SCG might have a greater oil adsorption performance than acid-treated SCG and raw SCG due to the elimination of hemicellulose, which consists of hydrophilic functional group.

3.2 Comparison on the Performance of Adsorbents

Figure 3 shows the effect of contact time on the oil adsorption capacity of the raw, acid-treated, and alkali-treated SCG samples. Minimum adsorptions were recorded at the earliest of 30 min of contact time for all samples (0.201 g/g for raw, 0.375 g/g for acid-treated, and 0.611 g/g for alkali-treated SCG). Then, a slow increment can be observed between 30 to 90 min of contact time. This observation could be linked

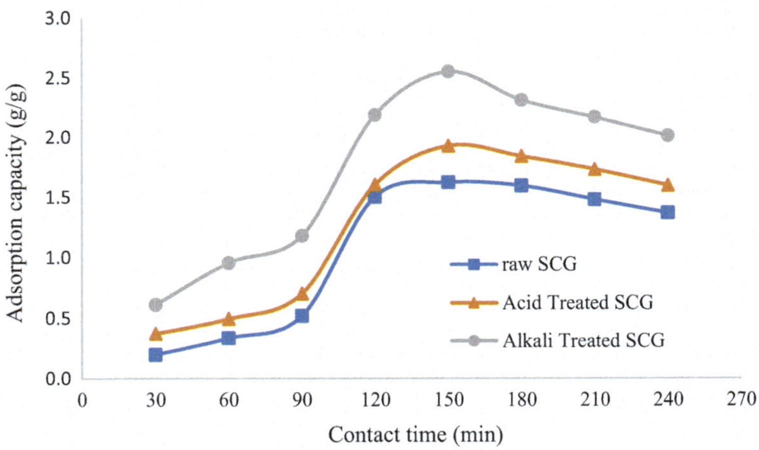

Fig. 3 The effect of adsorption capacity by manipulating the contact times (min)

to the external diffusion of palm oil on the surface of the sorbent, which was just starting to break through the pores [16].

Figure 3 shows that adsorption capacity has significantly increased between 90 to 120 min of contact time for all samples. Maximum adsorptions were recorded at 150 min for acid-treated (1.930 g/g), alkali-treated (2.549 g/g), and raw (1.625 g/g) SCG samples. This behaviour was progressively observed due to the internal diffusion of oil into the pores of the adsorbents, which became significantly enhanced once external diffusion was established within 30 to 90 min of contact time. At maximum adsorption capacity, available sites on the adsorbent surfaces would mainly be occupied by oil. However, starting from 150 min onwards, a slight decrement in adsorption capacity was observed for all samples. This is due to the desorption of oil from the binding sites. Previous studies have reported similar results for the effect of contact time on oil adsorption using *lansium parasiticum* shell and chicken feather [17, 18].

The results showed that alkali-treated SCG were the best adsorbents compared to the raw and acid-treated SCG. Based on the results obtained at 150 min of contact time, alkali-treated SCG showed 1.57 times higher improvement percentage compared to raw SCG. This conclusion was made based on the FTIR results that indicated the elimination of hydrophilic hemicellulose from the alkali-treated SCG, leading to a greater oil adsorption performance.

4 Conclusion

The potential use of SCG as adsorbent for oil removal was studied by treating the SCG with H_2SO_4 and NaOH. The SEM results showed that more pores had developed on acid- and alkali-treated SCG compared to on the raw SCG sample. The FTIR results indicated that the hydrophobic functional groups were enhanced following the acid treatment due to the reduction in hydrogen bonding in the cellulosic hydroxyl groups. Meanwhile, the alkali treatment has led to the elimination of hydrophilic hemicellulose, which then resulted in a decreased hydrophilic behaviour and increased hydrophobic behavior. Both acid- and alkali- treated SCG samples showed enhanced oil adsorption capacity when compared to raw SCG sample. Maximum oil adsorption capacity was obtained by the acid-treated SCG sample (1.930 g/g) and alkali-treated SCG sample (2.549 g/g) at 150 min of contact time. Therefore, it was concluded that both H_2SO_4 and NaOH pretreated SCG can potentially be used as oil removal sorbents.

Acknowledgements The authors would like to acknowledge National Collaborative Research Grant National (CRG) (4B514 and 9023-00014) for sponsoring research activities.

References

1. Rahim FA, Goh PJ, Cheah LF (2019) Malaysian Coffee Culture: Attributes considered to purchase coffee beverages. J. of Market. Adv. and Practice 1(1), 50–62. Retrieved from http://jmaap.org
2. Efthymiopoulos I, Hellier P, Ladommatos N, Kay A, Mills-Lamptey B (2019) Effect of solvent extraction parameters on the recovery of oil from spent coffee grounds for biofuel production. Waste Biomass Valor 10:253–264. https://doi.org/10.1007/s12649-017-0061-4
3. Doshi B, Sillanpaa M, Kolliola S (2018) A review of bio-based materials for oil spill treatment. Water Resour 135:262–277. https://doi.org/10.1016/j.watres.2018.02.034
4. Ballesteros LF, Teixeira JA, Mussatto SI (2014) Chemical, functional, and structural properties of spent coffee grounds and coffee silverskin. Food Bioprod Technol 7:3493–3503. https://doi.org/10.1007/s11947-014-1349-z
5. Gao L, Wen H, Tian Q, Wang Y Li G (2017) Influence of surface modification by sulfuric acid on coking coal's adsorption of coking wastewater. Water Sci Technol 76(3): 555–566. https://doi.org/10.2166/wst.2017.219
6. Liu M, Xiao C (2018) Research progress on modification of activated carbon. In: E3S Web of Conferences 38, 02005 EDP Sciences. https://doi.org/10.1051/e3sconf/20183802005
7. Hayati B, Mahmoodi NM (2012) Modification of activated carbon by the alkaline treatment to remove the dyes from the wastewater: mechanism, isotherm and kinetic. Desalinat Water Treat 47(1–3):322–333. https://doi.org/10.1080/19443994.2012.696429
8. Halim HNA, Yatim NSM (2011) Removal of acid green 25 from aqueous solution using coconut husk as adsorbent. Int Conf Environ Industr Innov 12: 268–72. http://ipcbee.com
9. Pauzan ASM, Ahad N (2018) Biomass modification using cationic surfactant Cetyltrimethylammonium Bromide (CTAB) to remove palm-based cooking oil. J Chem 2018:1–8. https://doi.org/10.1155/2018/5059791
10. Gopalakrishnan Y, Al-gheethi A, Malek MA, Azlan MM, Al-Sahari M, Mohamed RMSR, Alkhadher S, Noman E (2020) Removal of basic brown 16 from aqueous solution using Durian shell adsorbent, optimisation and techno-economic analysis. Sustainability 12(21): 1–22. https://doi.org/10.3390/su12218928
11. Ciufu AG, Raducanu C, Parvulescu OC, Cioroiu DR, Dobre T (2019) Natural wool for removal of oil spills from water surface. Rev Chim (Bucharest) 70(11): 3977–3980. http://revistadechimie.ro
12. Pirbazari AE, Saberikhah E, Badrouh M, Emami MS (2014) Alkali treated Foumanat tea waste as an efficient adsorbent for methylene blue adsorption from aqueous solution. Water Res Indust 6:64–80. https://doi.org/10.1016/j.wri.2014.07.003
13. Cerino-Cordova FJ, Davila-Guzman NE, Garcia-Leon AM, Salazar-Rabago JJ, Sot Regalado E (2020) Revalorization of coffee waste. Coffee-Product Res 1:1–26. https://doi.org/10.5772/intechopen.92303
14. Díaz-Muñoz LL, Bonilla-Petriciolet A, Reynel- Ávila HE, Mendoza-Castillo DI (2016) Sorption of heavy metal ions from aqueous solution using acid-treated avocado kernel seeds and its FTIR spectroscopy characterization. J Molec Liq 215:555–564. https://doi.org/10.1016/j.molliq.2016.01.022
15. Loganthan TM, Sultan MTH, Ahsan Q, Jawaid M, Naveen J, Shah AUM, Hua LS (2020) Characterization of alkali treated new cellulosic fibre from *Cyrtostachys renda*. J Mater Res Tech 9(3):3537–3546. https://doi.org/10.1016/j.jmrt.2020.01.091
16. Anastopoulos I, Karamesouti M, Mitropoulos AC, Kyzas GZ (2017) A review for coffee adsorbents. J Mol Liq 229:555–565. https://doi.org/10.1016/j.molliq.2016.12.096
17. Kurniawati D, Bahrizal STK, Adella F, Salmariza S (2021) Effect of contact time adsorption of Rhodamine B, Methyl Orange and Methylene blue colours on Langsat shell with batch methods. J Phys Conf Ser 1788:1–6. https://doi.org/10.1088/1742-6596/1788/1/012008
18. Okoya AA, Ochor NO, Akinyele AB and Olaiya OO (2020) The use of chicken feather as an adsorbent for crude oil clean up from polluted water. J Agri Econ Res Int 21(3): 43–53. https://doi.org/10.9734/jaeri/2020/v21i330136

Rainwater Harvesting System: Design Performances of Optimal Tank Size Using Simulation Software

Farisya Aliya Hilmi and Azianabiha A Halip Khalid

Abstract Rainwater harvesting (RWH) system has become a high potential source of water supply in urban areas. It is a practical solution to combat water crisis and can also contribute to reduce a storm's peak flow. This research aims to determine the feasibility and design performance of RWH for Kolej Melati, UiTM Shah Alam, Selangor. The performances of various storage tank sizes were calculated using Tanki NAHRIM 2.0. Finding from the software simulation was used to select the optimum tank size which provide highest water saving and reliability. The monthly analysis of the RWHS shows promising results of collectable water on the demand, in which the average reliability was higher than 50%. The result showed that the best tank size Kolej Melati is 100 m^3. About 46.6% of rainwater from roof can be used that provide 83.7% of water demand. This amount is equivalent to 2639.56 m^3 to fulfill the water requirement for 297 days per year. Therefore, the proposed implementation of RWH system in Kolej Melati is highly viable and the outcome could support green campus initiative by the university.

Keywords Rainwater harvesting · Optimum tank size · Tangki NAHRIM · Water saving efficiency

1 Introduction

Water availability is important for the health and productivity of the environment, ensuring the continuous supply of a variety of goods and services to support human well-being. Our earth is comprised of 71% covered in water which are mostly found in oceans and large water bodies. Water has been used by humans for various purposes either outdoor or indoor such as for drinking, watering garden and transportation. With such increasing in urbanization, water has been one of the highest demands and there is greater pressure to conserve existing water supplies [1]. According to

F. A. Hilmi · A. A H. Khalid (✉)
School of Civil Engineering, College of Engineering, Universiti Teknologi MARA, 40450 Shah Alam, Selangor, Malaysia
e-mail: azianabiha@uitm.edu.my

N. Mohamed Noor et al. (eds.), *Proceedings of the 3rd International Conference on Green Environmental Engineering and Technology*, Lecture Notes in Civil Engineering 214,
https://doi.org/10.1007/978-981-16-7920-9_52

Water World Council [2], the world population will rise by another 30–40% for the next fifty years and hence contribute to an increase of water demands which later will have significant outcome on the environment.

Malaysia is one of the countries that receive high rainfall in which the annual average rainfall for Peninsular Malaysia, Sabah and Sarawak are 2420 mm, 2630 mm and 3830 mm respectively. The high volume of rainwater in Malaysia has been taken as an alternative source to supply water by using rainwater harvesting system. However, the continuous growth of population and increasing in urbanization and industrialization have increase the demand of water. Therefore, in the planning and management of available water supplies, the principle of sustainability must be considered. In Sustainable Development Goal (SDG 6) has emphasizes the importance of conserving and handling water resources in a sustainable manner. The Malaysian government has taken many constructive steps to resolve the issue of water scarcity, including recycling and reuse of grey or waste water, increasing water tariffs, raising public awareness and encouraging the use of RWH system [3].

Besides, Malaysia has endured droughts and floods in various areas, so studying the feasibility of alternative water supplies in Malaysia to manage and preserve the sustainability of urban townships is important [4]. For the utilisation of rainwater as a water resource, the use of appropriate rainwater harvesting technology is significant. Rainwater harvesting system able to solve the issue of water shortage at the same time and minimise dependency on domestic water supply [5].

Rainwater Harvesting (RWH) system is a method to collect or store rainwater from roof as a catchment which later can be used as non-potable purpose such as landscape irrigation, toilet flushing and laundry [6]. It is also known as one of the sustainable methods as it emphasizes the using of natural resources. It can be used as potable usage but the microbial component needs to be reduced or removed with a proper treatment [7]. Even though it needs to be treated to satisfy legal requirements, rainwater has always been known to have considerable potential to add to the supply. This paper presents a study on the feasibility of RWH for Kolej Melati UiTM, Shah Alam and the design performance based on the optimum tank size by using Tangki NAHRIM 2.0.

2 Method

2.1 Sampling Location

UiTM campus Shah Alam was selected as the location of this study. A residential college of Block 4B of Kolej Melati (Fig. 1) was chosen to identify the optimum tank size. It is inevitable to develop RWH technology with one of its most important components, precipitation must be considered. A secondary data of hydrological data of rainfall was obtained from Drainage and Irrigation Department Malaysia (DID).

Fig. 1 Melati residential college, UiTM Shah Alam

The previous 15 years of rainfall data from the year 2003 to the year 2017 were used in the study.

2.2 *Tangki NAHRIM 2.0*

The latest Tangki NAHRIM 2.0 software was used to determine the optimum tank size for the selected location and to assess the viability of using rainwater tanks as a source of water addition. There are specific inputs such as rainfall data, roof information, water demand and the tank capacity needed to be filled in to get the optimum tank size for rainwater harvesting. The steps of the design are as the following:

2.2.1 Rainfall Data

To begin the analysis, UiTM Shah Alam rainfall station (Station No: 3014091) was selected. Table 1 shows the average annual and monthly rainfall. These rainfall data were used to analyse the potential of rainwater harvesting. Potential of rainwater harvesting can be calculated using the following equation [8]:

$$PRH = A_{RT} \times RC \times RI \tag{1}$$

Table 1 Average rainfall for station no: 3014091 per year and month

Year	Rainfall (mm)	Month	Rainfall (mm)
2003	1912.5	January	222.3
2004	2557.5	February	167.5
2005	1977.9	March	198.7
2006	3107.6	April	235.1
2007	2002.8	May	170.7
2008	3309.2	June	129.9
2009	2451.7	July	162.9
2010	2416.1	August	137.1
2011	2122.6	September	173.6
2012	2247.8	October	212.2
2013	1833.6	November	321.3
2014	2344.0	December	266.6
2015	2595.0		
2016	2319.6		
2017	2471.5		

Table 2 Runoff coefficient for roofing material [9]

Type of roof	Runoff coefficient
Galvanized iron sheet	>0.9
Tiles	0.8–0.9
Concrete	0.6–0.8

where, *PRH* is the potential rainwater harvesting (m^3), *A* is the area of the rooftop (m^2), *RC* is the runoff coefficient (−), and *RI* is the rainfall (m).

2.2.2 Roof Information

Block 4B Kolej Melati has approximately 1026 occupants. The roof measurements considered based on Google Earth were 104 m long and 31 m wide. Thus, resulting the catchment area to be about 3224 m^2. The rooftop runoff coefficient varies depending on the roof's material and slope as presented in Table 2. The building has runoff coefficient of 0.8 for concrete tiles material. First flush used was 1 mm as recommended by NAHRIM.

2.2.3 Water Demand

The required water demand for domestic use can be referred in guidelines provided by Urban Stormwater Management Manual (MSMA 2nd Edition) by Department of

Table 3 Rainwater consumption for domestic use [10]

Application	Average consumption	Average total rainwater demand
Toilet - Single flush - Dual flush	9 l/flush 6 or 3 l/flush	120 l/day 40 l/day
General cleaning	10–20 l/min	150 l/day
Sprinkler/ Handheld hose	10–20 l/min	1000 l/hour
Washing car	10–20 l/min	100–300 l/wash

Table 4 Water demand litres per day for block 4B Kolej Melati

Application	Unit	Average water used	Total water use (litres/day)
Single flush toilet with 9 l/flush	72 no	120 l/day	8640

Irrigation and Drainage (DID) as shown in Table 3. The toilet flushing was considered as the major non-potable used for the residential college. The total water demand is 8640 L/day. Therefore, the annual rainwater demand is 3153.6 m^3/year. The estimated water consumption was calculated in Table 4.

2.2.4 Tank Capacity

The maximum volume of water captured from the roof area to the rainwater harvesting system is used to construct the rainwater tank. The assumption on the tank size was based on previous research with almost similar amount of water demand. The smallest size to be about 10 m^3 and largest size 100 m^3 with an increment of 10 m^3 [11].

2.2.5 Reliability

Volumetric reliability is defined as the total volume of captured rainwater supplied divided by the total water demand and time reliability is the percentage of time when demand is fully met [12]. The following formula were used to determine the time and volumetric based reliability:

$$ReliabilityRatio = \frac{Deliverd\ Value}{Demand\ Value} \quad (2)$$

$$TR = 1 - \frac{df}{n} \times 100 \quad (3)$$

where *TR* is the time reliability, *df* is the number of failure days and *n* is the total number of days.

3 Result and Discussion

3.1 Feasibility of RWH Based on Hydrological Data

Before evaluating rainwater harvesting as a prospective source of water supply addition, it is critical to obtain and analyse rainfall data for at least the previous 10 years to obtain a good grasp of the rainfall pattern in a given location [13]. Precipitation data is required to calculate the volume of water that can be collected, considering rain depth and roof area, and then comparing that to the volume of demand needed. According to the rainfall data provided by Department of Irrigation and Drainage for year 2003 to 2017, the rainfall recorded are starting from January to December every year. From the data given, the pattern of the rainfall at station UiTM Shah Alam has be obtained.

The variation of historical data of annual rainfall for UiTM Shah Alam station is shown in (Fig. 2) where the mean annual rainfall is 2378 mm. The highest annual rainfall intensity recorded was in the year 2008, which is 3309.2 mm. The second highest rainfall is in 2006 with 3107.6 mm. Given the high seasonal variations, rainwater harvesting system will contribute to ensuring water supply for domestic and other needs [14]. All areas in Malaysia are subject to heavy rainfall with 80% a year which is between 2000 to 2500 mm of rain each year [15]. Hence, it is feasible for implementing RWH system as the intensity is more than 2000 mm. The result for the volume of the rainwater that can be harvest in a month using the Eq. (1)

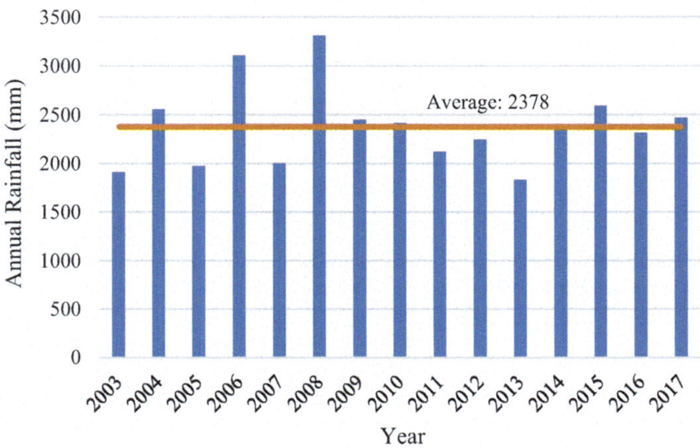

Fig. 2 Average annual rainfall for year 2003 to 2017 for UiTM Shah Alam station

Table 5 Potential of rainwater harvesting per month for catchment area of 3224 m^2

Month	Rainfall (m)	Coefficient	Catchment area (m^2)	Potential rainwater harvesting (m^3)
January	0.2223	0.8	3224	573.36
February	0.1475	0.8	3224	380.43
March	0.1987	0.8	3224	512.49
April	0.2351	0.8	3224	606.37
May	0.1707	0.8	3224	440.27
June	0.1299	0.8	3224	335.04
July	0.1629	0.8	3224	420.15
August	0.1371	0.8	3224	353.61
September	0.1736	0.8	3224	447.75
October	0.2122	0.8	3224	547.31
November	0.3213	0.8	3224	828.70
December	0.2666	0.8	3224	687.61

showed that November and December were the highest with 828.70 m^3 and 687.61 m^3 respectively (Table 5). This is because these months falls in northeast monsoon season.

When assessing the possibility of implementing rainwater harvesting systems, some factors such as local water demand, cost feasibility, and precipitation are considered to ensure the success of the system. The process for sizing a rainwater harvesting tank is based on historical rainfall data that may be used to estimate the amount of rainwater runoff gathered from the chosen roof [16]. Due to its location in the humid tropics, Malaysia has a high number of rainy days (138 to 181 days per year) thus the RWH system should be used to its full potential in order to save the most water in the reservoir that will be used during the dry season [8].

3.2 *Design Performance Analysis*

3.2.1 Water Saving and Storage

The storage and water saving efficiency for 8640 L per day of water demand for toilet flushing purpose is shown in Fig. 3. The water-saving and storage efficiency would grow with tank capacity for sizes ranging between 10 m^3 and 100 m^3 with the efficiency varied in the range from 35.9% to 83.7% and 19.9% to 46.6% respectively. Despite the small percentage variations, they resulted in water savings of several cubic meters each year and enhance the water supply network. In terms of cost, it was more cost effective to choose smaller tank considering economic purpose [13].

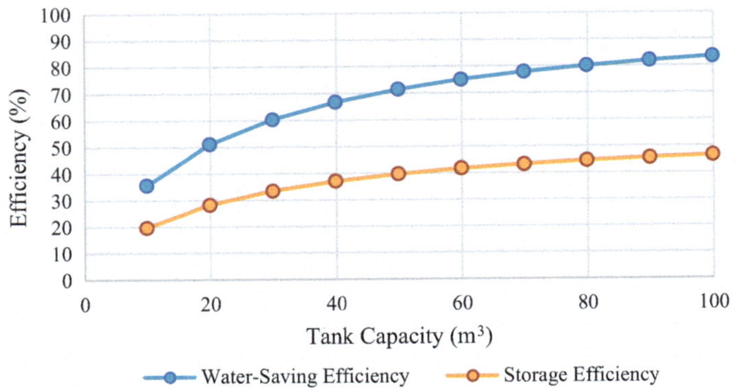

Fig. 3 Water saving and storage efficiency for tank size ranging from 10 m^3 to 100 m^3

Water saving efficiency is also known as volume reliability. It is calculated by referring on Eq. (2). The tank volumetric reliability (%) was used as a measure of the tank performance to assess the rainwater tank's water saving efficiency [17]. Based on water-saving efficiency curve, it shows that harvested rainwater can satisfied more than 80% of total water demand for tank capacity of 80 m^3 and larger. However, increasing the size of the tank by 10 m^3 for the given ranging capacity will only saves less than 5% of the total water consumption. A 50 m^3 storage capacity could provide 71.7% water demand which can save 1132.14 m^3 of water per year.

One of the key motivations for RWH is to save water, particularly in urban areas where fresh water is scarce [17]. Table 6 shows the volume of water that can be saved per year by implementing RWH system. It was found that for tank capacity of 20 m^3 and larger, it can provide more than half of the water demand which is 1576.80 m^3/year. Therefore, smaller tank capacity can be considered to be used as water saving efficiency is met nearly above half (50%) of the demand.

Table 6 Volume of water (m^3/year) that can be saved for different tank sizes

Tank capacity (m^3)	Water demand (m^3/year)	Water saving efficiency (%)	Water saved (m^3/year)
10	3153.6	35.9	1132.14
20	3153.6	51.3	1617.80
30	3153.6	60.6	1911.08
40	3153.6	67.0	2112.91
50	3153.6	71.7	2261.13
60	3153.6	75.3	2374.66
70	3153.6	78.1	2462.96
80	3153.6	80.3	2532.34
90	3153.6	82.2	2592.25
100	3153.6	83.7	2639.56

Storage efficiency curve is generally low compared to water saving efficiency curve. This is due to the amount of rainfall is quite high, so the low water demand causes spillage. It shows that almost half percent can be achieved if the tank capacity is 90 m^3 and above. To properly utilize the high amount of rainfall volume, the water demand must be increased to improve storage efficiency [13]. The tank size with reliability of more than 70% can be selected as optimum tank size [11]. Therefore, the tank capacity of 100 m^3 is chosen as an optimum size for Kolej Melati, UiTM Shah Alam. About 83.7% of water demand can be provided which equivalent to 2639.56 m^3 of water per year and 46.6% of rainwater from roof can be used.

3.2.2 Yield Spillage

A behavioural model was used to assess the effectiveness and performance of rainwater collection systems. This Tangki NAHRIM 2.0 software adopting a yield after spillage (YAS) model in which the concept is rainwater is collected until the storage is full. From the data provided, volume and time reliability can be analyse. Figures 4 and 5 show the volume and number of yield and spillage in each tank size for average annual and number of days. The reliability of the time is calculated using Eq. (3) and the result is tabulated in Table 7.

Based on Fig. 4, the volume of spill is higher than the volume of yield. However, the spill shows decreasing over the increasing of tank capacity while yield shows increasing in pattern. The YAS rule states that the yield is the lesser of the demand and supply [18]. From tank size 80 m^3 up to 100 m^3, the yield shows slightly consistent increase in which it matched with the water saving efficiency pattern. Whereas the spillage consistent with the trend storage efficiency. Thus, a larger tank should be provided so that the volume of spilled water can be minimized [19].

As shown in Fig. 5 is the time reliability. The pattern for yield and spill is contrast compared with (Fig. 4). The number of days rainwater yield satisfied the demand as

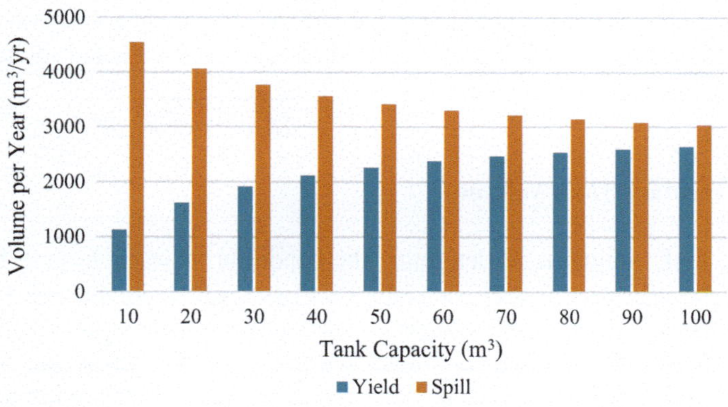

Fig. 4 Water saving and storage efficiency for tank size ranging from 10 m^3 to 100 m^3

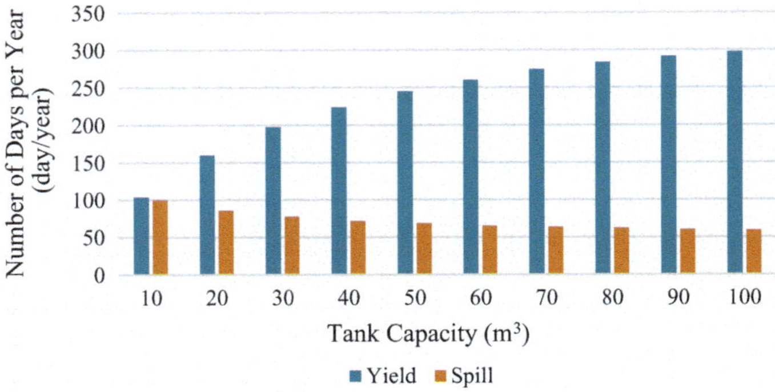

Fig. 5 Number of days yield and spillage in a year for each tank size

Table 7 Time reliability for water demand of 8640 L per day

Tank capacity (m^3)	Yield (day/year)	Time reliability (%)
10	104	28.5
20	160.5	44.0
30	197.9	54.2
40	224.1	61.4
50	244.9	67.1
60	260.6	71.4
70	274.7	75.3
80	284.1	77.8
90	291.7	79.9
100	297.9	81.6

tank size increases, roughly from 28.5% to 81.6% of days per year. On the other hand, the number of days per year for spillage decrease as tank size increases ranging from 99 to 59 days. For the tank capacity selected (100 m^3), the number of days rainwater yield fulfilled demand is 297 days per year which is about 81.6% and spillage occur at 59 days per year.

3.2.3 Percentage Tank Volume

Percentage tank volume is to determine the condition of water in tank based on number of days per year. It is divided into five classes which are the tank volume at 75% to 100%, 50% to 75%, 25% to 50%, less than 25%, and empty. The condition of water in tank within 15 years are illustrated in Fig. 6. From chart, the volume of water to full the tank increases as the tank size increase while the volume of water tank to empty decreases. The larger tank capacity will capture more rainwater and increase the period as it can store for a longer time compared with the small tank

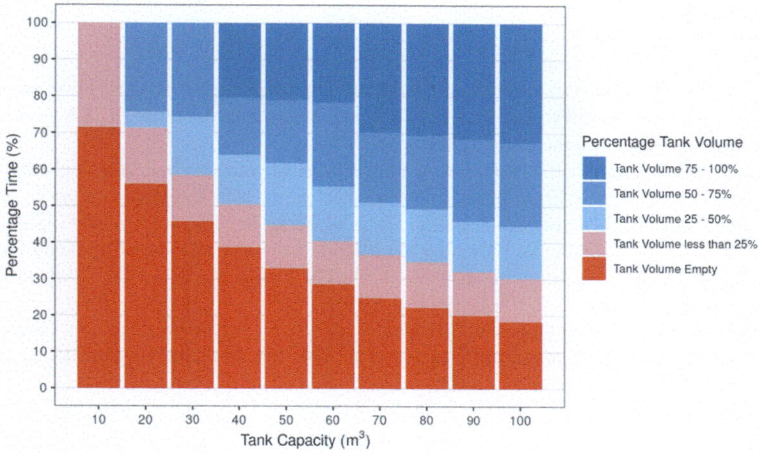

Fig. 6 Condition of water in tank within 15 years

capacity [13]. Furthermore, the smaller tank capacity will increase the spillage due to a large amount of rainwater enter the tank, hence, increase in the percentage time for tank volume to empty as it has higher water demand.

4 Conclusion

In conclusion, the rainfall data was chosen based on the nearest station from the study area which is UiTM Shah Alam station. The average annual rainfall is 2378 mm, thus, it is highly feasible for implementing rainwater harvesting system. The potential of rainwater that can be collected was identified based on the average monthly rainfall data. Next, a few inputs such as catchment area, roof information, water demand in litres per day and tank capacity were collected. From these, the optimum tank size for Block 4B Kolej Melati UiTM Shah Alam was determined. The tank size of 100 m^3 is selected as the optimum design. Based on this optimum size, about 46.6% of rainwater from roof can be used that provide 83.7% of water demand which equivalent to 2639.56 m^3 to fulfil the water requirement for 297 days per year. Most importantly, the system is expected to save water bill of approximately RM 4249.69 based on current tariff for government department.

Acknowledgements The authors appreciate the support from the College of Engineering, Universiti Teknologi MARA.

References

1. Che-Ani AI, Shaari N, Yahaya H, Jamil M, Mohd-Tawil N, Zain MZM (2009) Focused evaluation of rainwater harvesting: a case study in Sandakan, Sabah, Malaysia. IRCSA Conference, on Aug.3–6, 2009 at Kuala Lumpur, Malaysia, 1–8
2. Water World Council (2010) Water Crisis. https://www.worldwatercouncil.org/en/water-crisis
3. Kuok KK, Chiu PC (2020) Optimal rainwater harvesting tank sizing for different types of residential houses: Pilot study in Kuching, Sarawak. J Eng Sci Technol 15(1):541–554
4. Woon YB, Ling L, Lun Tan W, Fai Chow M (2019) Community rainwater harvesting financial payback analyses - case study in Malaysia. IOP Conf Series: Mater Sci Eng 636(1):012019. https://doi.org/10.1088/1757-899X/636/1/012019
5. Sample DJ, Liu J (2014) Optimizing rainwater harvesting systems for the dual purposes of water supply and runoff capture. J Clean Prod 75:174–194. https://doi.org/10.1016/j.jclepro.2014.03.075
6. Adnan A, Che Ahmad A, Teriman S (2020) Rainwater Harvesting (RWH) Installation for buildings: a systematic review and meta-analysis approach. Malaysian J Sustain Environ 6(1): 89. https://doi.org/10.24191/myse.v6i1.8681
7. Kasmin H, Bakar NH, Zubir MM (2016) Monitoring on the quality and quantity of DIY rainwater harvesting system. IOP Conf Series: Mater Sci Eng 136(1):012067. https://doi.org/10.1088/1757-899X/136/1/012067
8. Lani NHM, Syafiuddin A, Yusop Z, Adam U, Amin MZM (2018) Performance of small and large scales rainwater harvesting systems in commercial buildings under different reliability and future water tariff scenarios. Sci Total Environ 636:1171–1179. https://doi.org/10.1016/j.scitotenv.2018.04.418
9. Biswas BK, Mandal BH (2014) Construction and evaluation of rainwater harvesting system for domestic use in a remote and rural area of Khulna, Bangladesh. Int Scholarly Res Notices 2014:1–6. https://doi.org/10.1155/2014/751952
10. DID (2012) Urban stormwater management manual for Malaysia (MSMA 2nd Edition). DID, Kuala Lumpur
11. Man S, Hashim N, Ahmad AH, Thet KM, Sidek NS (2017) Kebolehupayaan Sistem Penuaian Hujan Sebagai Bekalan Air Alternatif di Malaysia: Suatu penelitian awal (The reliability of rainwater harvesting system as an alternative source of water supply in Malaysia: A preliminary study). Geografia: Malaysian J Soc Space 10(6): 97–104
12. Khan ST, Baksh AA, Papon MTI, Ali MA (2017). Rainwater harvesting system: an approach for optimum tank size design and assessment of efficiency. Int J Environ Sci Dev 8(1): 37–43. https://doi.org/10.18178/ijesd.2017.8.1.917
13. Goh YC, Ideris M (2021) Tangki NAHRIM 2.0: An R-based water balance model for rainwater harvesting tank sizing application. Water Pract Technol 16(1): 182–195. https://doi.org/10.2166/wpt.2020.106
14. Aladenola H, O. O., Adeboye, O. B. (2010) Assessing the potential for rainwater harvesting. Water Resour Manage 24(10):2129–2137. https://doi.org/10.1007/s11269-009-9542-y
15. Malaysia (2016) Malaysia Information. https://www.malaysia.gov.my/portal/content/144
16. Mohd Unaini A, Razali MR, Mohammed Ali TA, Alias AH, Mahsum E (2017) Assessment on the performance of a rainwater harvesting system. Sci Res 5(3): 36. https://doi.org/10.11648/j.sr.20170503.13
17. Campisano A, Modica C (2014) Selecting time scale resolution to evaluate water saving and retention potential of rainwater harvesting tanks. Procedia Eng 70:218–227. https://doi.org/10.1016/j.proeng.2014.02.025
18. Rebecca AO, Coker Akinwale O, Sridhar Mynepalli K, Adewole Esan M (2013) Examining the effectiveness of rainwater collection systems in a nigerian leper colony using the behavioural model. ARPN J Eng Appl Sci 8(1):1–8
19. Wong CY, Teo FY, Goh BH, Mah DYS (2018) Feasibility study of rainwater harvesting for a campus University in Semenyih, Malaysia. November

Study on Suspended Particles Velocity at Ogee Dam Spillway Using Discrete Particle Method (DPM) and Particle Image Velocimetry (PIV)

Nazirul Mubin Zahari, Mohd Hafiz Zawawi, Lariyah Mohd Sidek, Mohamad Aizat Abas, Farah Nurhikmah Ghazali, and Nurul Husna Hassan

Abstract Theories of dam construction that were created eventually laid the groundwork for dam safety, allowing for the design of higher-quality dams to be built. Dam engineers continued to experiment with new technology to build dams in a safer, more cost-effective, and environmentally friendly manner. The Discrete Particle Method (DPM) is commonly used in various study. Objective of this study is to identify a different velocity of different particles influence at ogee dam spillway using Discrete Particle Method (DPM). Simulation of 1:1 and 1:40 scaled model and experimental 1:40 scaled model had been done. This study presents a validation of particles velocity on 1:40 scaled ogee dam using Particle Image Velocimetry (PIV). Three different types of particles namely as silt, sand and gravel was use in DPM simulation. The results indicate a pattern of particle velocity peak at certain point. Percentage different for 1:40 scaled simulation and PIV experimental is less than 10%.

Keywords Discrete Particles Method · Ogee Dam · Spillway · Suspended Particles · Froude Numbers · Shield Numbers

---

N. M. Zahari (✉) · M. H. Zawawi · F. N. Ghazali · N. H. Hassan
Department of Civil Engineering, College of Engineering, Universiti Tenaga Nasional, 43000 Kajang, Selangor, Malaysia

M. H. Zawawi
e-mail: mhafiz@uniten.edu.my

L. M. Sidek
Institute of Energy Infrastructure, Universiti Tenaga Nasional, 43000 Kajang, Selangor, Malaysia
e-mail: lariyah@uniten.edu.my

M. A. Abas
School of Mechanical Engineering, Engineering Campus, Universiti Sains Malaysia, Nibong Tebal, 14300 Pulau Pinang, Malaysia
e-mail: aizatabas@usm.my

N. Mohamed Noor et al. (eds.), *Proceedings of the 3rd International Conference on Green Environmental Engineering and Technology*, Lecture Notes in Civil Engineering 214,
https://doi.org/10.1007/978-981-16-7920-9_53

1 Introduction

Dam safety theories evolved over time allowed for better dam design. Dam engineers continued to experiment with new technology to build dams in a safer, more cost-effective, and environmentally friendly manner. Discrete Particle Method (DPM) has been widely used in testing a variety application [1, 2]. Numerical software was utilized to simulate particle behavior in terms of velocity parameter. The software allowed the user to create a model, inject particles based on their attributes, and collect velocity information for the particles. Many investigations on numerical simulations based on ogee dam spillway have revealed a variety of issues [3–5].

Particle Image Velocimetry (PIV) validation was performed to confirm that the results are useable and acceptable [6–9]. PIV works by introducing seeding particles into the fluid flow that act as tracking particles and tracking the movement of tracking particles when illuminated. PIV is a non-intrusive experimental method to visualize the fluid flow velocity vector. The capability of PIV as a flow velocity visualization tool had been proved viable in various applications, such as water filtration unit [10], ogee spillway crest [6], encapsulation process [1], aeration study [8] and electrostatic precipitator [11]. Objective of this study is to identify different velocity of different particles influence to ogee dam spillway using Discrete Particle Method (DPM). This study also introduces a validation of particles velocity on 1:40 scaled ogee dam using PIV.

2 Materials and Method

2.1 Numerical Modelling of Suspended Particles

This section describes methods to use numerical simulation software to simulate suspended particles in a dam spillway. The fluid-particle interaction between the continuous phase fluid flow and dam structure, as well as the suspended particles, will be examined in this simulation. The Navier–Stokes equations, which include the continuity equation and momentum equation, were used to govern the continuous fluid phase [12]:

$$\frac{\partial \rho}{\partial t} + \nabla(\rho \mathbf{u}) = 0, \tag{1}$$

$$\frac{\partial}{\partial t}(\rho \mathbf{u}) + \nabla(\rho \mathbf{u} \cdot \mathbf{u}) = -\nabla P + \nabla \tau + \rho g, \tag{2}$$

where t is the time, ρ is the fluid density, $\mathbf{u}$ is the flow velocity, τ is the shear stress and gravitational acceleration, g.

The particulate phase of sediment presented in water was modelled by the discrete phase model (DPM) [1, 13]. By the forces balance, the trajectory of the sediment

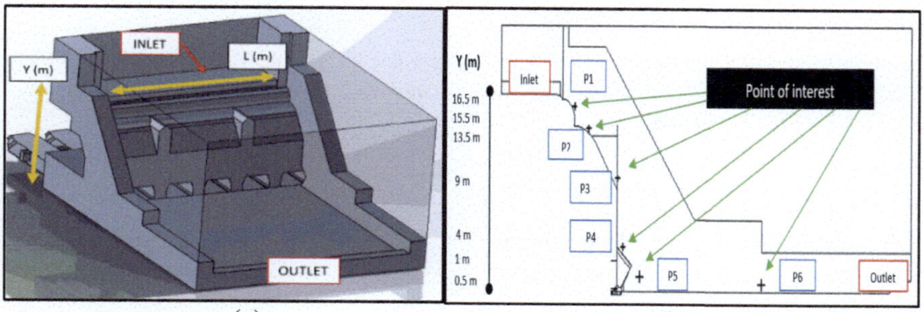

(a) (b)

Fig. 1 **a** The boundary conditions assigned on the fluid domains in the numerical simulation (ISO view) **b** Point of interest of 1:1 model (from side view)

particles as suspended in the water can be determined with known particle's velocity, $\mathbf{u}_p$:

$$\frac{\partial}{\partial t}\mathbf{u}_p = \frac{18\mu_f}{d_p \rho_p C_c}\left(\mathbf{u} - \mathbf{u}_p\right) + \overrightarrow{\mathbf{g}}\,\frac{\rho_p - \rho}{\rho_p} + \overrightarrow{F}_x, \tag{3}$$

where μ_f is the dynamic viscosity of the fluid, ρ_p is density of particle, d_p is the diameter of particle and C_c is the Cunningham correction factor.

Two models were drew using CAD software which are a scale 1:1 and 1:40. The characteristic length of the dam with scaled down dimensions were formed by considering the complexity of the geometry. Using two-way interactions of multiphase VOF and DPM model, the environment of a standard reflow process is recreated in numerical simulation. The discrete particle phase will allow for exact trajectory particle prediction. The schematic boundary diagram with inlet, outlet of water flow, L (characteristic length) and Y (height of the point location) are shown in Fig. 1. The mesh developed for the current model is based on tetrahedrons mesh. The fluid domain divided into three parts which are air, water and particles. The magnitude of the velocity of the water flow into the inlet is set at 0.032 m/s and 0.005 m/s for 1:1 and 1:40 scale respectively. Three different size of particle was used as input namely silt (1400 kg/m^3), sand (1800 kg/m^3) and gravel (2000 kg/m^3) and particle size 0.00125 m, 0.7 m and 4 mm respectively.

2.2 *Scaling Formulation*

The flow conditions in the physical model should be as close to those in the prototype as possible. This model is meant to show how geometrically similar, kinematically similar, and dynamically similar the prototype. Scaling down model to 1:40 scale size was done using similarity method. In this study, the scaled down dimension of the dam was expressed in terms of its length. The L length for 1:1 scale model is

Table 1 Scaling parameter

Parameter	Scale Relationship 1:1 model (p), 1:40 (m)	Ratio Scale Value of each parameter	1:1 Scale Value	1:40 Scale Value
Geometric (L)	S = Lp/Lm	1/40	31 m	0.78 m
Inlet Velocity (V)	Vp = Vm * S0.5	1/6.32	0.32 m/s	0.05 m/s
Scaled Gravel (D)	S = Dp/Dm	1/6.667	4 mm	0.6 mm

31 m and scale size for 1:40 is 0.78 m. The comparison of spillway length when scaled down is shown in Table 1. Shield Number was used to scale the particle size from 1:1 to 1:40. The Shield Number [14] general equation was used as follows:

$$\tau = \rho \times \sqrt{gyS} \tag{4}$$

$$\tau^{\circ} = \frac{\tau}{(\rho s - \rho)\ \times \mathrm{g} \times \mathrm{Dp}} \tag{5}$$

where, ρ is the fluid density, ρ_s is the particle density, S is the references slope, y is the references water depth, τ is the shear stress and gravitational acceleration, g and Dp is particle diameter.

For velocity similarity, Froude number was used to scale the velocity of the model from scale 1:1 to 1:40. Froude number should be constant between full scale model and scale down model [15]. The following formula was used, where, V is the velocity, g is gravitational constant, and L is the characteristic length.

$$\mathrm{Fr} = \frac{\mathrm{V}}{\sqrt{\mathrm{gL}}} \tag{6}$$

2.3 Particle Image Velocimetry (PIV) Experimental

Figure 2 presents the particle image velocimetry (PIV) experimental setup for conducting the validation process on ogee dam spillway 1:40 physical model. In setting up, a camera video is placed perpendicularly to the laser to capture the flow of water from the side. For fluid scenario, amount of polyamide seeding particles (same density of water) were added to the water as tracking water particles meanwhile for particle scaled gravel, 0.6 mm diameter were use as tracking particles velocity. Upon commencing the experiment, the ambient is kept dark to ensure the camera is capturing the thin illuminated by laser light. All images were analysed using PIVlab to obtain the flow velocity and flow vectors.

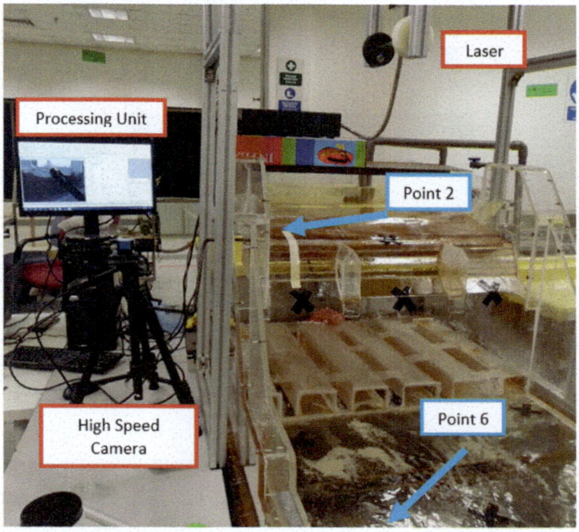

Fig. 2 Arrangement of PIV experimental and Point of interest location

3 Result and Discussion

3.1 Fluid and Particle Velocity

Figure 3 shows the results of numerical simulation based on fluid and 3 different particles as describe in Sect. 2.1. Figure 4 shows results of fluid velocity and particles velocity by point. On the right of the graph shows the Y (m) which is height of the point located. The pattern of velocity is different at each point and each case. For particle velocity, highest trend of velocity is sand particle with diameter of 0.7 mm and sand density is 1800 kg/m^3. Overall trend of highest velocity for water occurs at point 4 which is 13.93 m/s. Point 4 have a highest flow because of the downfall of water on that point toward settle basin area (point 6). Velocity from point 5 to point 6 is slowing in range 10.71 m/s to 6.43 m/s. This trend of velocity affect a velocity of particles since particles velocity depends on water velocity [5, 15].

4 Validation

The validation from the software has to be investigated to make sure the results obtained from the simulation were accurate. 1:40 model is simulated to validate a 1:1 scale and 1:40 model was validated by Particle Image Velocimetry (PIV) in experimental. Figure 5 presents the velocity of fluid and scaled gravel simulation 1:40 results comparison with PIV at two-point locations. Simulation 1:40 for particular point was used to validate with PIV. Validation at two locations namely Point 2 (P2) and Point 6 (P6) for fluid velocity and scaled gravel velocity. The

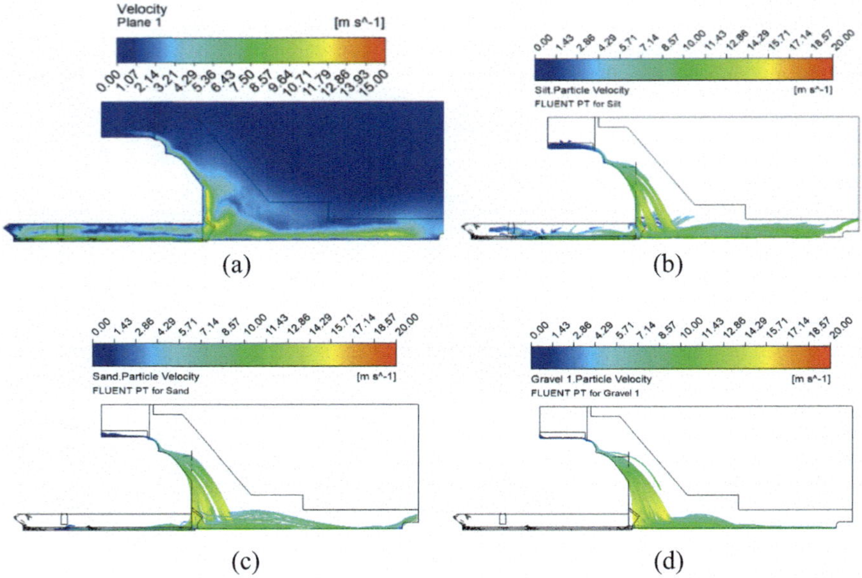

Fig. 3 Simulation Results for **a** fluid velocity, **b** Silt Particle Velocity, **c** Sand Particle Velocity, **d** Gravel Particle Velocity

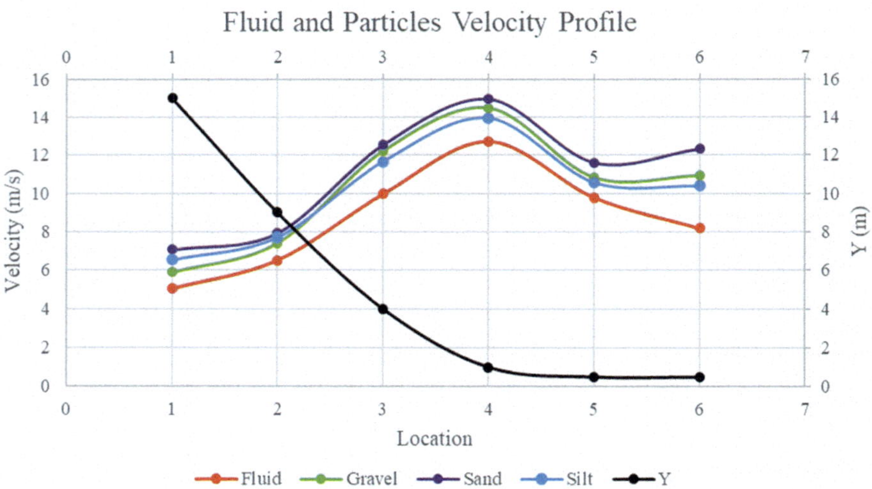

Fig. 4 Results Fluid and Particle Velocity of 1:1 Scale

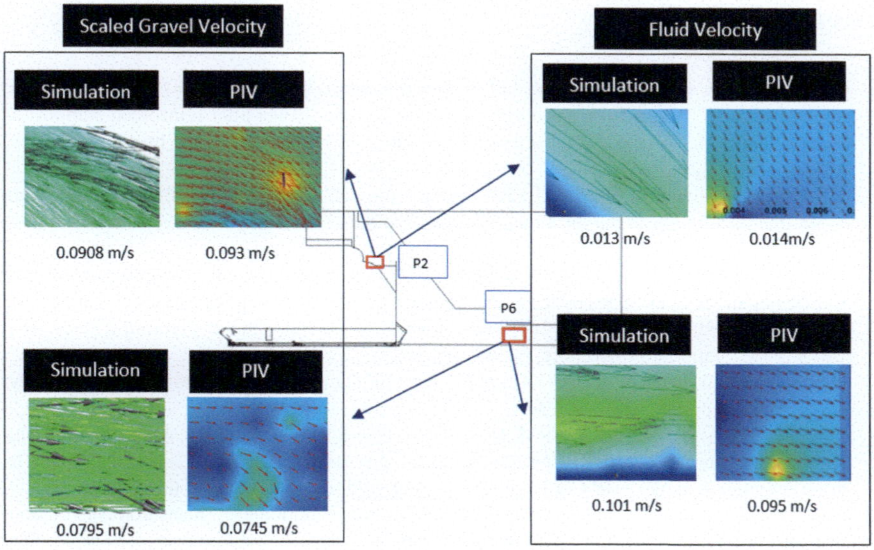

Fig. 5 Validation results for 1:40 Simulation and PIV

percentage difference at P2 for fluid and particles velocity respectively are 7.69% and 2.42% respectively. The percentage difference at P6 for fluid and scaled gravel velocity respectively are 5.94% and 6.29% respectively. According to the results, it is acceptable range of different percentage less than 15% [12, 15].

5 Conclusion

This study is helpful in identifying a part of the ogee dam spillway indicate high velocity. On simulation 1:1, general trend of highest velocity occurs at point 4 because of the downfall of water on that point and velocity is slowing toward to point 6. For simulation 1:40 is validated with PIV and shows the discrepancy percentage is less than 10%.

Acknowledgements The authors would like to acknowledge Universiti Tenaga Nasional for providing the facilities and financial assistance under BOLD Research Grant 2021.

References

1. Ng FC, Abas A, Gan ZL, Abdullah MZ, Che Ani F, Yusuf Tura Ali M (2017) Discrete phase method study of ball grid array underfill process using nano-silica filler-reinforced composite-encapsulant with varying filler loadings. Microelectron Reliab 72:45–64. https://doi.org/10.1016/j.microrel.2017.03.034
2. Kharoua N, AlShehhi M, Khezzar L (2015) Prediction of black powder distribution in junctions using the discrete phase model. Powder Technol 286:202–211. https://doi.org/10.1016/j.powtec.2015.07.045
3. Erpicum S, Blancher B, Peltier Y, Vermeulen J, Archambeau P, Dewals B, Pirotton M (2018) Experimental study of ogee crested weir operation above the design head and influence of the upstream quadrant geometry. In: 7th IAHR International Symposium on Hydraulic Structures, ISHS 2018. Utah State University, pp 99–108
4. Hinge GA, Balkrishna S, Khare KC (2010) Pawana dam energy dissipation—A case study. Aust J Basic Appl Sci 4:3261–3267
5. Zawawi MH, Aziz NA, Radzi MRM, Hassan NH, Ramli MZ, Zahari NM, Abbas MA, Saleha A, Salwa A, Muda ZC (2018) Computational fluid dynamic analysis at dam spillway due to different gate openings. AIP Conf Proc 2030: . https://doi.org/10.1063/1.5066886
6. Biegowski J, Paprota M, Sulisz W (2020) Particle image velocimetry measurements of flow over an Ogee-Type Weir in a hydraulic flume. Int J Civ Eng. https://doi.org/10.1007/s40999-020-00538-z
7. Zheng D, Lu G, Zhang J, Wang M (2016) Experimental method of three-dimensional velocity field measurement in circular pipe based on PIV. 1584–1589
8. Abas MA, Jamil R, Rozainy MR, Zainol MA, Adlan MN, Keong CW (2017) PIV Study of aeration efficient of stepped spillway system. IOP Conf Ser Mater Sci Eng 209. https://doi.org/10.1088/1757-899X/209/1/012090
9. Abdulwahab MR, Ali YH, Habeeb FJ, Borhana AA, Abdelrhman AM, Ali Al-Obaidi SM (2020) A review in particle image velocimetry techniques (developments and applications). J Adv Res Fluid Mech Therm Sci 65:213–229
10. Azman A, Zawawi MH, Hassan NH, Abas A, Razak NA, Mazlan AZA, Mohd Remy Rozainy MAZ (2018) Effect of step height on the aeration efficiency of cascade aerator system using particle image velocimetry. MATEC Web Conf 217:1–6. https://doi.org/10.1051/matecconf/201821704005
11. Miyashita H, Ehara Y, Thunoda T, Enomoto J, Inui T (2018) PIV analysis of particle behavior in electrostatic precipitator. Electron Commun Japan 101:16–23. https://doi.org/10.1002/ecj.12011
12. Zawawi MH, Hassan NH, Ramli MZ, Zahari NM, Radzi MRM, Saleha A, Salwa A, Sidek LM, Muda ZC, Kamaruddin MA (2018) Fluid-structure interactions study on hydraulic structures: a review. AIP Conf Proc 2030. https://doi.org/10.1063/1.5066885
13. Zahari NM, Zawawi MH, Sidek LM, Mohamad D, Itam Z, Ramli MZ, Syamsir A, Abas A, Rashid M (2018) Introduction of discrete phase model (DPM) in fluid flow: a review. AIP Conf Proc 2030. https://doi.org/10.1063/1.5066875
14. Mokashi AA, Hirpurkar PS (2019) Hydraulic scaling and similitude from model to prototype. Int J Recent Technol Eng 8, 390–392. https://doi.org/10.35940/ijrte.B1066.0982S1019
15. Hassan NH, Zawawi MH, Zainol MRRMA, Mazlan AZA, Abas MA, Azman MA (2020) Physical modelling analysis of Chenderoh spillway. AIP Conf Proc 2291. https://doi.org/10.1063/5.0023921

Synergistic Effect Between Iron and Food/Microorganism (F/M) Ratio in Biological Wastewater Treatment

Leela Sri Subramaniam, Nabilah Aminah Lutpi, Yee Shian Wong, Soon An Ong, Nazerry Rosmady Rahmat, and Chairat Siripatana

Abstract Biological wastewater treatment is mainly dependent on the actions of microorganisms that can be used to treat wastewater. Microorganisms will start to stick together when they degrade the organic matter in wastewater for food and flocculate to settle the pollutants. This study aimed to investigate the effect of food to microorganism (F/M) ratio and iron in a biological process using aerobic treatment. For this purpose, four aerobic tanks (A, B, C, D) were set up using activated sludge as the seed sludge, air pump as air diffuser to provide oxygen to the system, and three litres of synthetic medium as carbon source for each tank. A specific amount of iron (II) sulfate was added into tanks B, C, and D with the weight of 3 g, 6 g, and 9 g, respectively. Tank A act as a control, and no iron dosage was added. The F/M ratio for tanks A, B, C, and D were 0.8, 0.5, 0.4, and 0.3 mg BOD/mg MLVSS, respectively. The aerobic tanks were operated for 40 days in sequential batch mode and sampling was collected four times per week to observe the COD and MLVSS. This study has found that Tank D shows the best performance compared to all tanks

L. S. Subramaniam · N. A. Lutpi (✉) · Y. S. Wong · S. A. Ong · N. R. Rahmat
Faculty of Civil Engineering Technology, Universiti Malaysia Perlis, 02600 Arau, Perlis, Malaysia
e-mail: nabilah@unimap.edu.my

L. S. Subramaniam
e-mail: s171130584@studentmail.unimap.edu.my

Y. S. Wong
e-mail: yswong@unimap.edu.my

S. A. Ong
e-mail: saong@unimap.edu.my

N. R. Rahmat
e-mail: nazerry@unimap.edu.my

Centre of Excellence Water Research and Environmental Sustainability Growth (WAREG), Universiti Malaysia Perlis, Perlis, Malaysia

C. Siripatana
School of Engineering and Technology, Walailak University, Ta Sala, Nakhon Si Thammarat 80160, Thailand
e-mail: schairat@wu.ac.th

N. Mohamed Noor et al. (eds.), *Proceedings of the 3rd International Conference on Green Environmental Engineering and Technology*, Lecture Notes in Civil Engineering 214,
https://doi.org/10.1007/978-981-16-7920-9_54

with 84.71% COD removal efficiency and a fivefold increment of microorganism growth rate. These findings suggest that a relationship exists between the iron and F/M ratio to enhance the aerobic treatment process.

Keywords F/M ratio · Iron · Aerobic · Biological wastewater treatment

1 Introduction

Biological wastewater treatment is a method used to treat wastewater using microorganisms. Microorganisms consume pollutants or organic matter as nutrients in treating wastewater to grow, live and reproduce. Biological wastewater treatment or pollutant degradation process can be done with or without oxygen. Aerobic wastewater treatment is with the presence of oxygen, while anaerobic wastewater treatment is without oxygen [1].

A bioreactor for wastewater treatment should provide suitable operational conditions for microorganisms to breed and increase growth by using energy in organic matter. In aerobic wastewater treatment, for instance, the degradation of the wastewater process will be efficient with oxygen and a sufficient amount of food provided for the microorganism [2]. F/M ratio is significant because it ensures the food and microorganism amount are balanced in the aeration tank. The optimum and ideal F/M ratio ranges from 0.2 to 0.6 kg BOD/kg MLVSS in a complete mix system [3]. A perfect F/M ratio could treat wastewater efficiently. A high or low F/M ratio could not give an excellent dense floc because of the dispersion, resulting in an inadequate settlement of sludge [4].

Limitations of biological wastewater treatment are due to the slow biodegradation processes. Cell recovery is challenging because substrates could significantly impede the recovered cell's activity. On the other hand, iron can help the process by acting as a supplement for microorganisms [5]. The synergistic effects of microorganisms and iron can be explained by the fact that the presence of iron produces changes in microbial activity and community structure that are beneficial to the effective degradation of environmental pollutants. Microbes that are insufficient in iron dosage could be less in performance, leading to a poor disperse and poor dense floc [6]. Synergistic effects between iron and microbes on the contaminant degradation make the role of iron beyond that of a nutritional necessity [6]. Therefore, the study aims to investigate the synergistic effect between iron and F/M ratio in treating wastewater using the aerobic treatment process.

2 Materials and Method

2.1 *Activated Sludge and Synthetic Medium Composition*

The activated sludge used for the study was obtained from wastewater treatment plant in Wang Bintong, Perlis. The sample was preserved and refrigerated at 4 °C prior to use in the study to decrease the biological degradation and acidification.

The synthetic medium used in the aerobic wastewater treatment contained sucrose as the sole carbon and energy source and supplements, including ammonium chloride and potassium dihydrogen phosphate based on Carbon: Nitrogen: Phosphorus (C:N:P) ratio of 100:5:1. This medium composition was applied for all experiments investigating the effect of iron dosage and F/M ratio in treating wastewater using an aerobic process. The pH value of the synthetic medium was adjusted to be in a neutral range between 6.5 and 7.5 using sodium bicarbonate.

2.2 *Method*

Four aerobic tanks (A, B, C, D) were set up using activated sludge as the seed sludge, air pump as air diffuser to provide oxygen to the system, and three litres of synthetic medium for each tank. A specific amount of iron (II) sulfate was added into tanks B, C D with the weight of 3 g, 6 g, and 9 g, respectively. Tank A act as a control, and no iron dosage was added. The F/M ratio of each tank was calculated using Eq. 1, whereby the ratio for tanks A, B, C, and D were 0.8, 0.5, 0.4, and 0.3 mg BOD/mg MLVSS, respectively. The tanks were operated for 40 days in sequential batch mode. Sampling was done four times per week to observe the COD and MLVSS performance for each tank. All wastewater parameters were analysed using Standard Methods for the Examination of Water and Wastewater [7].

$$\frac{\mathrm{F}}{\mathrm{M}} = \frac{\mathrm{BOD5}\frac{\mathrm{mg}}{\mathrm{L}} \times (\text{Volume of wastewater, L})}{\mathrm{MLVSS}\frac{\mathrm{mg}}{\mathrm{L}} \times (\text{Volume of wastewater, L})} \tag{1}$$

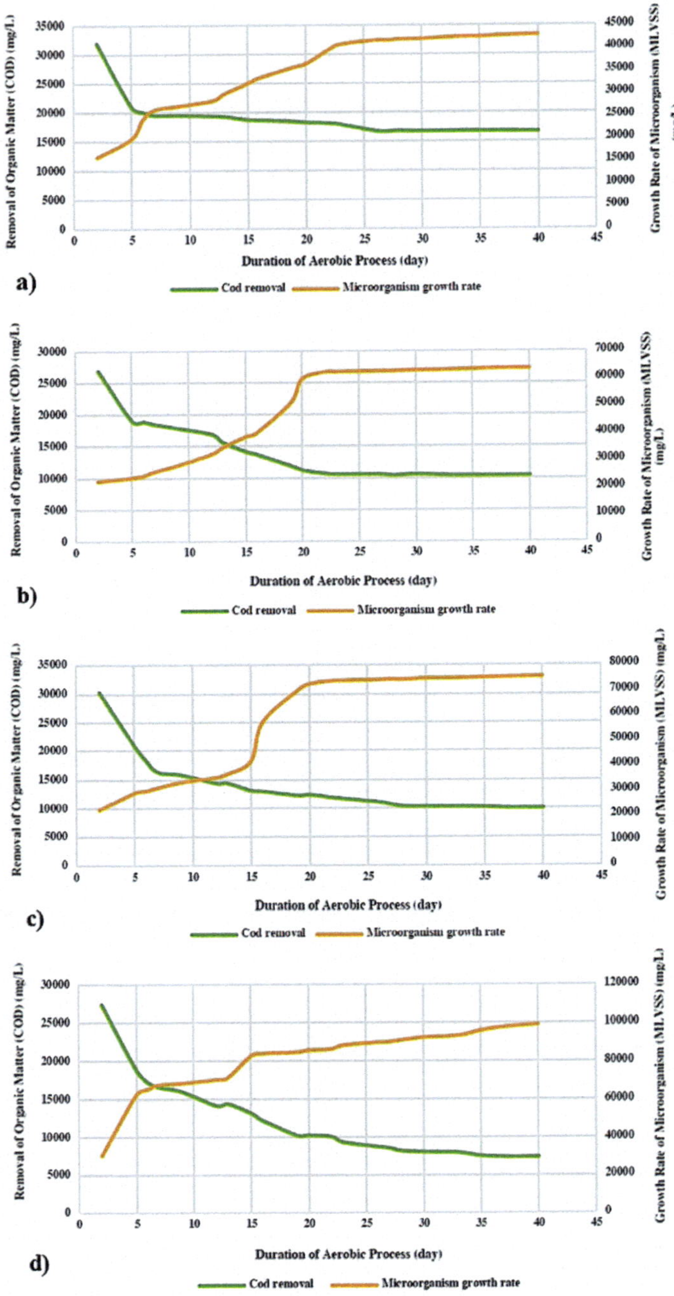

Fig. 1 Performance of **a** Tank A with F/M ratio 0.8 mg BOD/ mg MLVSS; **b** Tank B with F/M ratio 0.5 mg BOD/ mg MLVSS; **c** Tank C with F/M ratio 0.4 mg BOD/ mg MLVSS; **d** Tank D with F/M ratio 0.3 mg BOD/ mg MLVSS

3 Result and Discussion

3.1 Performance of Organic Matter (COD) Removal and Microorganism Growth Rate (MLVSS) at Different F/M Ratios and Iron Dosage.

Figure 1 shows the relationship between the F/M ratio and the performance of all tanks. Tank A with 0.8 F/M ratio has the most minor performance compared to the other three tanks. Tank A could only remove organic matter from 34,550 mg/L to 16,640 mg/L, 51.84% of COD removal. Nevertheless, the microorganism's growth rate for Tank A still shows almost threefold increment from the initial amount of 13,700 mg/L MLVSS. Tank B with a 0.5 F/M ratio has better performance than Tank A. Tank B could remove organic matter by 65.23%, from 26,910 mg/L to 10,360 mg/L of COD removal. The microorganism growth rate increased by three-fold, which was 17,500 mg/L MLVSS initially. Tank C with a 0.4 F/M ratio could remove 72.73% of COD, and the microorganism was increased by fourfold from 18,200 mg/L to 75,300 mg/L of MLVSS. However, Tank D depicted the highest performance compared to other tanks. Tank D with a 0.3 F/M ratio achieved the highest organic matter removal by 84.71% of COD removal, initially from 27,420 mg/L and decreased to 7340 mg/L of organic matter. Tank D also showing the highest growth rate of microorganisms which was almost fivefold. Overall in this research, Tank D shows the highest performance compared to all other tanks then followed by Tank C, Tank B, and Tank A (Tank: D > C > B > A). These results confirm the synergistic effect between iron and F/M ratio and in agreement with Eljamal et al. (2020) findings which showed an increment of 15% of COD removal and 25% of microorganism growth rate after adding the iron [5].

Research conducted by Wu et al. (2018) stated that the amount of food or organic matter and microorganism in the reactor should be stable and balanced for the aerobic treatment process to work efficiently [4]. The range of F/M ratio that works best in a system ranges from 0.2 to 0.6 kg BOD/ kg MLVSS [3]. A high or low F/M ratio could make microorganisms shocked because of the imbalance in the wastewater. Other than that, according to Hamza et al. (2018), the stable long-granule stability process appears to be favorable when the reactor's F/M ratio is 0.5 - 1.4 g COD/g SS [8]. However, granules cannot be observed in this research because the granulation process usually takes at least 90 days of observation; meanwhile, this research was only observed for 40 days because the system has achieved a steady-state COD removal period as depicted in Fig. 1.

4 Conclusion

Mixed liquor volatile suspended solids (MLVSS) represent the microbiological suspension composition changes in respect to variation of iron dosage in the aerobic system of the biological wastewater treatment process. In conclusion, Tank D shows the best performance compared to all tanks with an optimal F/M ratio of 0.3 and 9 g of iron dosage, resulting in 84.71% COD removal efficiency and a fivefold increment of microorganism growth rate. This study indicates that a relationship exists between the iron and F/M ratio for the enhancement of the aerobic treatment process. The experiments from the study confirmed that the F/M ratio and the amount of iron added influence the tanks' performance in COD removal and MLVSS increment.

References

1. Narayan CM, Narayan V (2019) Biological wastewater treatment and bioreactor design: a review. Sustain Environ Res 29:33. https://doi.org/10.1186/s42834-019-0036-1
2. Guo N, Zhang J, Xie H-J, Tan L-R, Luo J-N, Tao Z-Y, Wang S-G (2017) Effects of the food-to-microorganism (F/M) Ratio on N_2O emissions in aerobic granular sludge sequencing batch airlift reactors. Water 9:477. https://doi.org/10.3390/w9070477
3. Metcalf & Eddy (2013) Wastewater engineering: treatment and reuse, 5th Edition. McGraw-Hill Education
4. Wu D, Zhang Z, Yu Z, Zhu L (2018) Optimization of F/M ratio for stability of aerobic granular process via quantitative sludge discharge. Bioresour Technol 252:150–156. https://doi.org/10.1016/j.biortech.2017.12.094
5. Eljamal R, Kahraman I, Eljamal O, Thompson IP, Maamoun I, Yilmaz G (2020). Impact of nZVI on the formation of aerobic granules, bacterial growth and nutrient removal using aerobic sequencing batch reactor. Environ Technol Innov 19, 100911
6. Tian T, Yu HQ (2020) Iron-assisted biological wastewater treatment: synergistic effect between iron and microbes. Biotechnol Adv 44:107610. https://doi.org/10.1016/j.biotechadv.2020.107610
7. American Public Health Association (APHA) (2012) Standard methods for examination of water and waste water. 23th Ed.Washington DC, USA
8. Hamza RA, Sheng Z, Iorhemen OT, Zaghloul MS, Tay JH (2018) Impact of food-to-microorganisms ratio on the stability of aerobic granular sludge treating high-strength organic wastewater. Water Res 147:287–298. https://doi.org/10.1016/j.watres.2018.09.061

The Effect of NaOH-Na_2CO_3 Precipitation Bath Concentration on Adsorptive Performance of Crosslinked and Non-crosslinked Chitosan Beads for Azo Dye Removal

Lui-Ruen Irene Teo, Voon-Loong Wong, and Siew-Shee Lim

Abstract The newly formed chitosan (CS) gel beads were primarily synthesized in a precipitation bath which contains 1 M NaOH and Na_2CO_3 at varied concentration (0.2, 0.4 and 0.6 M). The CS gel beads formed with 0.2 M Na_2CO_3 exhibited the highest porosity (ε%) (97.10%) and dye uptake% (64.34%) after 72 h. Besides, the porosity and adsorptive performance for CS gel beads with 0.4 M and 0.6 M Na_2CO_3 were lower as compared to the pure CS gel beads due to low interconnection density caused by the deeper pores. To counteract the destabilizing effect, glutaraldehyde (GA) solution was added to the CS gel beads for chemically crosslinking. As result, all the GA-CS beads demonstrated noteworthy dye uptake% up to 99.6% after 4.5 h with disregards of precipitation bath concentration. This study had proven on the stabilising effect provided by GA crosslinking on the outstanding methyl orange (MO) adsorptive performance.

Keywords Adsorption · Azo dye · Chitosan · Precipitation · Porosity · Batch mode

---

L.-R. I. Teo · V.-L. Wong (✉)
School of Engineering and Physical Sciences, Heriot-Watt University Malaysia, 62200 Putrajaya, Wilayah Persekutuan Putrajaya, Malaysia
e-mail: v.wong@hw.ac.uk

L.-R. I. Teo
e-mail: it6@hw.ac.uk

S.-S. Lim
Department of Chemical and Environmental Engineering, Faculty of Science and Engineering, University of Nottingham Malaysia, 43500 Semenyih, Selangor, Malaysia
e-mail: siewshee.lim@nottingham.edu.my

N. Mohamed Noor et al. (eds.), *Proceedings of the 3rd International Conference on Green Environmental Engineering and Technology*, Lecture Notes in Civil Engineering 214,
https://doi.org/10.1007/978-981-16-7920-9_55

1 Introduction

Textile industry is the major contributor towards the hazardous dyes release due to incomplete dye exhaustion [1]. These dye effluents are subjected to high persistent colour and chemical oxygen demand (COD) which are aesthetically and environmentally unacceptable [2]. At present, adsorption by using CS, a natural biopolymer derived from chitin deacetylation, has been widely reported in the literature due to its noteworthy features of environment-friendly, outstanding dye removal performance, cost-effectiveness, ease of derivation, facile and controllable operation [3]. Besides, modifications onto CS adsorbents are highly recommended to improve its surface area and strength for effective adsorption at industrial scale [4]. Therefore, porogenic solvents such as Na_2CO_3 are utilised in CS gel beads synthesis to induce pores [5]. Additionally, chemically crosslinking agents such as GA are commonly introduced to produce a mechanically and chemically stable CS adsorbent [6].

In the present study, the effect of NaOH-Na_2CO_3 precipitation bath concentration on the structure of the newly formed CS gel beads was investigated. The CS gel beads were developed via phase inversion methods in a basic NaOH gelling solution supplemented with Na_2CO_3 at a varied concentration to enhance CS surface area. Furthermore, the effect of chemically crosslinking of CS gel beads with GA solution was studied. The key objective of this research is to elucidate the fundamental phenomenon of adsorption science in MO dye removal by using the newly formed CS gel beads and GA-CS beads.

2 Materials and Method

2.1 Material

CS powder (DD% > 90%) and acetic acid (HAc) (glacial 99.9%) were purchased from Fulltime Asia Sdn. Bhd. NaOH ($\geq$99%) and Na_2CO_3 ($\geq$99.9%) were purchased from Merck, Germany. GA (25.55%) was purchased from Calbiochem, United States. MO powder (85%) were purchased from QREC (Asia) Sdn. Bhd.

2.2 Synthesis of CS and GA-CS Beads

CS solution (3 wt%) was prepared by dissolving 6 g of CS powder into 194 g of HAc (3 wt%) under stirring for 3 h. A 200 mL precipitation bath solution of 1 M NaOH with various Na_2CO_3 concentrations (0.2, 0.4 and 0.6 M) was prepared at a fixed volume ratio of 1:1 for CS gel beads synthesis. Spherical CS gel beads were obtained by dropwise addition of CS solution into NaOH-Na_2CO_3 bath using a syringe pump (NE-1000, Netherlands) at 0.5 mL/min. The bead sizes were mainly controlled by

Table 1 Synthesis of CS gel beads and GA-CS beads for MO adsorption studies

Process conditions			CS gel beads		GA-CS beads	
CS (wt%)	HAc (wt%)	NaOH (M)	Set	Na_2CO_3 (M)	Set	Na_2CO_3 (M)
3	3	1	1	0.0	5	0.0
3	3	1	2	0.2	6	0.2
3	3	1	3	0.4	7	0.4
3	3	1	4	0.6	8	0.6

a 22-gauge needle tip. After solidification, the CS gel beads were thoroughly rinsed with distilled water to remove the alkaline residues until neutral pH. Next, some batches of newly formed CS gel beads were chemically crosslinked in 2 wt% GA solution for comparative studies. Similarly, the crosslinked GA-CS beads were also thoroughly rinsed. After rinsing, both CS gel beads and GA-CS beads were dried in an oven at 60 °C for 3 h. Table 1 tabulates the process conditions of synthesizing CS beads for MO adsorption studies.

2.3 *Porosity Analysis of CS Gel Beads and GA-CS Beads*

The porosity ($\varepsilon\%$) of CS beads were calculated based on the weight differences of CS beads before and after drying by using Eq. (1) [7]. All the experiments were conducted in duplicate to get the mean values for analysis purposes.

$$\varepsilon\% = \frac{\frac{(W_W - W_D)}{\rho_w}}{\frac{W_D}{\rho_{CS} + \frac{(W_W - W_D)}{\rho_w}}} \quad (1)$$

where, W_W (g) is the weight of wet CS bead before drying; W_D (g) is the weight of Batch dried CS bead; ρ_{CS} ($\rho_{CS} = 1.07$ g/cm^3) is the density of CS solution; and ρ_w ($\rho_w = 1.00$ g/cm^3) is the density of water.

2.4 *Batch Adsorption Studies*

Adsorption was performed based on experimental sets in Table 1. 0.35 g of dried CS and GA-CS beads were separately added into MO dye solution (30 ppm, 50 mL each) and agitated by using an incubator shaker at 200 rpm under ambient temperature. A constant amount of dye sample (1 mL) was eliminated from the solution at predetermined time intervals for spectrophotometric quantification of dye concentration. The adsorption capacity was calculated in term of uptake% based on Eq. (2).

All the experiments were conducted in triplicate to get the mean values for analysis purposes.

$$Uptake\% = \frac{C_0 - C_t}{C_0} \times 100 \quad (2)$$

where, uptake% is the MO dye uptake efficiency (%), C_0 is the initial concentration of dye solution (ppm) and C_t is the concentration of dye solution at time t (ppm).

3 Result and Discussion

Figure 1 illustrates the CS gel beads and GA-CS beads formed before drying and after drying, whereas Table 2 summarises the ε% of CS beads formed based on experimental sets in Table 1. Besides, MO uptake% was analysed and illustrated

Fig. 1 CS gel beads and GA-CS beads illustration. Both beads were measured in ImageJ software (v1.53e, Java 1.8.0, USA) to identify the bead dimension (diameter)

Table 2 Porosity for CS and GA-CS beads based on experimental sets in Table 1

Set	$W_{W, mean}$ (g)	$W_{D, mean}$ (g)	ε(%)	Set	$W_{W, mean}$ (g)	$W_{D, mean}$ (g)	ε(%)
1	14.02	0.44	97.05	5	9.66	0.75	92.68
2	14.33	0.44	97.10	6	10.12	0.75	93.00
3	13.46	0.45	96.89	7	9.37	0.73	92.72
4	11.60	0.45	96.38	8	10.08	0.74	93.08

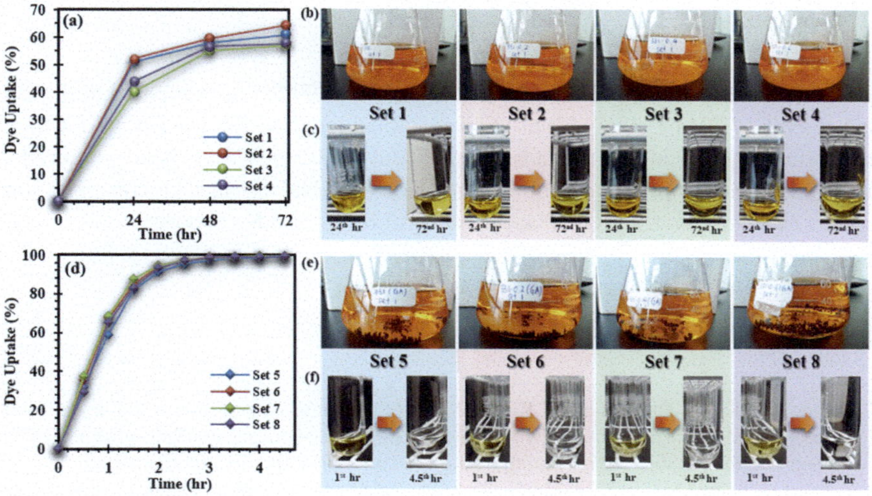

Fig. 2 The MO uptake% on the CS and GA-CS beads: **a** CS gel beads before **b** and after **c** the adsorption; and **d** GA-CS beads before **e** and after **f** the adsorption

in Fig. 2. Theoretically, CS gel beads with the highest porosity exhibit the highest adsorption capacity, resulting in highest uptake% [5]. Based on Table 2, CS gel beads showed the highest ε% under Set 2 > Set 1 > Set 3 > Set 4. Based on Fig. 2a, these CS gel beads demonstrated the highest uptake% under Set 2 > Set 1 > Set 4 > Set 3. The highest ε% and uptake% of Set 2 had shown that the presence of Na_2CO_3 porogenic solvent could aid in inducing pores to enhance the accessible adsorption sites for better dye adsorptive performance. Nevertheless, when the Na_2CO_3 concentration was further increased from 0.2 M to 0.6 M, the ε% and uptake% were reduced. This is because the deeper pores formed at higher Na_2CO_3 concentration will contribute towards low interconnection densities [8]. As a result, destabilising effect happen and cause the reduction of accessible adsorption sites. Moreover, Wahba [8] had studied on the properties of CS gel beads synthesized via ionotropic gelation with sodium tripolyphosphate (TPP) at varied Na_2CO_3 concentration. His result had proven that CS gel beads with 0.2 M Na_2CO_3 were too shallow to initiate the reduction in interconnection density. Thus, this has further supported the present work in which the highest uptake% was achieved for CS gel beads with 0.2 M Na_2CO_3.

The poor mechanical strength and stability of CS have restricted its practical application at larger scale. Chemically crosslinking of CS gel beads with GA serve as one of the possible approach while contemporaneously remain its original features [9]. As for GA-CS beads, the highest ε% were illustrated under Set 8 > Set 6 > Set 7 > Set 5. Based on Fig. 2d, the GA-CS beads demonstrated a remarkable uptake% up to 99.6% at the 4.5th hours. In addition, the effect of the NaOH-Na_2CO_3 bath concentration shown minimal effect on the adsorption capacity of GA-CS beads. The dye adsorptive performance of GA-CS beads had outstood the CS gel beads specifically because of the stabilising effect provided by GA crosslinking [8].

4 Conclusion

GA-CS beads exhibited an excellent MO uptake% of 99.6% as compared to the CS gel beads (57–64%). GA-CS beads showed a relatively faster removal rate in which equilibrium reached after 3 h of adsorption. Conversely, the CS gel beads required a longer adsorption time to reach equilibrium. However, NaOH-Na_2CO_3 concentration showed more significant effects on MO adsorption using CS gel beads.

References

1. Pandey AK, Sarada Dronamraju V, Kumar A (2016) Microbial decolorization and degradation of reactive red 198 azo dye by a newly isolated Alkaligenes species. Proc Natl Acad Sci, India Section B: Biol Sci 86(4):805–815. https://doi.org/10.1007/s40011-015-0497-x
2. Yaseen DA, Scholz M (2019) Textile dye wastewater characteristics and constituents of synthetic effluents: a critical review. Int J Environ Sci Technol 16:1193–1226. https://doi.org/10.1007/s13762-018-2130-z
3. Tay SY, Wong VL, Lim SS, Irene Teo LR (2021) Adsorption equilibrium, kinetics and thermodynamics studies of anionic methyl orange dye adsorption using chitosan-calcium chloride gel beads. Chem Eng Commun 208(5):708–726. https://doi.org/10.1080/00986445.2020.1722655
4. Vieira MLG, Pinheiro CP, Silva KA, Lutke SF, Cadaval TRSA, Dotto G, de Pinto LAA (2019) Chitosan and cyanoguanidine-crosslinked chitosan coated glass beads and its application in fixed bed adsorption. Chem Eng Commun 206:1474–1486. https://doi.org/10.1080/00986445.2019.1581618
5. Wong KT, Wong VL, Lim SS (2021) Bio-sorptive removal of methyl orange by Micro-Grooved Chitosan (GCS) beads: optimization of process variables using Taguchi L9 Orthogonal array. J Polym Environ 29:271–290. https://doi.org/10.1007/s10924-020-01878-6
6. Ramachandran S, Rajinipriya M, Soulestin J, Nagalakshmaiah M (2019) Recent developments in chitosan-based nanocomposites. In: Sanyang M, Jawaid M (eds) Bio-based Polymers and Nanocomposites. Springer, Cham. https://doi.org/10.1007/978-3-030-05825-8_9
7. Zhao F, Yu B, Yue Z, Wang T, Wen X, Liu Z, Zhao C (2007) Preparation of porous chitosan gel beads for copper (II) ion adsorption. J Hazard Mater 147(1–2):67–73. https://doi.org/10.1016/j.jhazmat.2006.12.045
8. Wahba MI (2017) Porous chitosan beads of superior mechanical properties for the covalent immobilization of enzymes. Int J Biol Macromol 105(1):894–904. https://doi.org/10.1016/j.ijbiomac.2017.07.102
9. López-Cervantes J, Sánchez-Machado DI, Sánchez-Duarte RG, Correa-Murrieta MA (2018) Study of a fixed-bed column in the adsorption of an azo dye from an aqueous medium using a chitosan–glutaraldehyde biosorbent. Adsorp Sci Technol. 36(1–2):215–232. https://doi.org/10.1177/0263617416688021

The manufacturer's authorised representative in the EU is Springer Nature Customer Service Centre GmbH, Europaplatz 3, 69115 Heidelberg, Germany. If you have any concerns regarding our products, please contact ProductSafety@springernature.com

Printed and bound by CPI Group (UK) Ltd, Croydon, CR0 4YY
30/09/2026
02225233-0014